Introduction to Composites and Polymer Material

Arvind Kaushal

BE. (Mechanical Engineering), ME. (Advance Production System),

Formerly Faculty of Mechanical Engineering,

Indira Gandhi Govt. Engineering College,

Sagar (MP), India.

Published by

Introduction to Composites and Polymer Material

Copyright © 2022 by Bonfring

ISBN 978-93-92537-04-2

Author
Arvind Kaushal

Bonfring

309, 5th Street Extension, Gandhipuram,
Coimbatore - 641 012,
Tamil Nadu, India.
E-mail: info@bonfring.org
Website: www.bonfring.org

This Book is Wholly Dedicated to

Lord Hanuman

&

My Parents, My Wife for their Love & Constant Supports

Preface

The suffix polymer 'mer' is originated from Greek word mers – which means part. The word polymer is thus recognised to mean material consisting of many part/mers. Polymer may either organic or inorganic compounds. Polymer showing remarkable properties and consequently used in many applications now a days. If we consider some historical aspect then our body is made up of polymer, for example – proteins, enzyme and other involves wood leather and silk these polymers are considered as a natural polymer. Further, its span of application are very wast and hence now a days it is used as synthetic polymer like – plastics, rubber, fiber, adhesives, adhesive tapes and resin etc. In this book I am going to focus on composites and polymer material. Composites material includes, type of composites, various matrix phases – like metal matrix composites, ceramic matrix composites, carbon matrix composites, properties of composites, application of composites, reinforcement of composites, manufacturing of composites, bonding of composites and etc. In case of polymer, it includes, type of polymers, properties of polymers, polymer manufacturing process and also describe product of composites in miscellaneous section. In present scenario, composites & polymer material & its product is widely used in so many applications because of their remarkable property. Now a days we have number of books on title **"Composites and Polymer Material"** but in this first edition of book, I am trying to arrange all the material is there specific sequence that is required for the student of material science. Furthermore, usefulness of this book is also augmented because it can not only applicable for material science & engineering but also equally applied for material & metallurgical engineering. A book that is being entitled as **"Introduction to Composites and Polymer Material"** completely dedicated to those students who is eager to achieve higher degree like **Master of Engineering / Master of Technology** in material science & metallurgy. Present study material is prepared according to the syllabus of **NIT'S & Other Technical Universities.** This book consists of two chapters & at the end of each chapter, a number of question have been provided for discussion which is not only helps to analyses the concept of the subject but also indicate how much he / she gained throughout the study.

Pedagogical Features:- Study material has following aspects.

1. Each chapter begins with basics of respective titles & elucidate each and every individual clause in easy & lucid language.
2. Whenever necessary suitable plots, graphs, curves are added with well supported theory for easy understanding.
3. Some advanced material is also provided to enhance the usefulness of the book.

Book / Course Outlines:- The overall course or syllabus are arranged into two chapters.

Chapter – 1 is dedicated to introduction, Introduction, Classification of composite material, Polymer matrix composites (PMC) or polymer matrix materials, Property of thermosets & thermo-plastic (Matrix phase), Polymer matrix composites (PMC) or polymer matrix materials, Metal matrix composites (MMCs), Property of metal matrix composites, Type of silicon carbide fibers, (CMCs), Carbon Matrix Composites (CAMCs), Applications of carbon–carbon composites, Function of matrix phase, Manufacturing of fibre reinforced composites, Discontinuous Fiber-Reinforced Metal Matrix Composites and Short fiber-Rubber Composites etc.

Chapter – 2 Deals with Introduction, Polymer a class of engineering materials, various field of application of polymer materials, various modes of classification of polymers, Condensation polymers, Addition polymerization technique, Property of polymer material, Polymer processing techniques and many more, in miscellaneous section, i am describing various types of products.

Arvind Kaushal

Acknowledgement

Any kind of manuscript or book preparation is very complicated and difficult task which could not be achieved by a single person. At very initial it is a constant supports & motivation of God for giving me the strength and thought for this task, my parents and others family members, teachers, colleagues, friends and many others individuals who supported & contributing to me for the preparation of this book. A special thanks is given by me to **Lard Hanuman** for their constant supports and encouragement. Last but not the least I am thanking to all teachers who have taught me thereby I have been able to achieve this state of study. Finally, this work is wholly dedicated to lord Hanuman.

TABLE OF CONTENTS

CHAPTER 1

COMPOSITES MATERIAL

1.1. Introduction

In present scenario diverse application & need, moves journey of human being toward advancement in the field of material science & technology which evolves number of materials, like – Metals, Alloys, Non-metals, Ceramics, Polymer & Composites etc. All those material falls under the category of engineering materials. Among those metals or materials here we will try to study about polymer material and composites material. Composites material is a recently newly developed, advanced promising & fascinating material because of their unique mechanical properties, physical properties and chemical properties. These properties are attributes as,

1. Mechanical properties: Superior Strength, Stiffness, Hardness, Outstanding strength to weight ration & modulus to weight ratio, Design flexibility, Higher wear resistance, Toughness and Durability etc.
2. Physical properties: Low density or Light weight, High temperature resistance.
3. Chemical properties: High corrosion resistance.

Furthermore, high performance rigid composites material that is manufacturing from other material like glass, graphite, Kevlar, boron and silicon carbide fibers in polymeric phase. Due to remarkable & outstanding property of composites it is extremely applied in Aero-space application, Space vehicle technology, and mechanical engineering application which includes internal combustion engines, Machine tool components, Automobile field, Rocket ships, Thermal engineering field, Trains, Brakes, Drive shafts, Flywheels, Tanks, Pressure vessels, Constructions, Oil & gas industries, Transportation, Sports, Electronic packaging and Medical applications & equipment's etc. Now a days we have diverse field of application of composites material that is not only famous due to superior mechanical, physical properties but also because of unique chemical property, electrical property and magnetic property as well as it has certain required, desirable property that can be created easily at minimum cost or less cost. Avery common & natural example of composites material is wood in which long cellulose fibers arrange & accumulated through a substance like lignin.

1.2. Definitions of Composites Materials

Composites material can be easily understood by the following definitions that has been given by material science researchers.

According to "**Jartiz**"

> Composites are multifunctional material systems that provide characteristics not obtainable from any discrete material. They are cohesive structures made by physically combining two or more compatible materials, different in composition and characteristics and sometimes in form.

According to "**Kelly**"

> He emphasis that composites should not be considered as a combination of two materials. In the broader significance, the combination has its own distinctive properties. In terms of strength to resistance to heat or some other desirable quality, it is better than either of the components alone or radically different from either of them.

According to "**Beghezan**"

> The composites are compound materials which differ from alloys by the fact that the individual components retain their characteristics but are so incorporated into the composite as to take advantage only of their attributes and not of their short comings

According to "**Van Suchetclan**"

> A composite material is a heterogeneous material consisting of two or more solid phases, which has an intimate contact with each other on a microscopic scale. They can be also considered as homogeneous materials on a microscopic scale in the sense that any portion of it will have the same physical property.

Except the above definitions, some other possible definitions are given below.

1. Composites material is defined as a material, consisting two or more than two material that has distinct phase which bound together so that we can achieve desired level of mechanical, physical properties, chemical properties & other properties to fulfill the desired purpose.

2. A composite material can be defined as a combination of two or more material and consequently we have better properties as a whole with respect to the properties of individual components.

3. A composites material is a structural material that is prepared by combination of two or more constituents at micro-scopic level but composition does not soluble in each other.

4. A composites material may be defined as a material that consists of two or more constituents of distinct physically & chemically characteristics which is distributed & systematically arranged that is separated by interface layer. Remember that matrix phase is continuous in nature while disperse phase or reinforcing phase is discontinuous in nature so therefore it will form a heterogeneous system.

5. Composites material is defined as, A heterogeneous material that is manufactured by combination of two or more constituents by using fillers (Reinforcing fibres) & compactable matrix.

6. Composites can be defined as, A material that consists of two or more different chemically & physically phases that is separated by a distinct interface.

7. A composites material is defined as, A material that is manufactured by combining two or more material that has quite different property which does not dissolve & blend to each other and final product give us desired level of property.

Whenever composites material has been prepared, we have two phases:

1. One constituent is called reinforcing phase or dispersed phase.

2. Another is, in which it is embedded is called matrix or in other word we can say that 1st phase is fixed firmly and deeply surrounding by the mass called compactable matrix. Remember that matrix phase material is always continuous.

Further, composites materials are heterogeneous in nature. Reinforcing material may be used in the form of fibers, particles or flakes and matrix phase may be metallic, ceramic or polymeric in origin. It has been observed that reinforcement is harder, stronger, stiffer than that of matrix which imparts desired level of strength & stiffness on the contrary matrix phase is generally ductile & less hard phase whose function is only to agglomerating the reinforcement or dispersed phase and cooperate to carry a load. Since dispersed phase or reinforced phase is embedded in the matrix and always in a discontinuous system hence for sake of simplification, matrix phase may be considered as a primary phase on the other hand dispersed phase / reinforcing phase is called secondary phase.

Some Most Common Example of Composites Material

1. Concreate reinforced with steel considered as a composites system.

2. Epoxy resin, reinforced with graphite fibres considered as a composites system.

3. Bone – Bones in which bone-salt plates made from calcium & phosphate ions reinforced with soft collagen considered as a composites system.

4. Wood – A natural composites, in which the lignin matrix is reinforced with cellulose fibres.

Note: In general, a composite consists of two components having two or more distinct phases i.e.

- The matrix which is a continuous phase.
- Reinforcements as the discontinuous or dispersed phase, including fibre and particle etc.
- A fine interphase region known as the interface.

1.3. Classification of Composite Material

We know that composites material usually composed of two or more material. it has two phases i.e. one is matrix phase that is known as primary phase and another is reinforced phase or dispersed phase known as secondary phase. Since the composite material has two phases so therefore, we can broadly classify the composite material in following two manners.

1. Classification of composites material on the basis of matrix phase.
2. Classification of composites material on the basis of reinforcement or dispersed phase.

Above both are briefly categorized as,

1.3.1. Classification of Composites Material on the Basis of Matrix Phase

On the basis of matrix, the composites material has following types.

1. Polymer Matrix Composites (PMC)
2. Metal Matrix Composites (MMC)
3. Ceramic Matrix Composites (CMC)
4. Carbon Matrix Composites (CAMC)

Consider following ray diagram which shows possible categorization of composite material.

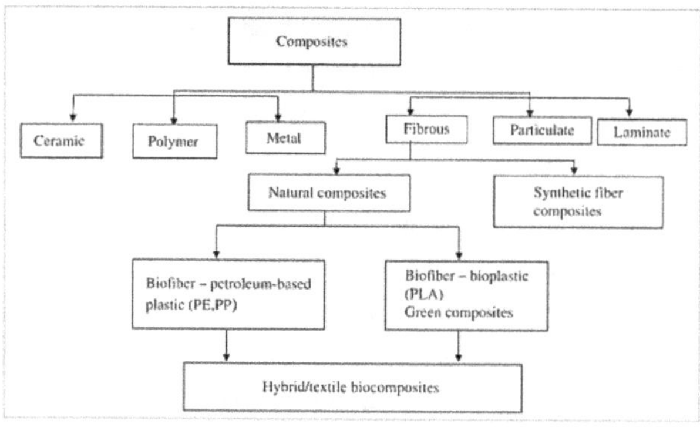

Diagram 1.1: Classification of Composites Material

1.3.2. *Classification of Composites Material on the Basis of Reinforcement or Dispersed Phase*

This type of classification can be performed on the basis of that, which kind of reinforcement is being used to manufacturing a composites material, usually it is achieved by using Particulate, fibre, laminate and hybrid. for broad categorization consider following ray diagram.

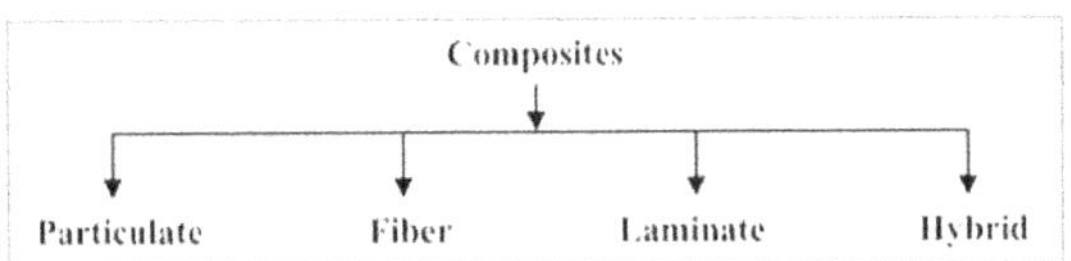

Diagram 1.2: Classification of Composites Material on the Basis of Reinforcement

Furthermore, if we want to deeply categorized the composite on the basis of reinforcing material i.e. Particulate, fibre, laminate and hybrid or structure we have following types.

1. **Particle-reinforced (Large-particle and Dispersion-strengthened):-** In this we will consider size of particle i.e. large particle which will enhance the dispersion strength and consequently take part in augmentation of mechanical properties of composites material.

2. **Fiber-reinforced [Continuous (Aligned) and Short Fibres (Aligned or Random)]:-** In this arrangement large size fibres are arranged & aligned in continuous manner in addition to it short fibres may either be arranged and oriented randomly inside the structure or it may align but in discontinuous short manner.

3. **Structural (laminates and Sandwich Panels):-** Structural reinforcement is depending upon the shape of reinforcement. for instance, reinforcement may either be transformed into laminates or in the form of sandwich panels. Consider the following ray diagram which shows complete information.

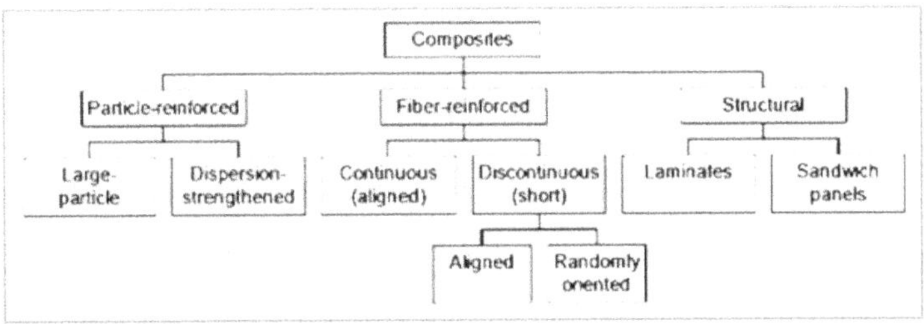

Diagram 1.3: Classification of Composites on the Basis of Basis of Reinforcement

Vividly, it is very clear that composite has two phases i.e. matrix phase & reinforcing phase or dispersed phase.

> **Composite material = Matrix Phase + Dispersed Phase (Reinforcement)**

As per the requirement, we will explain both the phases under here,

1.4.   Matrix phase (Primary Phase)

Usually it has been found that matrix material may be either polymers, metals, ceramics or carbon. Matrix phase or primary phase exhibits continuous nature which is ductile & showing less hardness. matrix phase plays a dominant role & co-operating to holding dispersed phase so that it can handle & share desired load further, from context point of view following phases are important which is being described below.

1.4.1. Polymer Matrix Composites (PMC) or Polymer Matrix Materials

Whenever composites material requires superior dimensional stability at elevated temperature, high resistance to solvent & also demanded resistance to corrosive environment then polymer matrix composites comes on the front line perspective to the context. Dimensional accuracy can be identified by glass transition temperature (T_g) that must be greater than that of maximum operating temperature. on the other hand, resistance to corrosion & solvent usually detected by either not dissolving characteristics, not swelling characteristics, not cracking characteristics or by not affecting characteristics of hot-wet environment. Furthermore, it has been observed that all the emphatic characteristics are included with the polymer matrix composites which always require for certain specific application. To fulfill the desired purpose, polymer composites are manufacture by using matrix phase such as thermosets or thermosetting plastic and finished product or final articles known as polymer matrix composites (PMC). Throughout the manufacturing process, following matrix phases or matrix material would be required.

- Thermo-sets matrix phase including, epoxy, polyester and urethane that is reinforced by thin diameter fibers like – Graphite, Aramids and Boron.
- Thermo-plastic matrix phase includes, vinyl ester, phenolics, polyamide, polypropylene, polyether-ketones in which reinforcement material are often fibers. Except the above poly-carbonates, poly-vinyl chloride, nylon, polystyrenes and embedded glass, carbon, steel or Kevlar fibers (dispersed phase) also considered in the cadre of thermo-plastic.

We know that in the class of polymer composites, two main types of polymers are very famous these are, thermosets and thermo-plastic. Thermosets showing, well bounded three-dimensional molecular structure after curing. The final product become decompose rather than melting. Thermosets are very popular matrix resin and usually employed for structural application and also best suitable for resistance to solvent & corrosion environment rather than thermo-plastic but the basic drawback is that after curing or fabrication it becomes rigid and further could not be reformed. Following types of thermo-setting resin are mostly used as matrix phases. These are epoxies, bismaleimides, thermo-setting polyimides, cyanate ester, benzoxazines, phenolics and so on. Consider the following ray diagram which shows broad classification of thermo-sets resins.

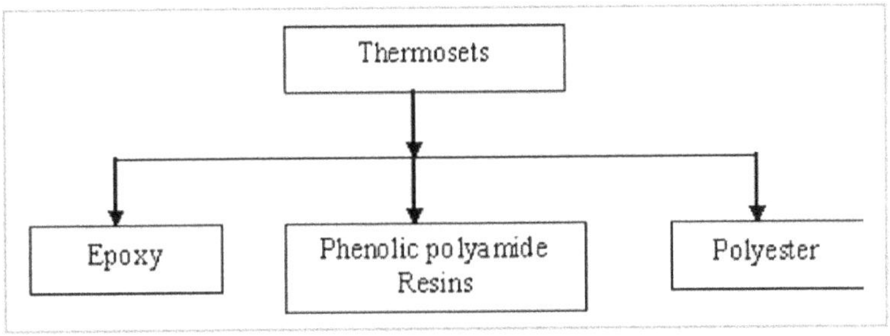

Diagram 1.4: Thermo-sets Resin

The brief description of some most common resin are given below.

- Epoxies are generally employed in areo-space application specially to manufacturing an air frame structure. It has maximum service or operating that is about 120°C. in addition to it, it has excellent structural characteristics and toughened composites showing enhance impact resistance but major drawback is that it indicates brittleness aswellas moisture sensitivity.

- Bismaleimide resins are also used to aero-space applications which indicates high bearing temperature capacity about 200°C. rather than epoxies.

- Thermo-setting polyimides are used to manufacturing a composite where temperature applications fall between the range of 250°C to 400°C. that is usually applied to aero-space applications.

- Cyanate ester does not show moisture sensitivity like epoxies but exhibits 250°C as operating temperature.

- Thermo-setting polyesters are widely used in commercial application because it is showing corrosion resistance and easily production at low cost.

- Vinyl esters matrix phase represent enhanced corrosion resistance in comparison to polyesters but having little expensive. It is widely used in commercial application, automobiles, marine sector, chemical & electrical applications.

- Phenolic resins matrix phases are generally employed in application of air-craft interiors, bulk molding compound and offshore oil platform structure because of due to high temperature resistance, fire resistance. Even though if burning is performed it release less smoke & toxic elements.

Structure of Thermosets (Matrix Phase) with Reinforcement Phase

According to previous literature it has been explicit that in the structure of thermosets polymer, molecules are chemically joined together through cross links as a result of which three dimensional network structure exists that indicate rigid structure to understand it consider following diagram.

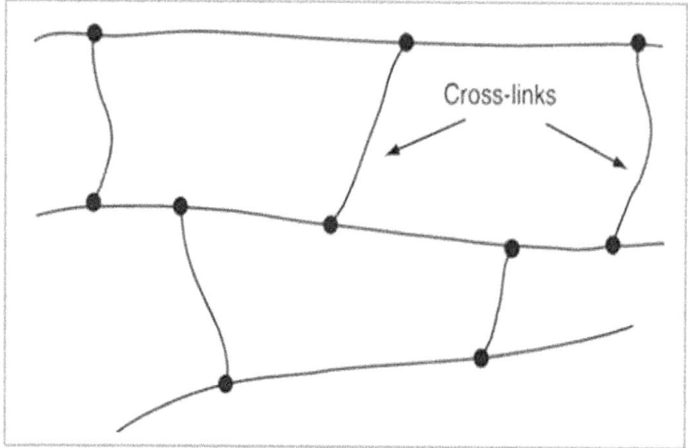

Diagram 1.5: Schematic Diagram of Thermoset Polymer Showing Cross Links

One thing must be kept in mind that during polymerization reaction or curing reaction, we have got cross links network throughout the structure that could not be melted even after the application of heat but it may become soften at elevated temperature if & only if it has a smaller number of cross links.

Note

- When thermosets material passes through the polymerization process / curing reaction or we can say normally curing then it would be transformed into solid from the liquid form. Further, in incurred form, it has small unlinked molecules known as monomers. during the polymerization / curing reaction we will have use following composition.

> **Thermosets (matrix phase) + Cross linkers / Curing agents / Catalysis**
>
> **+**
>
> **Application of heat**

Throughout the reaction, cross links arrangement of molecules will be formed and consequently solidify polymer composites material indicate longer molecular chain & cross link network in the structure.

- Thermosets are liquid resins at room temperature but when processed it will adopt cross linking of molecules so therefore final structure appear as cross link network structure. For instance – Epoxy, polyester, polyimides and phenolics. it has solid rigidity and could not be melted or re-processed.

- Epoxy resins (Matrix phase) which is also widely used to manufacturing an advanced composite. Now a day's it has application in plotting & encapsulating for environmental protection, tooling, adhesive bonding, laminated & filaments wound composites.

- Epoxy resins exhibits, better adhesion to fillers, fibers, corrosion protections, high strength, lower shrinking nature, good electrical & fatigue properties but on the other hand it has also a few unfavorable aspects like – higher costs, long curing temperature and poor / unpleasant appearance.

1.4.2. *Property of Thermosets & Thermo-Plastic (Matrix Phase)*

We have plenty of fascinating application of thermosets & thermo-plastic, if it used as matrix phase then it will showing some certain specific properties like mechanical properties, physical properties, chemical property & thermal property. Consider following table which shows some common & famous property of both the polymer matrix phase composites.

Table 1.1: Mechanical, Physical and Thermal Property of Thermosets & Thermo-plastics
(Polymer Matrix Composites)

Material phases	Density (g/cm^3)	Modulus (GPa)	Tensile strength (MPa)	Elongation up to break in %	Thermal conductivity (W/M-k)	Coff. Of thermal expansion (PPM/K)	Transition temperature (T_g) in ^{0}C	Maximum operating or service temperature in ^{0}C
Thermosets				Thermosets				Thermosets
Epoxy	1.1 to 1.4	3 to 6	35 to 100	1 to 6 %	0.1	60		
DGEBA Epoxy							180	125
TGDDM Epoxy							240 to 260	190
Thermosetting polyester	1.2 to 1.5	2 to 4.5	40 to 90	2	0.2	100 to 200		
Polyimide, polybenzimidazoles, poly-phenyl-auinoxaline								250 to 400
Acetylene – terminated polyimide (ACTP)							320	280
Bismaleimides							230 to 290	232
PMR – 15							340	316
Thermoplastics				Thermoplastics				Thermoplastics
Polypropylene	0.90	1 to 4	25 to 38	> 300	0.2	110		
Nylon (6-6)	1.14	1.4 to 2.8	60 to 75	40 to 80	0.2	90		
Polycarbonates	1.06 to 1.20	2.2 to 2.4	45 to 75	50 to 100	0.2	70		
Poly-sulfone	1.25	2.2	76	50 to 100		56	185	160
Polyetherimide (PEI)	1.27	3.3	110	60		62	217	267
Poly-ami-deimide (PAI)	1.4	4.8	190	17		63	280	230
Polyphenylene-sulfide (PPS)	1.36	3.8	65	4		54	85	240
Polyether-etherketones (PEEK)	1.26 to 1.31	3.6	93	50		47	143	250
K – III Polyimide							250	225
LARC – TPI Polyimide							265	300

Eventually, we can say that thermosets are very flexible and most favorable / suitable for matrix phase. Its advancement can be achieved whenever reinforcement is performed with fibers.

1.5. Thermo-plastic Polymer Matrix Phase

We know that polymer matrix composites have two phases i.e. thermosets & thermo-plastic polymer which work as matrix phase to manufacturing polymer matrix composites during the manufacturing process of polymer matrix composites, we will achieve three components.

1. Matrix phase (May be either thermosets or thermoplastic resins exhibits continuous phase.
2. Dispersed phase or reinforcement indicate discontinuous nature of phase.
3. Last components are a thin interface region which will differentiates / separate both matrix phase & dispersed phase or reinforcement.

All those three elements come on the front line throughout the composites phases these are thermosets & thermo-plastic polymer composites in which thermosets & thermoplastic behave as matrix phase. Whenever high impact strength, fracture resistance, higher strain to failure or superior resistance to matrix cracking are demanded or necessary then thermoplastic matrix phase would be used for manufacturing polymer matrix composites. Except the above property we can also include some other profit with thermo-plastic such as can be processed or recycle, high storage can be obtained at room temperature for long period of time, easily joining & handling etc. In general polymer material (Plastic) is a group of plenty of polymer molecules of similar chemical structure. Thermoplastic are resins usually divided in three category, such as amorphous, crystalline and liquid crystals. The diverse form of amorphous, crystalline and liquid crystals matrix phase of thermoplastic composites is given below.

- Amorphous matrix phase includes, Polycarbonate, acrylonitrile–butadiene–styrene (ABS), polystyrene, poly-sulfone, and polyetherimide.

- Crystalline thermoplastics include, nylon, polyethylene, polyphenylene sulphide, polypropylene, acetal, polyether sulfone, and polyether ether ketone (PEEK).

However thermo-plastic broad classification can be understood by the following ray diagram.

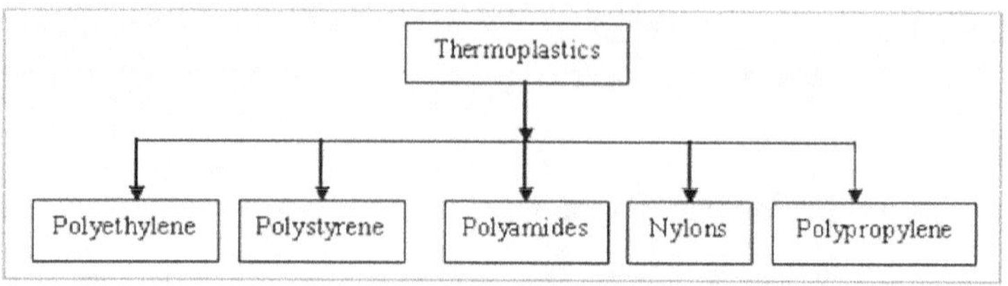

Diagram 1.6: Thermoplastic Resins (Matrix Phase)

In the above, matrix phase usually embedded glass, carbon, steel or Kevlar that work as dispersed phase. Moreover nylon (Matrix phase) used with chopped E-glass fibers reinforcement / dispersed phase. Thermoplastic exhibits one - or two-dimensional structure of molecules at elevated temperature and have exaggerated melting temperature. If we examine the structure of thermoplastic composites then it has been found that individual molecules are joined together either by secondary bonds like Vander walls bonds or hydrogen bonds as a result of which an inter-molecular force is setup amongs the molecules so that molecules may be hold together throughout the structure. Consider the following diagram, which shows a structure of thermoplastic polymer.

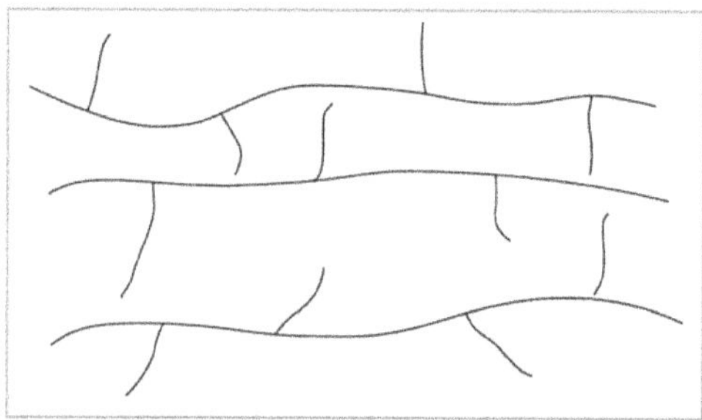

Diagram 1.7: Structure of Thermo-plastic Polymer

1.5.1. Effect of Heat on the Matrix of Thermoplastic Polymer Composites

Whenever heat is applied on thermoplastic then secondary bonds become weaken and it breaks and consequently molecules attain new position or configuration. Even though if this worse solid structure is being cooled then freezing takes place as a result of which secondary bonds again restored in new configuration. hence, we can say that thermoplastic polymer composites can be heated and hence it become soften, melted and then reshaped in desired shape.

Note

- Amorphous thermoplastics exhibits poor solvent resistance rather than crystalline materials.

- Thermoplastic polymer like nylons & polycarbonate represent low creep resistance & low thermal resistance so therefore unsuitable for structural application.

- Thermoplastic may be heated and reform in desired shape after cooling.

- Thermoplastic matrix phases are tougher, low brittleness, best chemical resistance, zero freezing capacity, easily recycle & re-transform into new shape.

1.6.　Metal Matrix Composites (MMCs)

Wherever, remarkable & fascinating properties like, high specific strength, low coefficient of thermal expansion, high strength at high temperature, high elastic properties, higher moisture resistance, high electric & thermal conductivity, better wear / fatigue and flaw resistance, specific stiffness, corrosion resistance, greater strength in shear & compression, non-in flammability, ductility, abrasion resistance, ability to coating, joining, forming and heat treatment, high toughness, yield strength and high transverse strength are demanded for any particular or specific application then name of metal matrix Phase always revived by the researchers, researcher scholars, material engineers and scientists. but except the above valuable quality its application is limited due to their high density & part processing. Metal matrix composites that is acronym as "MMCs" considered in the class of newer engineering material in which reinforcement or dispersed phase a metal matrix phases (second phase) are embedded and distributed evenly throughout the metallic matrix phase (primary phase) so that we can develop required property in composites material. Such type of composites is known as metal matrix composites and respective phase is considered as a metal matrix phase. Metal matrix composites (MMCs) may also be defined as, a compromising materials whose micro-structure has metallic continuous phase (metal matrix phase) into which a second phase (reinforced / dispersed phase) have been artificially introduced so as to achieve required physical, chemical, thermal and mechanical property throughout the structure that is always demanded for some specific application. Usually it has been observed that metal matrix composites composed of a metallic matrix phase like aluminum, magnesium, iron, cobalt, copper in association with dispersed ceramic phase such as oxides and carbides or other metallic phases like, lead, tungsten and molybdenum. Sometimes other metallic matrix phases have also used to fulfillment of desired purpose, these are lead magnesium, cobalt, silver aluminum – lithium and super alloys.

Consider following example of composites in which metal matrix phase is used.

- Composites of aluminium and silicon carbide fibres. In which aluminium work as metal matrix phase on the other hand silicon carbide fibres behave as dispersed or reinforced phase.
- Aluminium & graphite fibres composites.

Metal matrix composites	**= Aluminium**	**+**	**Graphite Phase**
	(Metal Matrix Phase)		(Dispersed phase / Reinforcement)

- Metal matrix composites (MMCs) that has composition of following.

Metal matrix composites	**= Titanium**	**+**	**Silicon carbide**
	(Metal Matrix Phase)		(Dispersed phase / Reinforcement)

Generally, apply for skin stiffeners, beams, frames of hyper sonic aircraft, bicycle frames and sporting goods.

- Metal matrix composites (MMCs) that has composition of following.

Metal matrix composites =	**Aluminium & magnesium**	**+**	**Graphite fibers**
	(Metal Matrix Phase)		(Dispersed phase / Reinforcement)

Usually this composite is used in missiles, satellites and in helicopter structure.

- Metal matrix composites (MMCs) that has composition of following.

Metal matrix composites =	**Lead**	**+**	**Graphite fibers**
	(Metal Matrix Phase)		(Dispersed phase / Reinforcement)

Usually this composite is used to manufacturing storage battery plates.

- Metal matrix composites (MMCs) that has composition of following.

Metal matrix composites =	**Copper**	**+**	**Graphite fibers**
	(Metal Matrix Phase)		(Dispersed phase / Reinforcement)

Usually this composite is used to fabricate electrical contacts and bearings.

- Metal matrix composites (MMCs) that has composition of following.

Metal matrix composites =   Aluminium   +   **Boron fibers**

                           (Metal Matrix Phase)   (Dispersed phase / Reinforcement)

Usually this composite is used to manufacturing compressor blades and structural supports.

- Metal matrix composites (MMCs) that has composition of following.

Metal matrix composites =   **Magnesium**   +   **Boron fibers**

                           (Metal Matrix Phase)   (Dispersed phase / Reinforcement)

Usually this composite is used to making antenna structure.

- Metal matrix composites (MMCs) that has composition of following.

Metal matrix composites =   **Titanium**   +   **Boron fibers**

                           (Metal Matrix Phase)   (Dispersed phase / Reinforcement)

Usually this composite is used to manufacturing jet – engine fan blades.

- Metal matrix composites (MMCs) that has composition of following.

Metal matrix composites = Cobalt base super alloys + Molybdenum / Tungsten fibers

                           (Metal Matrix Phase)   (Dispersed phase / Reinforcement)

Usually this composite is used manufacturing high temperature engine components.

We know that metal matrix composites (MMCs) has various kind of composition in which metal matrix phase play a dominant role among those some are given below.

1. Matrix phase of either aluminium, magnesium or titanium into which secondary phase of fibres like carbon and silicon carbide are embedded.

2. Whenever fibre of silicon carbide is being introduce into the metal matrix phase then we have got increased elastic stiffness, strength of metallic phase on the contrary coefficient of thermal expansion, thermal conductivity & electrical conductivity become reduces.

3. Metal matrix phase of (Aluminium, magnesium, aluminium lithium copper and super alloys) used with fibres material (secondary phase) like aluminium oxides, graphite, silicon carbide, boron tungsten and molybdenum are mostly used.

4. Aluminium matrix phase with boron fibres dispersed phase used to produce MMCs which is being used in space shuttle orbiter application for structural tubular supports.

Note

- Aluminium & titanium are two most important & common metal matrix phase. Now a days nickel & cobalt super alloys, aluminium & its alloys say Al – 201, Al – 6061 & Al – 1100 and titanium & its alloy say (Ti – 6 Al – 9 V) & (Ti – 10 V – 2 Fe – 3 A) are also used as a metal matrix phase for metal matrix composites.

Throughout the manufacturing of metal matrix composites, one of the most difficult tasks is a selection of metal matrix phase & dispersed phase because both are equally responsible and have great impact for alteration of properties of metal matrix composites. Usually it has been found that various kind of oxides are used as dispersed phase or reinforcement in the form of particulates, fibres and whiskers. For example – Aluminium, magnesium and other metallic matrix phases are generally dispersed with oxides of alumina, zirconium and thorium oxides particulates for manufacturing a metal matrix composites (MMCs).

1.6.1. *Property of Metal Matrix Composites*

As it has been previously introduce the name of distinct remarkable property of metal matrix composites and we have clearly understand that metal matrix composites has metal matrix phase as a primary phase. For example, aluminium & its alloys, titanium & its alloys, magnesium alloys, copper-based alloys, nickel based super alloys, stainless steel has been used as matrix phase since long period of time to produce MMCs material. These matrix phase materials can be employed between temperature range of 300°C to 500°C. But these operating temperature range may be varied which is completely depending upon the composition of both i.e. quantity of matrix phase (primary phase) & dispersed phase / reinforcement (secondary phase) and sometimes also coincide to shape & size of composites. MMCs possesses various advantage but its application limits due to lower thermal fatigue, thermo-chemical compatibility and transverse creep resistance. To overcome & maintain such type of property aluminium – based dispersion is to be performed by applying powder metallurgy, direct casting, rolling, forging and extraction and consequently we can have development in desirable property throughout the metal matrix composites. Consider the following table which shows some mechanical properties of metal matrix composites.

Table 1.2: Mechanical Properties of Metal Matrix Composites

Metal matrix composites	Specific gravity	Young modulus (GPa)	Ultimate tensile strength (MPa)	Coefficient of thermal expansion μ / M / $^\circ$C
Aluminium / SiC	2.6	117.2	1206	12.4
Aluminium / Graphite	2.2	124.1	448.2	18
Steel	7.8	206.8	648.1	11.7
Aluminium	2.6	68.95	234.4	23

Metal matrix composites has certain advantages over polymer matrix composites for instance, resistance to environment, temperature resistance, high transverse strength, high modulus, high compressive strength and easily plastically deformed & strengthening may be obtained by applying diverse thermal & mechanical treatments. Consider following example.

1. Two most common metal matrix phases are aluminium & titanium. Both the metals indicate low density & also available in alloys form.

2. Magnesium is light in nature but due to corrosion it become unsuitable for many applications.

3. Beryllium is light in nature which indicate higher tensile modulus than that of steel but could not be used in the form of metal matrix phase due to extreme brittleness.

1.6.2. Application of Metal Matrix Composites

Unique mechanical property, thermal property, physical property and chemical property with respect to polymer matrix composite, enhance the span of application of metal matrix composites. Some most important & common application of metal matrix composites are given below.

1. **Space Application:-** The space shuttle uses boron / aluminium tubes to support its fuselage frame. In addition to decreasing the mass of the space shuttle by more than 145 kg, boron / aluminium also reduced the thermal insulation requirements because of its low thermal conductivity. The mast of the Hubble Telescope uses carbon-reinforced aluminium.

2. **Military Application:-** Precision components of missile guidance systems demand dimensional stability that is, the geometries of the components cannot change during use. Metal matrix composites such as SiC/ Aluminium composites satisfy this requirement because they have high micro yield strength. In addition, the volume fraction of SiC can be varied to have a coefficient of thermal expansion compatible with other parts of the system assembly.

3. **Transportation Application:-** Metal matrix composites are finding use now in automotive engines that are lighter than their metal counterparts. Also, because of their high strength and low weight, metal matrix composites are the material of choice for gas turbine engines.

4. Used to manufacturing gas turbine engine components which requires high strength, low weight or density and high operating temperature.

5. MMCs is usually applied for Aero-plane structure and such a candidate of MMCs is known as structural composites. This structural composite is most favourable / feasible to this application due to light weight or lower weight and exhibits suitability at higher operating temperature.

6. MMCs is also promising material which is used to manufacturing a spare parts of advanced fighter planes such as air fuse Lage structure, nose landing gears, arriving gears, drag brakes and torque tubes because due to advantage of life cycle, cost saving and weight saving.

7. Superior characteristics of MMCs like light weight and high operating temperature is also utilized to manufacturing a components of gas turbine engines like exhaust nozzles lines, vanes blade, cases shaft and rings.

8. One of the most important application of MMCs comes on the front line due to its fabrication from complex shapes to simple shape transformation. For example – Flat panels, net shapes, cylinders, I- Beams, channels and complex shape of gas engine shaft manufacturing.

1.7. Ceramic Matrix Composites (CMCs)

Basically we know that composites material has number of property or characteristics, such kind of inherent property are desirable for many specific application or say general application but in case of ceramic matrix composites (CMCs), researchers mostly emphasis to augment the fracture toughness rather than the strength or other remaining properties. For ceramic matrix composites, composition is given below.

Composition

=

Ceramic matrix phase (Primary phase)

+

Fiber phase of other ceramic material (Secondary phase)

Ceramic matrix composites (CMCs) have ceramic matrix phase such as alumina, calcium-alumino-silicates (CAS), lithium – alumino – silicates (LAS) that reinforced by fiber of carbon or silicon carbide. The basic function of reinforcing ceramic fibers such as silicon carbide or carbon is to enhance the fracture toughness of composition. In addition to fracture toughness, inclusion of fibre reinforced phase also develops mechanical properties and increasing useful life or service temperature of composites material. Usually four categories of ceramic matrices are used to manufacturing ceramic matrix composites. These are:

1.   Glass, boro-silicate and alumino-silicate.
2.   Conventional ceramic materials that includes silicon – carbide, silicon – nitrides, aluminium oxides and zirconium oxides.
3.   Cements.
4.   Concreates carbon composites.

Furthermore, whenever ceramic reinforces with continuous fibres then ceramic matrix composites would be formed which is used to manufacturing structural material. Further addition of whiskers with discontinuous fibers somewhat improve fracture toughness. More over for manufacturing structural ceramic, matrix of ceramic material may be either oxides or non-oxides, in case of oxides ceramics, alumina (Al_2O_3), and mullite ($Al_2O_3 – SiO_2$) is preferred which is mostly used due to their remarkable thermal & chemical stability On contrary in case of non-oxides ceramic silicon carbide (SiC), Silicon nitride (Si_3N_4), boron carbide (B_4C) and aluminium nitrides (AlN) are most common. SiC (Silicon carbide) is used when high modulus & high temperature resistance, Si_3N_4 (Silicon nitride) for high strength and AlN (Aluminium nitride) impart high thermal conductivity throughout the ceramic matrix composites. In addition to ceramic matrix phase there is also requirement of reinforcement or dispersed phase to manufacturing ceramic matrix composites. Usually in order to fulfil the desired purpose we will use fibres of SiC, Carbon, Si_3N_4 and AlN or fibres of other ceramics. Reinforcement of SiC impart thermal stability & compatibility with oxides & non-oxides ceramic matrices. Some other kind of reinforcement are whiskers, platelets, particulates and mon-filaments & multi-filament continuous fibres.

Consider following highlighted points regarding to fibres.

- Silicon carbide (SiC) is a ceramic fibre which is known for high temperature resistance, exhibits melting point temperature about 2830°C and maintain their strength up to 650°C. **[Suitable for reinforcement / Dispersed phase]**.

- Aluminium oxides (Al_2O_3) is also a ceramic fibre apply for high temperature application. It representing melting point temperature about 2045°C and showing excellent strength sustainability up to 1370°C. **[Suitable for reinforcement / Dispersed phase].**

1.7.1. Type of Silicon Carbide Fibers

Usually three type of silicon fibres are used to manufacturing, ceramic matrix composites (CMCs). These are:

1. **Mono-filaments:-** Mono-filaments fibre is produced by depositing a layer of β – SiC over the substrates of carbon mono-filaments that has diameter of 10 to 25 mm. The average diameter of substrate maybe increased up to 140 mm after depositing a layer of β–SiC.

   - For manufacturing of mono-filaments fibres, Consider the following ray diagram.

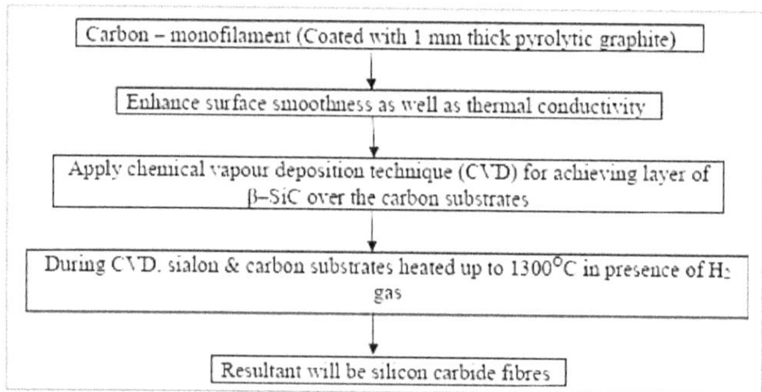

2. **Multi – Filament Yarn:-** Multi-filament yarn has been also used as fibres. The average diameter of yarn is about 14.5 mm, for commercial purpose yarn consisting 500 fibres but the basic drawback of yarn fibre have lower strength rather than mon-filaments.

   - For manufacturing of multi-filaments fibres, Consider the following ray diagram.

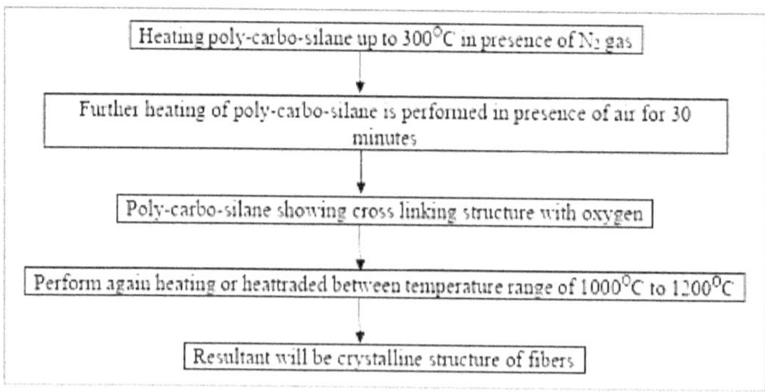

3. Whisker of SiC also used as dispersed phase or reinforcement phase, such a SiC whisker contains following compositions.

SiC (whisker) composition = [10 % weight of SiO_2 + 10 % weight of Si_3N_4]

Manufacturing process, consider following ray diagram.

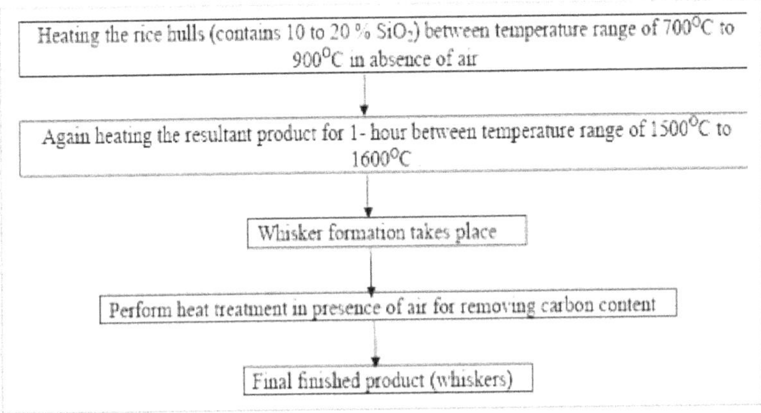

1.7.2. Aluminum Oxides Fiber

In order to fulfill the purpose of fibers, aluminum oxides fibers may also be used as dispersed phase / reinforcement. It has two types and manufactured by sol gel process. These are:

1. **Nextel – 610:-** It contains 99 % of Al_2O_3 which includes single phase structure of $\alpha - Al_2O_3$. It has grain size of about 0.1 mm & average filament diameter is about 14 mm. Indicate high tensile strength at room temperature.

2. **Nextel – 720:-** It contains following composition.

Composition = 85 % (Al_2O_3) + 15 % (SiO_2)

It indicates low tensile strength at room temperature, low creep rate than that of Nextel – 610. Nextel – 720 has capability to retain their tensile strength up to 1400°C that is about 85%. if we consider the structure then we have got that it contains ($\alpha - Al_2O_3$) which embedded in mullite grains. Another ceramic fiber known as "Fiber frax" that contains following.

Composition = Al_2O_3 (50%) + SiO_2 (50%)

It available in short, discontinuous length which has fiber diameter that is about 2 to 12 mm. In addition to it we have also a new fiber that is known as discontinuous alumino-silicate fiber having following composition.

$$\text{Composition} = Al_2O_3\ (95\%) + SiO_2\ (5\%)$$

This is available in diameter which ranges between 1 to 5 mm.

1.8. Property of Ceramic Matrix Composites (CMCs)

The basic & important property of ceramic matrix composites (CMCs) is higher toughness which valued its dignity amongst all other composites materials. Further, rest of the property are high-temperature and corrosion-resistant high stiffness, hardness, resistance to wear, corrosion, oxidation, poor resistance to thermal and mechanical shock, modest tensile stresses, chemical inertness, low density, Low tensile or impact loading (fail catastrophically), high fracture toughness, higher ductility, better transfer of load. Since ceramic matrix composites material recognized for higher toughness then consider the following which shows fracture toughness of some common ceramic matrix composites.

Table 1.3: Fracture Toughness of Some Common Ceramic Matrix Composites

Ceramic matrix composites	Fracture toughness In ($MPa\text{-}m^{1/2}$)
Alumina (1) – zirconia particle (2)	6 to 15
Alumina (1) – Silicon carbide whiskers (2)	5 to 10
Silicon carbide (1) – continuous Silicon carbide fibbers (2)	25 to 30
Alumina (1) – Silicon carbide fibres (2)	27
Silicon carbide (1) – Silicon carbide fibres (2)	30

Remark – (1) – Stand for Matrix Phase & (2) – Stand for Dispersed Phase

Further, also consider the following table which shows the mechanical property of ceramic matrix composites (CMCs).

Table 1.4: Indicate Mechanical Property of Some Common Ceramic Matrix Composites

Ceramic matrix composites	Specific gravity	Young modulus in MSi	Ultimate tensile strength in KSi	Coefficient of thermal expansion in $\mu/M/^oC$
Lithium - Alumino – Silicate (LAS) [1] – SiC (Fibre) [2]	2.1	13	72	2
Calcium - Alumino – Silicate (CAS) [1] – SiC (Fibre) [2]	2.5	17.55	58	2.5

Remark – (1) – Stand for matrix phase & (2) – Stand for dispersed phase

1.8.1. Advantages & Application of Ceramic Matrix Composites

Advantages of CMCs include high strength, hardness, high service temperature limit, chemical inertness, and low density. However, ceramics by themselves have low fracture toughness. Low tensile or impact loading, they fail catastrophically. Reinforcing ceramics with fibers, such as silicon carbide or carbon, increases their fracture toughness because it causes gradual failure of the composite. This combination of a fibres and ceramic matrix makes CMCs more attractive for applications in which high mechanical properties and extreme service temperatures are desired. The advantages of CMC include high strength, hardness, high service temperature limits for ceramics, chemical inertness and low density. Naturally resistant to high temperature, ceramic materials have a tendency to become brittle and to fracture. Ceramic matrix composites are finding increased application in high-temperature areas in which metal and polymer matrix composites cannot be used. Typical applications include cutting tool inserts in oxidizing and high-temperature environments. Other applications include heat exchangers for corrosive and high-temperature environments. Silicon carbide particle-reinforced alumina has been used to make parts for abrasive slurry pumps. C/ SiC brakes are being used in high-end automobiles. Further in lightly loaded structures, concrete reinforced with dispersed asbestos, steel, glass, and carbon fibres allows modest tensile stresses to support is also considered as ceramic matrix composites.

Note

- Throughout the manufacturing of ceramic matrix composites, we will use raw material in the form of powder.

- Aluminium oxide fibres have low thermal & electrical conductivity & high coefficient of thermal expansion rather than silicon carbide fibres.

- Whisker has length of 50 mm & diameter falls between range of 0.1 to 1 mm.

1.9. Carbon Matrix Composites (CAMCs)

The carbon matrix composites consist of carbon fibres in carbon matrix and it is usually known as carbon–carbon (C–C) composites. It gives superior service up to $3100^\circ C$ in absence of oxidation environment because they exhibit oxidation in presence of oxygen and start to degrade. The protection of C–C composites against oxidation at high temperatures can be obtained by providing either an external or internal coating. Both oxide coatings, such as SiO_2 and Al_2O_3, and non-oxide coatings, such as SiC, Si_3N_4, and HFB_2, have been used. It has been found that after coating, maximum service of C–C composites can be attained up to 1700°C. Fibres in the C–C composites can be either continuous or discontinuous. Continuous fibres are selected for structural applications. Further graphite may also use for high temperature applications in which C–C composites. The microstructure of carbon matrix in C–C composites may vary from small, randomly oriented crystallites of turbo-stratic carbon to large, highly oriented, and graphitized crystallites. The carbon matrix may also contain large amount of porosity, which causes the density to be lower. Carbon – carbon composites are available in various form among those, C – C with ID – unidirectional continuous, C – C with 2D fabric, C – C with 3D woven orthogonal fibres etc.

1.9.1. Advantage of Carbon – Carbon Composites

Carbon is brittle in nature and flaw sensitive like ceramics. Reinforcement of a carbon matrix allows the composite to fail gradually and also gives advantages such as ability to withstand high temperatures, low creep at high temperatures, low density, good tensile and compressive strengths, good thermal shock resistance, abrasion resistance, fracture toughness, excellent high temperature durability in inert or vacuum environment, good corrosion resistance, high fatigue resistance, high thermal conductivity, and high coefficient of friction. Drawbacks include high cost, low shear strength, and susceptibility to oxidations at high temperatures. Consider following table which shows mechanical property of C – C composites.

Table 1.5: Mechanical Property of C – C Composites

Composites material	Specific gravity	Young modulus (GPa)	Ultimate strength (MPa)	Coefficient of thermal expansion in μm / M / °C
C – C Composites	1.68	13.5	35.7	2.0

1.9.2. Applications of Carbon–Carbon Composites

Following are some important application of carbon–carbon composites material. These are:

1. **Space Shuttle Nose Cones:-** As the shuttle enters Earth's atmosphere, temperatures as high as 1700°C are experienced. Carbon–carbon composite is a material of choice for the nose cone because it has the lowest overall weight material, high thermal conductivity to prevent surface cracking, high specific heat to absorb large heat flux and high thermal shock resistance to low temperatures in space of –150°C to 1700°C and hence the carbon–carbon nose remains undamaged and can be reused many times.

2. **Aircraft Brakes:-** The carbon–carbon brakes are used in commercial aircraft such as Airbus due its high durability, high specific heat, low braking distances, low braking time and large weight savings. Air brakes made from carbon – carbon composites material exhibits 2 to 4 times more durability with respect to steel as well as 2 to 5 times specific heat than that of steel. Further it has been also noticed that fuel saving identified about 1360 L / year due to weight saving of full service commercial air craft.

3. **Mechanical Fasteners:-** Fasteners needed for high temperature applications are made from carbon–carbon composites because it loose their strength slightly at high temperatures.

1.10. Function of Matrix Phase

We know that composite material composed of matrix phase & reinforced / dispersed phase. Both the phase has its own signification and play a dominating role in the composites. The basic function of matrix phase is given below.

1. To bind together the reinforcement Particles or particulates / fibers / flakes by virtue of its cohesive and adhesive characteristics.

2. To protect reinforcement / dispersed phase from environments and handling.

3. To maintain the desired level of orientation and spacing of dispersed phase say Particles or particulates / fibers / flakes throughout the matrix.

4. To transfer stresses equally to Particles or particulates / fibers / flakes by adhesion and / or friction across the matrix & dispersed phase interface when the composite is under load, and consequently we can prevent propagation of cracks and subsequent failure of the composites.

5. To sustain compatibility with chemically and thermally atmosphere.

6. To be adopt easiness in manufacturing or fabrication.

1.11. Classification of Composites Material on the Basis of Reinforcement

Under this classification, we will concentrate upon dispersed phase or reinforcement rather than matrix phase. Usually reinforcing material either used in the form of particles (Particulates), fibres, laminate or hybrid with the matrix phase. Basically, the function of reinforcing phase or dispersed phase is not only to impart strength and stiffness but also develop resistance to heat resistance to conduction, resistance to corrosion and rigidity. Usually, the reinforcement is harder, stronger, and stiffer than the matrix. Reinforcements for any kind of composites can be performed either by fibers, fabrics particles or whiskers. Fibres are generally attributed by one very long axis with other two axes either may be circular or approximately circular. Particles have no preferred orientation therefore it adopts the position according to the shape of composites. Whiskers have a preferred shape and exists or available in small diameter and length.

Further broad classification can be easily assimilated by the following diagram.

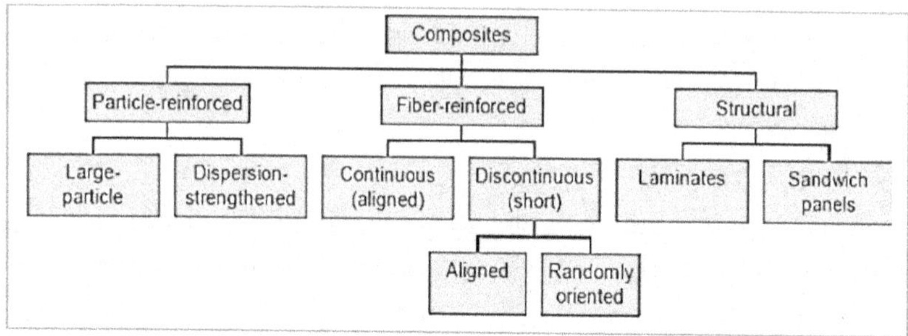

Diagram 1.8: Classification of Composites on the Basis of Reinforcement

Particles / Particulates dispersed phase may be of two types, These are:

1. Composite having random orientation of particles
2. Composite having preferred or specific orientation of particles

A composite that is generated from particulate in conjunction with matrix phase known as particulate composites. The composition of such kind of composites is given below.

Composition	=	Matrix phase	+	Dispersed phase / Reinforced phase
				in the form of particulates
		[Primary phase]		**[Secondary phase]**

In 1st case we will have matrix phase, in which particles or particulates showing random orientation and work as dispersed phase.

In 2nd type we will have matrix phase and dispersed phase that contains two dimensional flat platelets (flaks) which arrange parallel to each other in overlap form.

1.12. Fiber Reinforcing Material

In this case we will use either matrix phase or fiber phase as a Reinforcing material / Dispersed phase, and such kind of composites material known as fibrous composites.

$$\text{Composition} \quad = \quad \text{Matrix phase} \quad + \quad \text{Fibers (Dispersed phase)}$$
$$[\text{Primary phase}] \qquad\qquad\qquad [\text{Secondary phase}]$$

Reinforcing of the fiber may be performed in following manners.

1. **Short Fiber Reinforcing Composites:-** In this case reinforcement is performed by discontinuous fibers (Short length fibers). Composition is given below.

$$\text{Composition} = \quad \text{Matrix phase} \quad + \quad \text{Short Fibers (Dispersed phase) [discontinuous fibers]}$$
$$[\text{Primary phase}] \qquad\qquad\qquad [\text{Secondary phase}]$$

    a)   Composites having random orientation of fibers.

    b)   Composites having preferred orientation of fibers.

2. **Long – fibers Reinforced Composites:-** In this case we will have matrix phase and fiber used as a reinforcement / dispersed phase which would be a long fibers or continuous fibers and the respective composites known as long – fibers reinforced composites. In this category we have two types.

    a)   Uni-directional orientation of fibers.

    b)   Bi – directional orientation of fibers (woven).

3. **Laminate Composites:-** Whenever composites, dispersed phase consists of several layer in different fibers orientation then such kind of composites known as multi-layer composites in which a whole mass arranged in lamina.

A possible explanation of all those reinforcement or dispersed phase are given.

1.13. Reinforcement by Particles or Particulates

We know that composites material have two phases i.e. matrix phase & dispersed phase. Particulates composites may be defined as a composite in which reinforcement is performed by using particles or particulates in such a manner that all the particles evenly distributed throughout the structure in all directions. These kinds of composites are known as particulates reinforced composites. The basic function of particulate dispersed phase is to imparts high temperature resistance, wear resistance, reduce friction & shrinkage, enhance stiffness, sharing a load equally with matrix phase, improve strength and oxidation resistance. Basically, in particulate composites, particles usually immersed in matrix of either alloys or ceramic or polymer. For example – aluminum particle in rubber, silicon carbide particles in aluminum and a composition of gravel, sand and cement known as concrete. Consider the following diagram which shows particulate reinforcement throughout the composites structure.

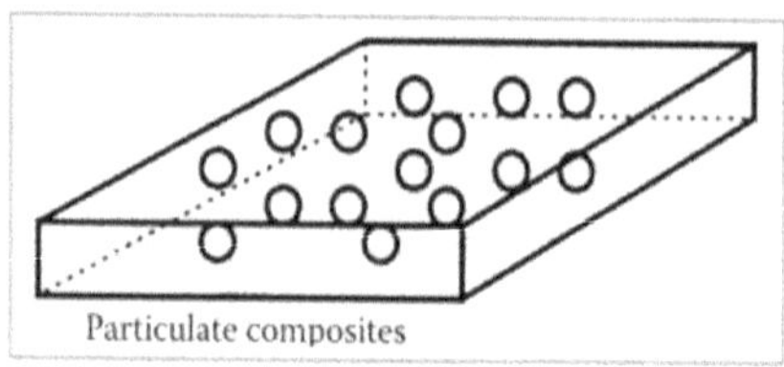

Diagram 1.9: Composites Material Reinforced by Particulates or Particles

Moreover, particulate composites have some certain specific dimensions that always equal in all directions. Particulate composites are either available in spherical, platelets, or any other regular or say in irregular geometry. Particulate composites are weak and less stiff in nature rather than continuous fibre composites.

1.13.1. Shape and Orientation of Dispersed Phase Inclusions

Particles have no preferred or specific directions but they improve properties at lower cost of isotropic materials. The shape of the reinforcing particles can be spherical, cubic, platelet, or regular or irregular geometry. Particulate reinforcements are equally or evenly distributed throughout the composite's material. Particle-reinforced composites has two types, known as sub class of particle-reinforced composites. These are large particle and dispersion-strengthened composites.

1.14. Particle-Reinforced Metal Matrix Composites

Whenever metallic particles are reinforced with metallic matrix then we have metal particle-reinforced metal matrix composites. They play a dominant role in the cadre of engineering material, in this class following are important.

1. Tungsten carbide particles embedded in a cobalt matrix which is used extensively in cutting tools and dies. This composite is known as a cermet, that has superior fracture toughness, high hardness and exhibits brittle nature.

2. When diamond particles embedded in a cobalt matrix, then polycrystalline diamond is formed which is used in rock drill bits and cutting tools.

3. Tungsten carbide particle-reinforced with silver is also form a composite considered as a commercial metal matrix composite which is being used as a circuit breaker contact pad material. Such a composites indicates electrical conductivity, high hardness and wear resistance than that of monolithic silver.

4. Ferrous alloys reinforced with titanium carbide particles. It indicates high wear resistance, stiffness and lower density.

1.14.1. *Titanium Carbide Particle-Reinforced with Steel*

A composite consisting of austenitic stainless steel, which reinforced with 45% by volume of titanium carbide particles then we have titanium carbide particle-reinforced steel composites. The modulus of this composite is about 304 GPa. The specific stiffness of the composite is almost double than that of the unreinforced metal in addition to it, lower density and higher wear resistance also achieved.

1.14.2. *Silicon Carbide Particle-Reinforced with Aluminium*

When aluminium metal reinforced with silicon carbide particles then we a metal matrix composites. Properties of such a composites material depending up on the type of particle, particle volume fraction, matrix alloy, and on manufacturing process. Usually it has been found that, as particle volume fraction increases, modulus and yield strength increase and fracture toughness and tensile ultimate strain decrease. Particle reinforcement also responsible for enhancement of little elevated-temperature strength properties and fatigue resistance. Consider the following table which represent mechanical properties of aluminium, titanium, and steel alloys when reinforcement is performed with silicon carbide particles.

Table 1.6: Mechanical Properties of Silicon - Carbide Particles - Reinforced Aluminium, 6061 - T6 Aluminium, 61 - 4V Titanium and 4340 Steel

	Aluminum	Titanium	Steel	Silicon Carbide Particle-Reinforced Aluminum		
	(6061-T6)	(6Al–4V)	(4340)	Particle Volume Fraction		
				25	55	70
Modulus, GPa	69	113	200	114	186	265
Tensile yield strength, MPa	275	1000	1480	400	495	225
Tensile ultimate strength, MPa	310	1100	1790	485	530	225
Elongation (%)	15	5	10	3.8	0.6	0.1
Density, g/cm³	2.77	4.43	7.76	2.88	2.96	3.00
Specific modulus, GPa	25	26	26	40	63	88

Note

- Silicon carbide particle-reinforced aluminium composites are widely used in thermal management and packaging of electronics and photonics.
- Alumina Particle-Reinforced Aluminium is also a composite.
- Silicon carbide particle-reinforced aluminium has been used in aerospace, commercial, automobile engine cylinder liners for wear resistance and optical structural applications.
- Boron carbide particle-reinforced aluminium has been used for spent nuclear fuel containers.

1.15. Fibre Reinforced Composites

As we have seen that in case of particulates / particles dispersed phase, inclusion of particulate with the matrix phase cause of development of diverse characteristics that is attributed as high temperature resistance, wear resistance, stiffness, strength and oxidation resistance & so on in the composites materials. But whenever fibres are reinforced or it involves in the form of dispersed phase in matrix phase, we can augment the mechanical property, physical property as well as chemical property etc. For example – we can attain, high strength, high flexibility, high ductility, high toughness, modulus, high temperature resistance, high impact resistance, chemical resistance, develop electrical insulation characteristic, creep & fatigue resistance, low density, tensile strength, corrosion resistance, low & high thermal conductivities, axial stiffness, tensile & compressive strength, manufacturing a low & high density composites, high melting point temperature composites, high transverse strength, high wear resistance, high strength & high fatigue strength at elevated temperature, to produce high structural composites, high specific strength & high specific stiffness, high green strength, high dimension stability, lower shrinkage, [Resistance to solvent, swelling, abrasion, tear & creep], Improve rheological properties, low cost creation, high electrical resistance [E - glass], high strength [S - glass], low weight or low density, electrical conductivity [Carbon composites], Nylon fibres [High durability, excellent physical property, very high strength, high elasticity, abrasion resistance, resistance to oil & chemical, high resilience, light weight, high temperature performance etc.], Green composites [Fibre + Bio-degradable resins] Eco-friendly characteristics, SiC / SiC composites [Excellent thermal shock resistance], Improve thermal & mechanical shock resistance, high fracture resistance, high temperature stability, high thermal conductivity & high temperature erosion resistance [C – C composites], high structural efficiency, high specific strength & high specific stiffness [Alumina fibre – reinforced with aluminium], low coefficient of friction & high wear resistance [Ceramic- carbon fibres], high fracture toughness [For discontinuous fibres & whiskers], high specific strength & modulus [Fibres reinforced with polymer composites], erosion & corrosion resistance [PMCs], enhance load transfer capacity, low compressive strength [E - glass], high compressive strength [Boron – fibres], polymer composites reinforced with glass fibres [High tensile strength, high chemical resistance, excellent insulating property, high hardness, low fatigue, abrasion sensitivity & low cost], high corrosion resistance to acid [C - glass] and enhancing many more property of composites materials.

1.15.1. *Fibre Reinforced or Fibre Dispersed Phase*

Initially let be define the term "FIBER" Fibre is defined as a single unit of any matter that has inherent characteristics which is attributed by flexibility, fineness and aspect ratio. Furthermore, aspect ratio is defined as the ratio between length of fibre to diameter of fibre. It has been found that governing property of any composite that is reinforced by fibre always depending upon the aspect ratio. Hence,

$$\text{Aspect ratio} = [\,(\text{Length of fibre}) / (\text{Diameter of fibre})] = [L / D]$$

Aspect Ratio is Dimension-less Parameter

The basic significance of aspect ratio is that it plays a dominant role to sustain the orientation of fibres. For instance, continuous fibre has high aspect ratio as a result of which such kind of fibre responsible for imparting high strength & stiffness to matrix phase eventually we have fibres composite of high strength & high stiffness with other desirable property on the contrary discontinuous fibre has same property but their value become lowers. The property of composites always coincides to orientation of fibre throughout the composites. For example, continuous fibres have specific or preferred orientation of fibre while discontinuous fibres showing random orientation of fibres. Usually fibre characterized by filament like structure of that has length about 100 µm and aspect ratio greater than 10. Fibre are classified into two group these are natural fibres & manmade fibres. Consider the following diagram which shows both the types.

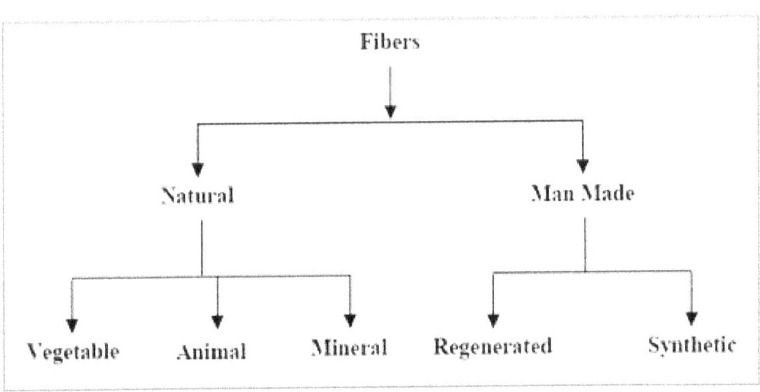

Diagram 1.10: Broad Classification of Fibres

For any kind of composites which is being manufactured synthetically by human being by using fibre reinforcement with the matrix phase, apply for diverse field of application which has distinct characteristics, known as fibre reinforced composites. From the context point of view, we will elaborate here only man-made fibre composites rather than natural, consider following diagram which shows broad classification of man-made composites.

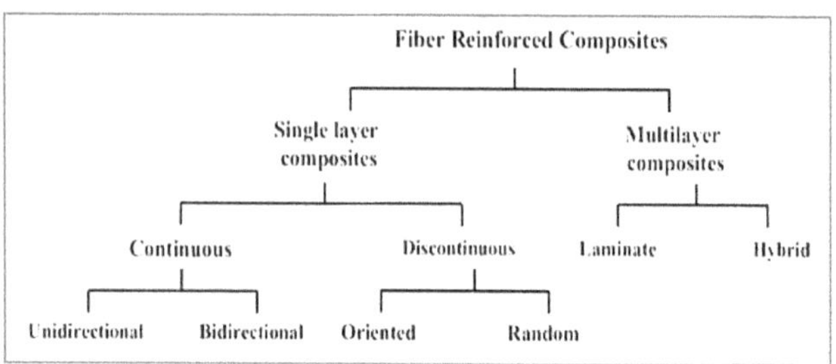

Diagram 1.11: Classification of Fiber Reinforced Composites

A prominent function of reinforcing fibre is to enhance the strength and stiffness of a matrix material. Major constituents in a fibre-reinforced composite material are the reinforcing fibres and a matrix, which acts as a binder for the fibres. Other constituents that may also be found are coupling agents, coatings, and fillers. Consider the following fibres that take parts in manufacturing of most fibre composites material. These are synthetic fibre Nylon 66, Nylon 6 and polyester, PPTA (Para-phenylene terephthal-amide), gel spun poly ethylene, Kevlar fibres, carbon fibres, yarn, For reinforcing polymer matrix composites, carbon fibres (AS4, IM7, etc.), glass fibre (E-glass, S-glass, C-glass etc.), aramid fibres (Kevlar and Twaron), and boron fibres are generally used. Consider the following table which shows the characteristics & application of some common glass fibres for manufacturing of composite materials are given below.

Table 1.7: Glass Fibres Grades & their Applications

Glass fibre grades	Characteristics & application
Grade A	High alkali grade glass, originally made from window glass
Grade C	Chemical-resistant grade glass for acid environments or corrosion
Grade D	Low dielectric grade glass, good transparency to radar (quartz glass)
Grade E	Electrical insulation grade; this is the most common reinforcement grade
Grade M	High modulus grade glass
Grade R	Reinforcement grade glass; this is the European equivalent of S-glass
Grade S	High strength grade glass has higher Young's modulus and temperature resistance than E-glass

Cotton fibre, silk fibre, rayon fibre, jute and Nylon, short carbon fibre, short Kevlar fibre, cellulosic fibre-reinforced green composites, Short Nylon fibre reinforced polypropylene, short glass fibre, short coir fibre reinforced natural rubber, cellulose fibre, Short jute fibre reinforced NR composites, short jute fibre reinforced carboxylates nitrile rubber [66], Nylon 6 fibre nitrile rubber composites, Natural rubber coir fibre composite, short Nylon fibre-natural rubber composites, short Kevlar aramid fibre-thermoplastic PU composite, short Kevlar fibre, Carbon fibers (graphite fiber), basalt fibers, HS glass fibre, Carbon fibers are made from three precursor materials: PAN, petroleum pitch, and coal tar pitch, PAN-based fibers, pitch-based carbon fibers, Boron fibers, Silicon-carbide-based fibers, silicon carbide fibre (Monofilament), Multifilament silicon-carbide-based fibers, Alumina-based fibers, aramid fibers (Kevlar - 49 and Twaron), High-density polyethylene fibers, Basalt fibers, SiC fiber-reinforced titanium, alumina–silica fibres, alumina fiber-reinforced aluminium, polyester fiber, aluminium oxide (Al_2O_3) fibres (Nextel 610 and Nextel 720), Silicon carbide fibers (Monofilaments & Multifilament yarn, SC- Whiskers), Fiber-frax (50 % Al_2O_3 and 50 % silica (SiO_2) and many more. Fibers are the principal constituents of fiber-reinforced composite. They have largest volume fraction in a composite laminate and share the major portion of the load acting on a composite structure. Inherent property of fiber-reinforced composite always coincide to type of fiber, fiber volume fraction, fiber length, and fiber orientation throughout the composites structure. These fiber used in the form of filament (small diameter) which facilitate easily handling they are accumulated in the form of bunch called bundle of fibers, which is a collection of large number of continuous filaments, either in untwisted or twisted form. The untwisted form is called strand and twisted form is called yarn.

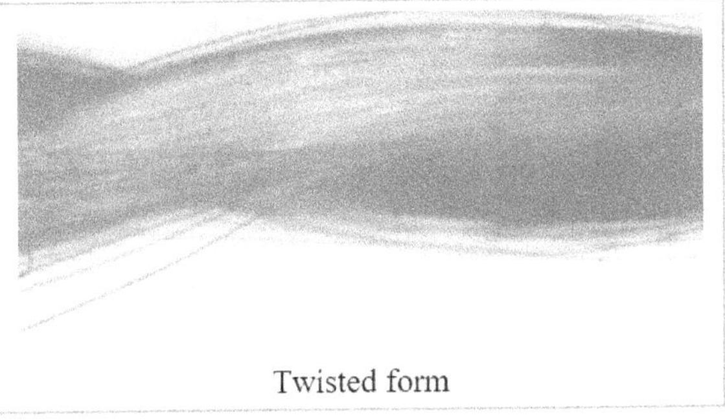

Twisted form

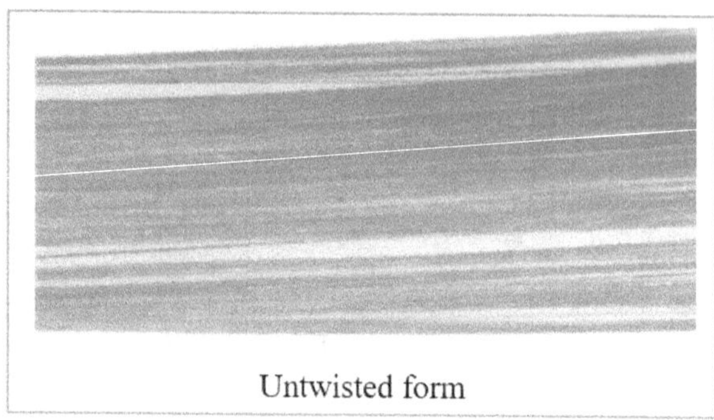

Except the Twisted form & Untwisted form, fibers are available in other form, consider the following diagram which shows various form of fibers.

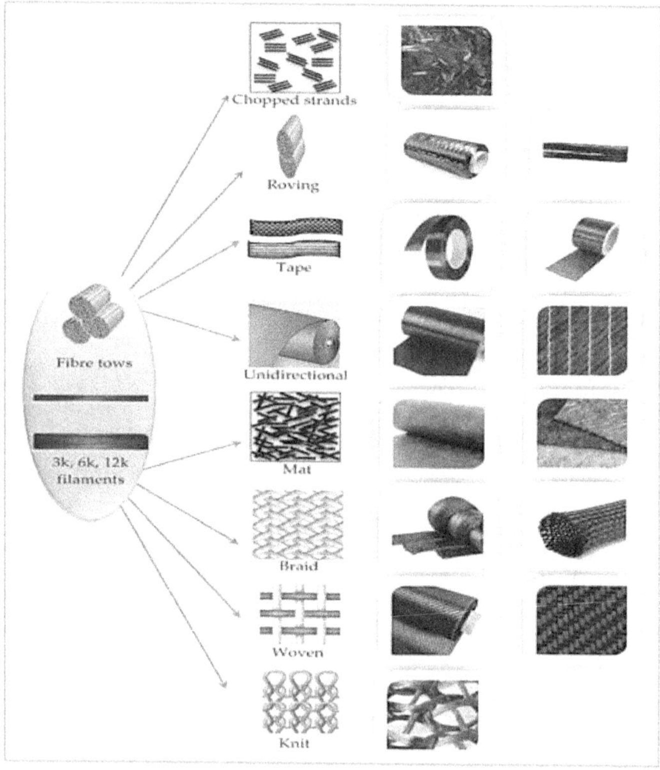

Diagram 1.12: Different Type of Fibers

Moreover, consider the following table which shows some most common commercial fibre that has been used since long period of time.

Table 1.8: Property of Some Common Commercial Reinforced Fibre

Fiber	Diameter In µm	Density g/cm^3	Tensile modulus In GPa (Msi)	Tensile modulus In GPa (Ksi)	Strain to failure In %	Coefficient of thermal expansion [10^{-6}/°C]	Poisson ratio
Glass							
E-glass	10 (round)	2.54	72.4 (10.5)	3.45 (500)	4.8	5	0.2
S-glass	10 (round)	2.49	86.9 (12.6)	4.30 (625)	5.0	2.9	0.22
PAN carbon							
T-300[c]	7 (round)	1.76	231 (33.5)	3.65 (530)	1.4	−0.6 (longitudinal) 7–12 (radial)	0.2
AS-1[d]	8 (round)	1.80	228 (33)	3.10 (450)	1.32		
AS-4[d]	7 (round)	1.80	248 (36)	4.07 (590)	1.65		
T-40[c]	5.1 (round)	1.81	290 (42)	5.65 (820)	1.8	−0.75 (longitudinal)	
IM-7[d]	5 (round)	1.78	301 (43.6)	5.31 (770)	1.81		
HMS-4[d]	8 (round)	1.80	345 (50)	2.48 (360)	0.7		
GY-70[e]	8.4 (bilobal)	1.96	483 (70)	1.52 (220)	0.38		
Pitch carbon							
P-55[c]	10	2.0	380 (55)	1.90 (275)	0.5	−1.3 (longitudinal)	
P-100[e]	10	2.15	758 (110)	2.41 (350)	0.32	−1.45 (longitudinal)	
Kevlar 49[f]	11.9 (round)	1.45	131 (19)	3.62 (525)	2.8	−2 (longitudinal) 59 (radial)	0.35
Kevlar 149[f]		1.47	179 (26)	3.45 (500)	1.9		
Technora[g]		1.39	70 (10.1)	3.0 (435)	4.6	−6 (longitudinal)	
Extended chain polyethylene							
Spectra 900[h]	38	0.97	117 (17)	2.59 (375)	3.5		
Spectra 1000[h]	27	0.97	172 (25)	3.0 (435)	2.7		
Boron	140 (round)	2.7	393 (57)	3.1 (450)	0.79	5	0.2
SiC							
Monofilament	140 (round)	3.08	400 (58)	3.44 (499)	0.86	1.5	
Nicalon[i] (multifilament)	14.5 (round)	2.55	196 (28.4)	2.75 (399)	1.4		
Al$_2$O$_3$							
Nextel 610[j]	10–12 (round)	3.9	380 (55)	3.1 (450)		8	
Nextel 720[j]	10–12	3.4	260 (38)	2.1 (300)		6	
Al$_2$O$_3$-SiO$_2$							
Fiberfrax (discontinuous)	2–12	2.73	103 (15)	1.03–1.72 (150–250)			

[a] 1 µm = 0.0000393 in.
[b] m/m per °C = 0.556 in./in. per °F.
[c] Amoco.
[d] Hercules.
[e] BASF.
[f] DuPont.
[g] Teijin.
[h] Honeywell.
[i] Nippon carbon.
[j] 3-M.

1.15.2. Fibrous Composites

Fiber composites may be further classified as, Long-fiber reinforced composites & Short-fiber reinforced composites.

1. **Long-fiber Reinforced Composites:-** Long-fiber reinforced composites consist of a matrix reinforced by a dispersed phase in form of continuous fibers.

    *   Unidirectional orientation of fibers.

    *   Bidirectional orientation of fibers (woven).

2. **Short Fiber Reinforced Composites:** Short-fiber reinforced composites consist of a matrix reinforced by a dispersed phase in form of discontinuous fibers. They are classified as,

    *   Composites with random orientation of fibers.

    *   Composites with preferred orientation of fibers.

1.15.3. Long-fiber Reinforced Composites

Long-fiber reinforced composites sometimes referred as continuous fiber in which fiber has a length greater that than that of its diameter. The length-to-diameter (l/d) ratio is called *aspect ratio*. Continuous fibers have long aspect ratios and consequently continuous-fiber composites normally have a preferred orientation, continuous fibre reinforcements include unidirectional, woven cloth, and helical winding like structure. Continuous-fiber composites usually fabricated into laminates form by stacking single sheets of continuous fibres in different orientations so that desired level of strength and stiffness can be sustained. In composites material continuous phase is considered as a matrix phase that may be either polymer, metal, or ceramic. Polymers have low strength and stiffness, metals have intermediate strength and stiffness but high ductility, and ceramics have high strength and stiffness but are brittle. The basic function of matrix phase (continuous phase) is to maintaining the position of fibers in the proper orientation and spacing so that protection of fibre-composites can be attained from abrasion and the environment. Furthermore, a composite, reinforcing the long fiber in a single layer in which fibres are oriented in one direction or fiber aligned to each other so that we can achieved enhanced properties like strength & stiffness are called continuous fiber reinforcement composites. These long & continuous, aligned fibers are mostly used in high performance applications. Continuous fiber – reinforced composites fibre may be oriented either in uni-directional, bi-direction or, filament (helical direction). Consider following diagram.

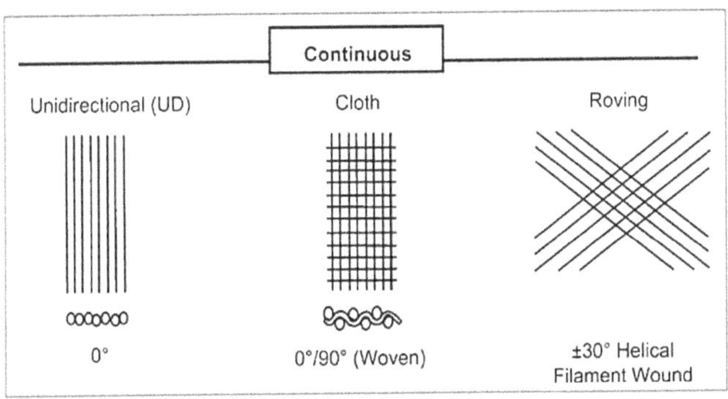

Diagram 1.13: Orientation of Fibre in Composites Material (Continuous Fibre)

Continuous fiber reinforced composites are employed for high performance structural applications. In carbon fiber reinforced composites, specific strength and specific stiffness of composites are superior than that of conventional metallic alloys. But structural properties always depending upon how fibres are oriented within the matrix during the fabricated of composites. However, carbon matrices are weak, brittle and has low-stiffness materials. As a result, properties in transverse and elastic moduli and strength throughout the thickness of unidirectional carbon–carbon composites remain low. Consider following table which shows a mechanical property of continuous carbon fibre reinforced with carbon matrix.

Table 1.9: Mechanical Property of Most Common C – C Carbon Composites

Material	RCC	ACC-4
Fiber	WCA (rayon)	T300 (PAN)
Fabric	Plain weave	Eight-harness satin
Density, g/cc	1.4–1.7	1.8
In-Plane		
Tensile modulus, GPa	14	110
Compression modulus, GPa	17	110
Tensile strength, MPa	120	330
Compression strength, MPa	88	180
Shear strength, MPa	—	6.9
Through Thickness		
Tensile strength, MPa	5.5	3.4

RCC - Stand for Reinforced C - C Carbon Composites (Reusable)

ACC - Stand for Advanced C - C Carbon Composites

Carbon fibers are commercially available in three basic forms. These are long & continuous tow, chopped (6–50 mm long), and milled (30–3000 mm long). The long and continuous tow are generally composed of untwisted bundle of 1,000–160,000 parallel filaments which is used for high-performance applications. Further Carbon fiber tows can also be weaved into two-dimensional fabrics of various styles. Hybrid fabrics consisting co-weaved carbon and other fibres for instance E-glass, Kevlar, PEEK, PPS, and so on. The modulus of carbon fibres coincides to the accuracy & correct perfection of the alignment. In class of high modulus fibres, we can consider those composites whose heat-treatment cab be performed up to 1650°C, which possess three-dimensional atomic arrangement & containing about carbon contents above 99% and have a tensile modulus about 350 GPa. However, Imperfections in alignment results in complex shaped voids that would be elongated parallel to the fibre axis, which act as stress raisers and points of weakness. In case of ceramic matrix composites attention is to be drawn to increase the toughness rather than the strength and stiffness further addition of continuous fibers to ceramics is not only increase the fracture toughness but also inhibit the changes of failure up to greater extent of ceramics. Consider the following table which shows fracture toughness of structure alloys, monolithic ceramic & ceramic matrix composites.

Table 1.10: Fracture Toughness of Structure Alloys, Monolithic Ceramic & Ceramic Matrix Composites

Matrix	Reinforcement	Fracture Toughness MPa·m^{1}
Aluminum	None	30–45
Steel	None	40–65[a]
Alumina	None	3–5
Silicon carbide	None	3–4
Alumina	Zirconia particles[b]	6–15
Alumina	Silicon carbide whiskers	5–10
Silicon carbide	Continuous silicon carbide fibers	25–30

Continuous fibers may also be fabricate with the metal matrix phase for enhancing the mechanical property of metal matrix composites. Consider the following table in which continuous fiber is reinforced with metal matrix phase (7075–T6 aluminium & Ti-6A-4V). Table - 1.11, Table - 1.12, Table - 1.13 & Table - 1.14, Table - 1.15, which showing elastic properties, specific axial modulus, strength, specific strength properties and physical properties respectively.

Table 1.11: Elastic Property of Unidirectional Metal Matrix Composites (7075-T6 Aluminium & Ti-6Al-4V Titanium)

Fiber	Matrix	Density g/cm^3	Axial Modulus GPa	Transverse Modulus GPa
—	7075-T6	2.8	70	70
—	Ti–6Al–4V	4.4	114	114
Boron	Al	2.6	210	140
Alumina	Al	3.2	240	130
SiC	Ti	3.6	260	170

Table 1.12: Specific Axial Modulus of Unidirectional Metal Matrix Composites (7075-T6 Aluminium & Ti-6Al-4V Titanium)

Fiber	Matrix	Specific Axial Modulus
—	7075-T6	26
—	Ti–6Al–4V	26
Boron	Al	81
Alumina	Al	75
SiC	Ti	72

Table 1.13: Strength property of Unidirectional Metal Matrix Composites (7075-T6 Aluminium & Ti-6Al-4V Titanium)

Fiber	Matrix	Density g/cm^3	Axial Tension MPa	Transverse Tension MPa	Axial Compression MPa
—	7075-T6	2.8	551[a]	551[a]	517[a]
—	Ti–6Al–4V	4.4	950[a]	950[a]	880[a]
Boron	Al	2.6	1240	140	1720
Alumina	Al	3.2	1700	120	1800
SiC	Ti	3.6	1700	340	2760

Table 1.14: Specific Strength Property of Unidirectional Metal Matrix Composites (7075-T6 Aluminium & Ti-6Al-4V Titanium)

Fiber	Matrix	Axial Tension MPa	Axial Compression MPa
—	7075-T6	200[a]	190[a]
—	Ti–6Al–4V	220[a]	200[a]
Boron	Al	480	660
Alumina	Al	530	560
SiC	Ti	470	780

Table 1.15: Physical Property of Some Common Unidirectional Metal Matrix Composites

Fiber	Matrix	Density g/cm^3	Axial CTE 10^{-6}/K	Transverse CTE 10^{-6}/K	Axial Thermal Conductivity W/m·K	Transverse Thermal Conductivity W/m·K
—	7075-T6	2.8	24	24	130	130
—	Ti–6Al–4V	4.4	8.6	8.6	6.7	6.7
Boron	Al	2.6	6	—	—	—
Alumina	Al	3.4	7	16	80	45
SiC	Ti	3.9	5.9	—	—	—

Metal matrix composites reinforced with continuous fibers has wide range of application but it is relatively expensive rather than other composites. Among those, most common metal matrix composite which is commercially manufactured by continuous alumina is fiber-reinforced aluminium. It has high specific stiffness and specific strength and widely used in construction (Support towers) & in transmission lines. Fabric-reinforced materials are weak in nature throughout the direction of thickness so therefore three-dimensional reinforcement forms are always preferred. It has maximum elastic modulus 340 GPa, tensile strengths about 700 MPa and compressive strengths approximately about 800 MPa in the direction of fibrous reinforcement on the contrary it showing maximum elastic modulus 10 GPa, tensile strengths about 14 MPa and compressive strengths approximately about 34 MPa in perpendicular direction of fibre directions. It has been found that mostly fibrous reinforcements are made up of brittle glasses, ceramics, or carbon. A composite that is manufactured by using GFRPs reinforcing with continuous fibres and fabrics are usually applied in commercial, consumer, and aerospace/defence structural applications. Aerospace/defence applications include solid fuel rocket motor cases, rocket launchers, aircraft fairings flooring, interiors, and cargo containers, helicopter rotor blades, patrol boats, mine hunter ships and ballistic armour. Such a composites has continuous unidirectional fibers or fabrics, in which reinforcement imparts superior structural performance than that of discontinuous reinforcement.

1.16. Manufacturing of Fibre Reinforced Composites

Manufacturing of a composite structure initiates with fibres in which large number of fibers are arranged inside a thin layer of matrix to form a lamina (ply) in which thickness of a lamina ranges between 0.1 to 1 mm (0.004 to 0.04 in.). If continuous (long) fibers are used in making the lamina, they may be arranged either in a unidirectional orientation (All fibres are arranged in one direction), in a bidirectional orientation (fibers are arranged in two directions) or in a multidirectional orientation (fibers are arranged in more than two directions). The bi - or multidirectional orientation of fibers is obtained by weaving or other processes used in the textile industry. For a lamina containing unidirectional fibers, the composite material has the highest strength and modulus in the longitudinal direction of the fibers. However, in the transverse direction, its strength and modulus are very low. For a lamina containing bidirectional fibers, the strength and modulus can be varied using different amounts of fibers in the longitudinal and transverse directions. For a balanced lamina, these properties are the same in both directions. The commercial form of continuous glass fibers is a strand, which is a collection of parallel filaments numbering 204 or more. Glass fibers are also available in woven form, such as woven roving or woven cloth. Woven roving fabric is coarse in nature in which continuous roving's are woven in two mutually perpendicular directions. Woven cloth is weaved using twisted continuous strands, called yarns. Both woven roving and cloth provide bidirectional properties that depend on the style of weaving as well as relative fiber counts in the length (warp) and crosswise (fill) directions. All of these forms of glass fibers are suitable for hand layup molding and liquid composite molding.

Note

- Fibers produce high-strength composites because of their small diameter.

- Continuous fibre imparts highest strength and modulus to continuous-fibre composites.

- It has been found that if smaller will be the diameter of the fiber then strength, flexibility would be higher in addition, it facilitates easily fabrication but cost become high. Typical fibres include glass, aramid, and carbon, which may be continuous or discontinuous.

- In polymer and metal matrix composites that form a strong bond between the fiber and the matrix, the matrix transmits loads from the matrix to the fibers through shear loading at the interface.

- Continuous-fiber composites are used where higher strength and stiffness are required.

- From the mechanical engineering application point of view dominating role is performed by the following fibre to manufacturing composites material these are glasses, carbons (also called graphite's), basalt fibres, several types of ceramics, and high-modulus organics. Most fibres are produced in the form of multifilament bundles, usually called strands or ends in their untwisted forms or yarns in twisted.

1.17. Discontinuous - Short Fiber Reinforced Composites

Short-fiber reinforced composites consist of a matrix phase that has a dispersed phase in form of discontinuous fibers. Discontinuous fibers has short aspect ratio and it is defined as the ratio between length of fibre-to-diameter of fibre say (L/D) is known as the aspect ratio. The discontinuous fibers may be arranged in uni-direction orientation but mostly has random orientation of fibre. For examples discontinuous reinforcements are chopped fibres and random mat. Consider following diagram which shows discontinuous fibres reinforcement.

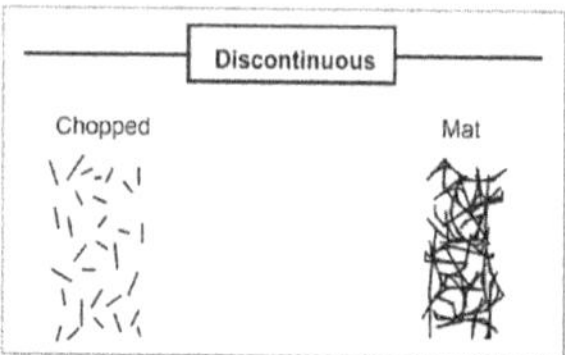

Diagram 1.14: Discontinuous Fibre Reinforcement in Composites

Discontinuous fiber-reinforced composites have lower strength, stiffness and modulus than that of continuous fiber composites due to random orientation of fibres throughout the composites.

1.18. Discontinuous Fiber-Reinforced Metal Matrix Composites

Discontinuous fiber-reinforced metal matrix composites are widely used in mechanical engineering applications like in manufacturing of internal combustion engine components. The basic function of fibre reinforcement is to enhance the wear resistance, fatigue, high thermal conductivities and strength at elevated-temperature, usually fiber-reinforced aluminium composites is used to fulfil the desired purpose. It also eliminates the requirement of cast iron sleeves in engine blocks and cast-iron inserts in pistons. If low volume fraction of fibres may use then we can reduce the chance of thermal stresses. Automobile and truck engines have been used different variety of discontinuous ceramic fibers in order to augment the capacity of cylinder bore and piston wear resistance and high-temperature strength cast iron inserts has replaced by fiber-reinforced aluminium composites. To full fill the desired purpose now a day's ceramic and carbon fibers composites are also in fashion which also not improve wear resistance but also reducing the coefficient of friction. Further titanium boride particle-reinforced titanium has been used in engine valves.

Note

- To increase wear resistance of internal engine components discontinuous alumina and alumina–silica fibres are used.

- It has been observed that when carbon fibres combine with alumina then wear resistance improves and coefficient of friction of cylinder walls reduces.
- It was also found that wear resistance of ring groove wear for an alumina fibre-reinforced aluminium piston was significantly reduces than that of cast iron insert
- Discontinuous fibers and whiskers (elongated single crystals) also somewhat improve fracture toughness.
- Discontinuous fibers have short aspect ratio.
- Discontinuous-fiber composites are normally somewhat random in alignment, which dramatically reduces their strength and modulus.

1.19. Short Fibre Composites

A composite in which short or staple fibers (Dispersed phase) are embedded or firmly fixed in the matrix phase then it is referred as discontinuous fiber reinforcement that is sometimes also known as short fiber composites. In short fiber composites, the length of short fiber should not be high so that individual fibers remain unmixed to each other and should not be very small to sustain their fibrous nature. Remember here reinforcement must be uniformed for proper dispersion of short fibers. There are various types of short fiber which may be reinforced with metal matrix, ceramic matrix, polymer matrix and carbon matrix. Short fibre has remarkable characteristics like high specific strength, stiffness, reduced shrinkage during fabrication, resistance to solvent, swelling, resistance to abrasion, tear and creep resistance and cheaper than long fibers, Let be consider the example of Short fiber-Rubber Composites.

1.19.1. Short Fiber - Rubber Composites

The term 'short fiber' means the fibers in the composites have a critical length which is neither too high to allow individual fibers to entangle with each other, nor too low for the fibers to lose their fibrous characteristics. A short fiber consisting two main constituents, i.e., the short fibers and the rubber matrix phase.

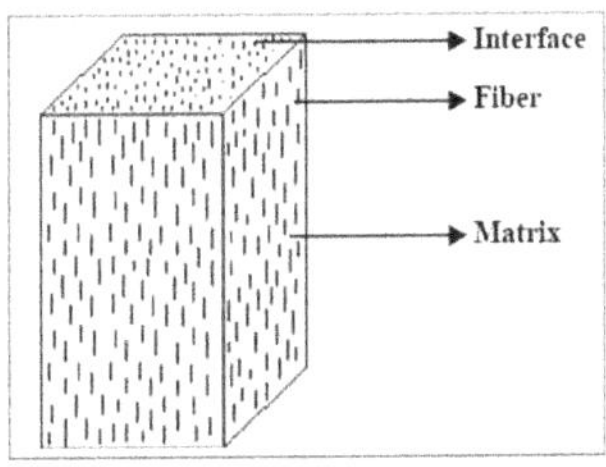

Diagram 1.15: Short Fiber Composites

Composites in which the short fibres are oriented uniaxially like in an elastomer then we have got superior strength and stiffness than that of fibers and good elasticity than that of rubber. Now a days such type of composites is being used for the fabrication of V-belts, hoses and complex shapes articles. Furthermore, Reinforced / dispersed phase using as short fibres such as cotton, silk, rayon, jute and Nylon are usually employed for natural rubber matrix or elastomeric matrix. These natural rubber matrix phases are styrene butadiene rubber (SBR), chloroprene rubber (CR), nitrile rubber (NBR) and ethylene propylene rubber (EPDM). Consider the following example in which short fibre that is used with matrix phase.

Table 1.16: Short Fibre with Matrix Phase

Short carbon fiber/ poly (trim ethylene terephthalate) composites.	Reinforcement of millable PU with short Kevlar fibre.
Poly (ethylene terephthalate) / short glass fiber composites	Natural rubber coir fiber composite
Reinforced Bromo butyl rubber using short Kevlar fiber.	Short Nylon 6 fibre nitrile rubber composites
Fiber-filled rubber composites	Short jute fiber reinforced carboxy-lated nitrile rubber
Cellulosic fiber-reinforced green composites	Short jute fiber reinforced NR composites
Short Nylon fiber reinforced polypropylene	Short melamine fiber reinforced EPDM rubber
Thermoplastic natural rubber reinforced short glass fiber	Short coir fiber reinforced natural rubber composites
Fiber reinforced plastic and rubber composites	EPDM/ Nylon-6 short fiber composites
Short Kevlar aramid fiber- thermoplastic PU composite	Short polyester fiber- polyurethane elastomer composite
Natural rubber short Nylon fiber composites	Short Nylon fiber reinforced SBR composite
Short fiber reinforced NBR composites	Short Nylon fiber reinforced NBR composite
Short aramid fiber reinforced chloroprene rubber composite	Short cellulose fiber reinforced with NR-SBR blends matrix
Short carbon fibers reinforced with Ethylene vinyl acetate (50%) and EPDM (50%)	Short fiber rubber composites [Jute / silk fibers reinforced with rubber matrices]
Short glass fiber reinforced NR and SBR composites	Carbon - short fiber-polymer composite
Randomly oriented carbon fiber-polymer composites	Short carbon fiber reinforced nitrile rubber composites
Short Nylon fiber reinforced NBR and CR composite	Nylon 6 short fiber composite
Short Nylon 6 fiber reinforced NR, NBR, SBR composite	Short Nylon fiber reinforced SBR composites [V-belt applications]
Short Nylon 6 fiber reinforced NBR and SBR composites	Short Nylon fiber and vinylon fiber reinforced nitrile rubber and SBR
Short Nylon fiber-rubber composites	Nylon fiber in NR and polyester fiber in CR
Short Nylon 6 fiber reinforced SBR composites	Carbon fiber-reinforced polymers (CFRPs) [high-performance applications]
Glass fiber-reinforced polymers (GFRPs)	PAN carbon fibers, [T-300 and AS-4]
Pitch carbon fibers & poly-carbo-silane fiber	Silicon carbide (SiC) and aluminum oxide (Al$_2$O$_3$) fibers

1.20. Fibre Orientation

Fibre orientation play a prominent role throughout the structure to develop an imperative property in the fibre composites. These orientations may be either continuous (Align) orientation of fibre or discontinuous (Random) orientation of fibre. Continuous fibres have long aspect ratios in addition to continuous-fiber composites normally have a preferred or specific orientation or sometimes we can say it has fibre flow alignment either in unidirectional, bi-direction or multi direction on the contrary discontinuous fibers has short aspect ratio and discontinuous fibers usually arranged either in uni-direction or in a random orientation. Consider following schematic diagram in which alignment of fibre are given for both continuous, discontinuous and laminates composites.

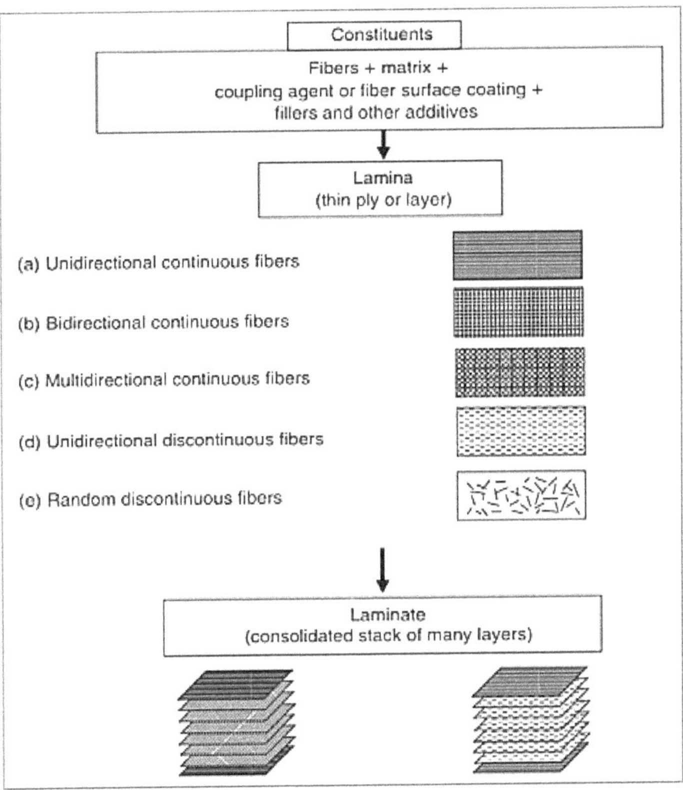

Diagram 1.16: Position of Fibres in Fibre – Reinforced Composites Material

Continuous-fiber composites are generally made into laminates by stacking single sheets of continuous fibers in different orientations to obtain the desired strength and stiffness properties with fiber. For a lamina containing unidirectional fibers, the composite material has the highest strength and modulus in the longitudinal direction of the fibers. However, in the transverse direction, its strength and modulus are very low. For a lamina containing bidirectional fibers, the strength and modulus can be varied using different amounts of fibers in the longitudinal and transverse directions. For a balanced lamina, these properties are the same in both directions. A lamina can also be constructed by using discontinuous (short) fibers in a matrix. The discontinuous fibers can be arranged either in unidirectional orientation or in random orientation. Discontinuous fiber-reinforced composites have lower strength and modulus than continuous fiber composites. However, in case of random orientation of fibers we can obtain same mechanical and physical properties in all directions. The thickness required to support a given load or to maintain a given deflection in a fiber-reinforced composite structure is obtained by stacking several laminas in a specified sequence and then consolidating them to form a laminate. Furthermore, properties of fibre – reinforced composites depending upon the type of fiber, fiber volume fraction, fiber length, and fiber orientation because they not only influence properties like Density, Tensile strength and modulus, Compressive strength and modulus, Fatigue strength, Electrical and thermal conductivities but also Cost. The basic function of a reinforcing fiber is to increase the strength and stiffness of a matrix material. For instance, mechanical properties like specific strength and stiffness, reduced shrinkage in moulded products, resistance to solvent swelling, abrasion, tear and creep resistance are greatly improved in the case of short fibre composites. Moreover, short fibres are cheaper than long fibers. By using continuous fibres, we can attain highest strength, improved impact resistance and modulus but subjected to higher cost on the other hand Discontinuous-fibre composites showing reduced strength, stiffness, modulus and less costly. In ceramic matrix composites, the objective is often to increase the toughness rather than the strength and stiffness; therefore, a low interfacial strength bond is desirable. Further, I shall describe here some common synthetic fibre that has been used for manufacturing / fabrication of composites material since long period of time,

1.21. Glass Fiber

We know that any composites material consists of two phase i.e.

$$\text{Composition} = \text{Matrix phase} + \text{Dispersed phase}$$

Glass fibre works as dispersed phase or reinforcement which is generally employed to manufacturing / fabrication of polymer matrix composites (CMCs) due to their fascinating characteristics that attributed as high strength, high chemical resistance, low cost and excellent insulating properties but except the above appealing property it has also some draw back or negative responses like low elastic modulus, poor adhesion, high specific gravity, sensitivity to abrasion, low fatigue strength, low tensile strength and high hardness are most common. From the context point of view, following types of glass fibres are used to manufacturing or for fabrication of composites materials. These are E-glass, S-glass, C-glass (CR-glass), S-2-glass, D-glass, A-glass and HS glass etc. Further, possible brief description of all those mention glass fibres are given below.

1.22. E – Glass

The primary functions of E – Glass is to reinforcing the polymer matrix composites (PMCs). Basically, the letter "E" recognized for electrical applications or in other word we can say that it is known as electrical fibre reinforced glass. It developed for electrical insulation application showing high electrical resistance, decoration and structural application however these fibres glass may also be employed for mechanical engineering application. E-glass is most widely used of all kind of fibrous reinforcements. This glass fibres have low cost as compare to other fibres and hence widely used in the FRP industry. Glass fibres are arranging in a bunch of glass fibres or multifilament bundles in which filament diameters ranges between 3 to 20 μm. Furthermore if we consider the characteristics of E – glass then it has low elastic moduli compared to other reinforcements, susceptible to creep rupture, low thermal and electrical conductivities (Thus used as thermal and electrical insulators), Low coefficient of thermal expansion of E - glass fibres then that of metals.

Composition of E-glass:- E– glass has following composition in % weight, i.e.

$$E\text{-glass composition} = 54.5\%(SiO_2) + 14.5\ \%\ (Al_2O_3) + CaO\ (17\%) + MgO\ (4.5\%) +$$
$$B_2O_3(8.5\%) + Na_2O\ (0.5\ \%)\,.$$

In the above composition, Na_2O and K_2O content in E - is very low therefore it does not mention in compositional formula but this little amount imparts better corrosion resistance to water as well as higher surface resistivity.

1.23. S – Glass

S – glass fibre has higher content of silica. Consider the chemical composition of s – glass which shows all the ingredient in % weight.

Chemical composition

64% Silicon(SiO_2) + 25% Aluminium(Al_2O_3) + 10%Magnesium oxides(MgO) + 0.01% Calcium oxides (CaO) + 0.01% Boron oxides (B_2O_3) + 0.8%Other elements

'S-glass' fibre recognize for strength. This glass fibres are generally used for reinforcing the plastics and sometimes for rubbers. A composite that is manufactured by S – glass showing capacity to sustain its strength at high temperatures than that of E-glass and also has higher fatigue strength and highest tensile strength. Basically, due to its high temperature resistance it is widely used in aerospace applications or manufacturing aircraft components and missile casings. Now a day's fibres of S-2-glass is also available which is advance version of S-glass, called S-2-glass that can be produced at low cost which showing similar tensile strength and modulus like a S-glass.

1.24. HS Glass Fibers

HS glass fibres usually employed for reinforcing polymers composites which is being used for mechanical engineering applications. HS glass is stiffer and stronger than E-glass and has better resistance to fatigue and creep.

1.25. CR – Glass Fibers

CR – glass fibers are recognize as corrosion-resistant glass fibres and widely used in the chemical process industries.

In addition to it we also have some other glass fibres which is being used to manufacturing a fibre reinforced composite, these are:

1.26. C-glass Fibre

"C" known for corrosion and respective c - fibre reinforced composites used in chemical environments, such as storage tanks.

1.27. R-glass Fibre

"R" glass famous for structural applications and respective R - glass - fibre reinforced composites mostly used in construction application.

1.28. D-glass Fibre

"D" known for dielectric characteristics, it has low dielectric constants used for low dielectric application for example radomes.

Moreover, Glass fibers are known for reinforcement of composites that is being used in high performance composite due to some remarkable properties & low manufacturing. The major element of glass fiber is silica which is mixed with other oxides. Glass fibres are mostly employed for reinforcing the plastics. High initial aspect ratio can be obtained with glass fibres, but brittleness causes breakage of fibres during processing. The tensile strength of glass fibre-rubber composite decreases with increasing effect of atmospheric humidity. Glass fibre-elastomer composites showing enhanced strength and wear resistance whose property can be used in auto tires, V-belts and conveyor belts. Glass fibres are the mostly used for reinforcing polymeric matrix composites (PMC). Glass fibre-reinforced PMCs are often used as thermal and electrical insulators. Because it represents low thermal and electrical conductivities than that of metals. The favourable characteristics of such a glass fibre are low cost, high tensile strength, high chemical resistance, and excellent insulating properties on the contrary it has low tensile modulus and high density, sensitive to abrasion, low fatigue resistance, and high hardness. Consider the following table which show comparative study of property of E – Glass & S- Glass fiber.

Table 1.17: Comparisons of Property between E – Glass & S - Glass Fiber

Glass - fibers	Property			
	Specific gravity	Young modulus (GPa)	Ultimate tensile strength (MPa)	Coefficient of thermal expansion (μm/m/$^{\circ}$C)
E – Glass	2.54	72.40	3447	5.04
G – Glass	2.49	85.50	4585	5.58

One of the most important aspects of glass-fibre is a chemical composition that always affects the inherent property of composites when it would be dispersed with matrix phase hence we can say composition coincide to property, consider following table which shows composition of E – Glass & S – Glass.

Table 1.18: Composition of E – Glass & S - Glass Fiber

Glass - fibers	Compositional elements					
	SiO_2	Al_2O_3	CaO	MgO	B_2O_3	Others elements
E – Glass	54	15	17	4.5	8	1.5
G – Glass	64	25	0.01	10	0.01	0.8

Glass – fibers has larger amount of silica & the other elements included in declining orders. Other oxides, such as B_2O_3 and Al_2O_3, are added whose function is to modify the network structure of SiO_2 as well as augmenting its workability. Further the addition of Na_2O and K_2O content in E- and S-glass fibres is very low say 1.5 % of weight & 0.8 % of weight respectively whose function is to imparts protection against corrosion resistance to water as well as higher surface resistivity. The internal structure of glass fibres is a three-dimensional, long network of silicon, oxygen, and other atoms arranged in a random fashion. Thus, glass fibres are amorphous i.e. non-crystalline and isotropic in nature.

Note

- The basic commercial form of continuous glass fibres is a strand, which is a collection of parallel filaments numbering 204 or more. A roving is a group of untwisted parallel strands (also called ends) wound on a cylindrical forming package.
- Chopped strands are produced by cutting continuous strands into short lengths.
- Strands of high integrity are called "hard" and those that separate more readily are called "soft." Chopped strands ranging in length from 3.2 to 12.7 mm, are used in injection-molding operations. Longer strands, up to 50.8 mm, in length, are mixed with a resinous binder and spread in a two-dimensional random fashion to form chopped strand mats (CSMs).
- Milled glass fibers are produced by grinding continuous strands in a hammer mill into lengths ranging from 0.79 to 3.2 mm. They are primarily used as a filler in the plastics industry and do not possess any significant reinforcement value.
- Glass fibres are also available in woven form, such as woven roving or woven cloth. Woven roving is a coarse fabric in which continuous rovings are woven in two mutually perpendicular directions. Woven cloth is weaved using twisted continuous strands, called yarns. Both woven roving and cloth provide bidirectional properties that depend on the style of weaving as well as relative fiber counts in the length (warp) and crosswise (fill) directions.
- The average tensile strength of freshly drawn glass fibres may exceed 3.45 GPa. However, surface damage due to abrasion become reduces. The strength of glass – fibres is reduced due to cyclic loading.
- The tensile strength of glass fibres is decreased with loading – times. Further it has been observed that the tensile strength of glass fibres is also reduces in the presence of water and under static fatigue condition. When glass fibres interact with water bleaches out the alkalis from the surface and deepens the surface flaws already present in fibres. Under static loading condition, the growth of surface flaws is increased & tensile strength become lowers.

1.29. Carbon Fibers

Carbon fibres exhibits fascinating property which become a responsible factor for application in high performance application. For instance, high stiffness, strength, low density, excellent resistance to creep, stress rupture, fatigue, low weight and resistance to corrosive environments and outstanding high thermal conductivities but the negative aspect of such type of carbon fibres is that its oxidization takes place at high temperatures. Carbon fibres sometimes also referred as a graphite fibre that can be used as reinforcing / reinforced phase for polymers composites, metals composites, ceramics composites, carbon composites including carbon – carbon composites. Now a day's carbon fibres are extensively used in polymer composites. Carbon fibres are available in three forms, these are long and continuous tow form, chopped form (6–50 mm long) and milled form (30–3000 mm long). The long and continuous tow carbon fibres which is a bundle of untwisted fibres of 1,000–160,000 parallel filaments, is mostly used for high-performance applications. Furthermore, Carbon fibres can also be classified on the basis of values of tensile modulus i.e. low tensile modulus carbon fibres & high tensile modulus carbon fibres. The value of tensile modulus ranges from 207 GPa to 1035 GPa respectively. It has been found that the low-modulus fibres exhibit lower density, lower cost, higher tensile and compressive strengths, and higher tensile strains-to-failure in comparison to high-modulus fibres. Consider the following table in which tensile strength & compressive strength are given for some common carbon fibers.

Table 1.19: Tensile Strength & Compressive Strength of Carbon Fibers

Mechanical properties	Type of carbon fibers			
	T – 300 Carbon fibers	AS -4 Carbon fibers	GY – 70 Carbon fibers	P – 100 Carbon fibers
Tensile strength in GPa	3.2	3.6	1.86	2.2
Compressive strength in GPa	2.7 to 3.2	2.7	1.06	0.5

1.29.1. Manufacturing of Carbon Fibers

Commercially these carbon fibres are manufactured from rayon, polyacrylonitrile (PAN), coal tar pitch or petroleum pitch.

1. **Rayon-based fibers** represent poor strength. It is widely used in space Shuttle Orbiter CCCs that was made from rayon-based carbon fibres.

2. **PAN-based fibers** are manufactured from carbon fibers. The composites that is manufactured from PAN-based fibers represent superior tensile and compressive strength but its compressive strength decreases as modulus increases. PAN-based fibers has maximum tensile strengths up to 7 GPa. PAN carbon fibers can be classified into high tensile strength (HT), high modulus (HM), and ultrahigh modulus (UHM) types. The high tensile strength PAN carbon fibers like T-300 and AS-4 has lowest modulus, while the ultrahigh-modulus PAN carbon fibers for example GY-70, have the lowest tensile strength as well as the lowest tensile strain to-failure. Intermediate modulus (IM) is a high-strength PAN carbon fibers, such as T-40 and IM-7 that possess the highest strain-to-failure among the carbon fibers. The PAN carbon fibers have higher compressive strength than pitch carbon fibers. It is also observed that higher the modulus of a carbon fiber then lower would be its compressive strength. The PAN carbon fibers have lower thermal conductivity and electrical conductivity than pitch carbon fibers. For both types of carbon fibers i.e. PAN carbon fibers & pitch carbon fibers, the higher the tensile modulus, higher thermal and electrical conductivities. Consider the following table which shows some common mechanical property of carbon fibres (PAN, Pitch, Rayon).

Table 1.20: Manufacturing Properties of Carbon Fibres

Precursor	PAN	PAN	Pitch	Pitch	Rayon	Pitch (K13D2U)
Modulus	Low	High	Low	High	Low	Ultrahigh
Tensile modulus (GPa)	231	392	161	385	41	931
Tensile strength (GPa)	3.4	2.5	1.4	1.8	1.1	3.7
Strain to failure (%)	1.4	0.6	0.9	0.4	2.5	0.4
Relative density	1.8	1.9	1.9	2.0	1.6	2.2
Carbon assay (%)	94	100	97	99	99	>99

3. **Pitch-based fibers** indicate higher axial moduli rather than those made from PAN precursors. UHM pitch fibres indicate maximum moduli value approximately about 895 GPa. In addition, UHK pitch fibres, has extremely high axial thermal conductivities, say 1000 W/m·K. However, the composites manufactured from pitch-based carbon fibres are somewhat weaker in tension and shear and much weaker in compression than those using PAN-based reinforcements. Carbon fiber is considered in a category of high-performance fibre which is mostly used to manufacturing short fiber-polymer composite or polymer composites. The composites made from fibre carbon has light weight as well as superior mechanical properties and consequently apply in aerospace application. It has been observed that carbon fibres are highly anisotropic. Axial stiffness,

tension, and compression strength and thermal conductivity are always greater than that of radial direction. Carbon fibers generally have small, negative axial coefficient of thermal expansion (Means they get shorter when heated) and positive coefficient of thermal expansion in radial direction (Means they get lengthen when heated). Diameters of common reinforcing fibres which transform into multifilament bundles ranges between 4 to 10 μm.

Note

- Carbon fibers with tensile moduli as high as 895 GP and with tensile strengths of 7000 MPa are commercially available.
- Now a day's characteristic of high thermal conductivities of carbon fibre composites employed for electronic packaging and other applications in which heat dissipation and thermal control does matter.
- Carbon fibers are used in high-performance aerospace and commercial PMCs and some CMCs.
- Carbon fibers are also made from CVD technique. These fibres has maximum axial thermal conductivities about 2000 W/m·K.
- Carbon fibres contain a blend of amorphous carbon and graphitic carbon. Their high tensile modulus results from the graphitic form, in which carbon atoms are arranged in a crystallographic structure of parallel planes or layers. The carbon atoms in each plane are arranged at the corners of interconnecting regular hexagons as a result of which carbon fibre has highly anisotropic physical and mechanical properties.
- Pitch carbon fibers have very high modulus values, but their tensile strength and strain-to-failure are lower than those of the PAN carbon fibers. It has high modulus.
- The axial compressive strength of carbon fibers is lower than their tensile strength.
- The PAN-based fibres are SM, UHS and UHM. SM PAN based fibers are very common type of carbon fiber reinforcement. UHS PAN based carbon fibers are also a carbon fibre which is strong one. Another type of carbon fibre is known as intermediate modulus (IM) carbon fibres because axial modulus falls between range of SM and high-modulus carbon fibres.
- Carbon fibres have several advantages including high stiffness, high tensile strength, low weight, high chemical resistance, and high temperature. These carbon fibres can be utilised in various applications such as aerospace, automotive, sporting goods, and consumer goods.

1.30. Aramid Fibres

Aramid fibers sometimes also referred as aromatic polyamide fiber which is a high-modulus organic reinforcement. Such kind of fibre are used to reinforcing polymer composites & cement composites in which the function of reinforcement is to imparts protection against ballistic effects. It is well known for its promising property such as lowest density, highest tensile strength-to-weight ratio, lightweight, high tensile strength, and resistance to impact damage, negative coefficient of thermal expansion in the longitudinal direction (As temperature increases its dimension become shorten and used to manufacturing low thermal expansion composite panels), soft lightweight, high temperature stability, very low thermal conductivity, better chemical resistance, highest tensile modulus, low creep rate. Except the above fascinating characteristics, they also impose some disadvantages like, low compressive strengths, very high vibration damping coefficient, complication in cutting or machining operation. Carbonization takes place at about 427°C, Sensitivity to ultraviolet lights (Indicate discoloration and loss in tensile strength). A very first aramid fibers was comes on the front line in year 1971 that was used to reinforcing of elastomeric matrices and thermoplastic elastomers. There are three kinds of aramid fibers which has been used to reinforcing the composites material these are Kevlar – 49, Kevlar – 149, Kevlar – 29 and Twaron. A possible brief description of all those are given below.

1.31. Kevlar 49

First aramid fiber is available in the market which is recognized as Kevlar 49. This aramid fiber is used to reinforcing the polymer composites. Such type of composites has lightweight, high temperature resistance, high tensile strength, high chemical resistance and superior resistance to impact damage due to these characteristics it is employed in marine and aerospace applications, soft lightweight body armors and helmets where weight saving always desirable.

1.32. Kevlar - 149

A newly developed aramid fiber is known as Kevlar-149. It has the high tensile modulus amongs all other aramid fibers. The tensile modulus of Kevlar 149 is 40% higher than that of Kevlar 49; however, its strain-to-failure is lower. Further, Kevlar 149 also has a lower creep rate than Kevlar 49. Furthermore, consider the following table which shows some common property of aramid fibers.

Table 1.21: Property of Aramid Fibers

Fibre type		E (GPa)	σ' (GPa)	ε' (%)
Kevlar 29	High-toughness, high-strength, intermediate modulus for tyre cord reinforcements	83	3.6	4.0
Kevlar 49	High modulus high-strength for composite reinforcement	131	3.6	2.8
Kevlar 149	Ultra-high modulus recently introduced	186	3.4	2.0

Note

- Kevlar 49 fibers do not melt up to 427°C but if temperature rise above it carbonization starts.
- The recommended maximum operating temperature for Kevlar-49 is 160°C.
- Kevlar 49 fibers exhibits sensitivity to ultraviolet lights. Prolonged direct exposure to sunlight causes discoloration and significant loss in tensile strength. For reducing the detrimental effect of ultra violet-light, ultraviolet light-absorbing fillers is added with the matrix phase.
- Kevlar 49 fibers are hygroscopic and can absorb up to 6% moisture at 100% relative humidity. The equilibrium moisture content (i.e., maximum moisture absorption) is directly proportional to relative humidity and is attained in 16–36 h. Moisture somewhat effect the tensile strength of Kevlar 49 fibers but when moisture content is high it cracks internally due to presence of pre-existing micro-voids, consequently responsible for longitudinal splitting.
- An aramid fibre also known as Kevlar. it has lightweight, stronger-than-steel fibre and used to manufacturing bullet proof vests and other body armour.

1.33. Nylon Fibers

Nylon fiber is a synthetic fiber comes on the front line during the year of 1939 which is an aliphatic polyamides. Usually these aliphatic polyamides are produced & prepared from diamine and a dicarboxylic acid. Nylon fibers are generally used for reinforcing – Polymer Matrix Composites that is generally consist of synthetic nylon fibers embedded in a polymer matrix, which surrounds and tightly binds the fibers and rubber matrix. But now a day's short nylon fibers can also be used as reinforced or reinforcing, nitrile butadiene rubber composites and chloroprene rubber. Short nylon – 6 fibers reinforcing natural rubber and SBR composites. It has been found that when nylon-6 fiber reinforced to (Natural rubber) NR, (Nitrile butadiene rubber) NBR and (Styrene butadiene rubber) SBR then mechanical properties were improved. Furthermore, dispersion of nylon fiber are responsible for augmenting, durability, physical property, strength, high elastic recovery, abrasion resistance, lustre, resistance to chemicals, resistance to oil, long life of composites due to long life of composites, light weight and high melting temperature stability. Nylon fibers are available in two forms i.e. Nylon – 66 & Nylon – 6. Short nylon fibers reinforced SBR composites is generally applied for manufacturing V–belt. Consider the following tables which shows physical property of nylon fibers & mechanical property of some synthetic fibers.

Table 1.22: Physical Property of Nylon Fibers

Property	Continuous filament	Staple
Tenacity at break N / tex, 65 % RH, 21 °C	0.40 - 0.71	0.35 - 0.44
Extension at break, %, 65 % RH, 21 °C	15 - 30	30 - 45
Elastic modulus N / tex, 65 % RH, 21 °C	3.5	3.5
Moisture regain (%) 65 % RH, 21 °C	4.0 - 4.5	4.0 - 4.5
Specific Gravity	1.14	1.14
Approx.volumetric swelling in water,%	2 - 10	2 - 10

Table 1.23: Mechanical Property of Some Synthetic Fibers

Fiber	Density (10^3 kg/m^3)	Elongation (%)	Tensile strength (MPa)	Young's modulus (GPa)
Aramid	1.4	3.3–3.7	3000–3450	63–67
Carbon	1.4	1.4–1.8	4000	230–240
Kelvar 49	1.45	2.0	2800	124
E-glass	2.5	2.5	2000–3500	70.0
S-glass	2.5	2.8	4570	86.0
SiC	3.08	0.8	3440	400
Alumina	3.95	0.4	1900	379

Usually following rubber composites are made by reinforcing nylon fibers, These are short Nylon 6 fiber nitrile rubber composites, short Nylon fiber-natural rubber composites, short Nylon fiber reinforced SBR composite, short Nylon 6 fiber reinforced SBR, short Nylon fiber reinforced NBR composite.

1.34. Ceramic Fibers

Ceramic fibers can be dispersed / reinforced or embedded (Fixed firmly) with either metal matrix phase or ceramic matrix phase, for manufacturing ceramic fiber based metal composites & ceramic fiber based ceramic composites material. In such type of ceramic fibers, fiber of Silicon carbide (SiC) and Aluminium oxides (Al_2O_3) are very famous. The mechanical, physical and chemical property could not define in accumulated form due to their diverse form, here I will try to evolve some properties with the name of ceramic fibers, Ceramic fiber composites exhibits, high temperature stability, high temperature strength (Up to 650°C in case of SiC - fibers), Excellent strength (Up to 1370°C in case of Al_2O_3 oxides fibers), Lower thermal & electrical conductivities (Aluminium oxide fibers), High coefficient of thermal expansion (Al_2O_3 - fibers), Low coefficient of thermal expansion (SiC - fibers), high tensile strength (Aluminium oxide fibers like Nextel – 610 at room temperature but decreases as temperature increases say up to 1100°C), Lower tensile strength [(Al_2O_3 fibers – Nextel – 720 (85% Al_2O_3 + 15% SiO_2) at room temperature and maintains its tensile strength up to 1400°C], Lower creep rate { Al_2O_3 – Fibers (Nextel - 720)}, Higher creep rate {Al_2O_3 – Fibers (Nextel - 610)}, high temperature insulation {Discontinuous alumino-silicate fibers (95% Al_2O_3 + 5 % SiO_2)}, high temperature insulation {Discontinuous Ceramic fibers (Fiber - frax), composition = 50% Al_2O_3 + 50% SiO_2}, good tensile strength & tensile modulus {SiC whisker – fibres (10%SiO_2 + 10%Si_3N_4 + 80%SiC)}. Except the above we also have so many other properties of ceramic fibers.

Silicon carbide fibers may be of three types:

1. **Monofilaments silicon carbide fibers** are made from chemical vapor deposition of β-SiC on a 10–25 mm diameter carbon monofilament substrate. The carbon monofilament is previously coated with 1 mm thick pyrolitic graphite so that its surfaces become smooth and also enhance thermal conductivity.

2. **Multifilament yarn** produced by melt spinning of a polymeric precursor, such as poly-carbo-silane, at 350°C in presence of nitrogen gas. To full fill the desired purpose initially polycarbosilane fiber is heated in presence of air up to 190°C for about 30 min and then heat-treatment to 1000°C–1200°C to form a crystalline structure. The fiber diameter of yarn is about 14.5 mm, further commercial yarn contains 500 fibers. Yarn fibers exhibits lower strength than that of monofilaments.

3. **Whiskers** has length about 50 mm that has 0.1 to 1 mm in diameter. They are produced from rice hulls, which contain 10–20 wt% SiO_2. Rice hulls are first heated in an oxygen-free atmosphere to 700°C–900°C and then to 1500°C–1600°C for 1 h to produce SC whiskers. The final heat treatment is performed in presence of air. The resulting SiC whiskers contain 10 wt% of SiO_2 and up to 10 wt% Si_3N_4. The tensile modulus and tensile strength of these whiskers are reported as 700 GPa and 13 GPa respectively.

Furthermore, plenty of aluminium oxide fibers are available like Fiber FP {polycrystalline α-Al_2O_3 fiber}. Its tensile modulus and tensile strength is about 379 GPa and 1.9 GPa respectively. Fiber FP also have high compressive strength which is about 6.9 GPa. Further we also have a two other ceramic fibers known as Nextel 610 and Nextel 720 which also falls under the category of aluminium oxide fibers. These fiber are exists in continuous multifilament form. Nextel-610 (Contains about 99% Al_2O_3) and available in filament diameter of 14 mm. It has a high tensile strength at room temperature and reduces with temperature increment. Nextel 720 (85% Al_2O_3 + 15% SiO_2) has a lower tensile strength at room temperature, but it may maintain 85% of its tensile strength up to 1400°C. Nextel 720 also has a much lower creep rate than that of Nextel 610. Another ceramic fiber is recognize as Fiber-frax that has 50% Al_2O_3 and 50% silica (SiO_2) and available in short, discontinuous fibers. The diameter of these fiber ranges between 2 to 12 mm and aspect ratio showing greater than 200. Further another variety of ceramic fiber is discontinuous alumino-silicate fiber which has composition of 95% Al_2O_3 and 5% SiO_2. It represents high temperature insulation and available in 1 to 5 mm diameter range.

1.35. Silicon Carbide based Fibers

Silicon-carbide-based fibers belongs to ceramic fiber cadre which is used to reinforcing metals and ceramics matrix. For example, monofilament boron fibers which is produced by high-purity silicon carbide with the help of CVD technique. The substrate of such a fiber is carbon - monofilament. There are number of multifilament silicon-carbide-based fibers. Some silicon carbide (SiC) fiber contains few amounts of silicon, carbon and oxygen, titanium, nitrogen, zirconium, and hydrogen. Consider the following table which shows mechanical properties of silicon carbide-based fibers at various temperature ranges.

Table 1.24: Mechanical Property of Silicon Carbide Fiber – Reinforced Silicon Carbide

Property	816°C	1038°C	1315°C
Warp tensile modulus (GPa)	208	209	158
Tensile strength (MPa)	362	325	295
Tensile proportional limit (MPa)	177	168	163
In-plane shear strength (MPa)	47.2	—	—

1.36. Alumina based Fibers

Alumina-based fibers are primarily used to reinforce metals and ceramics composites, although they have been used to produce polymer matrix composite aircraft firewalls. Silicon-carbide-based fibers have different chemical formulations. The primary constituents, in addition to alumina, are boria, silica, and zirconia.

1.37. Boron Fibers

Boron fibers are used to reinforcing polymers matrix and metals matrix both. Boron fiber is known for outstanding property like excellent resistance to buckling and extremely high tensile modulus which ranges between 379 to 414 GPa. Boron fiber composites also represent high compressive strength but it's extremely high cost inhibit its application however due to their remarkable property it is widely used in aerospace applications.

1.38. Manufacturing of Boron Fiber

Boron fibers are manufactured by chemical vapor deposition (CVD) of boron onto a heated substrate. Substrates may either be tungsten wire or a carbon monofilament. For deposition of boron particles over the substrate, boron vapor is prepare by the reaction of boron chloride with hydrogen.

$$2BCl_3 + 3H_2 = 2B + 6HCl$$

Diameter of tungsten wire substrate is taken which is about 0.0127 mm. This tungsten wire is continuously supply through a reaction chamber at a temperature of $1300^{\circ}C$. Throughout the chamber boron atoms are deposited on the tungsten substrate and then pull-out slowly, wire from chamber. After deposition of boron atoms, core diameter increases from 0.0127 mm to 0.0165 mm. Now a days, boron fibers are available in various diameters such as 0.1, 0.142, and 0.203 mm. The properties of boron fibers always depending upon the ratio of overall fiber diameter to that of the tungsten core.

Note

- Boron fibers diameters available between range of 100–140 µm. which is greater than that of E-glass and carbon fibers.

- Properties of boron fibers affected by the ratio of overall fiber diameter to that of the tungsten core. For example, specific gravity is 2.57 for fiber diameter of 100-µm and 2.49 for 140-µm diameter of boron fibers.

- Boron fibers usually employed for reinforcing the polymers matrix composites and metal matrix composites.

- Boron fibers always manufacturing in the form of monofilaments (single filaments) by using CVD technique, in this technique boron atoms deposited over the tungsten wire.

1.39. High-Density Polyethylene Fibers

High-density polyethylene fibers also falls cadre of synthetic fiber which is always used to reinforcing polymers and for ballistic protection. Further high-density of polyethylene decreases drastically with temperature rise as a result of which it subjected to creep under given applied load, even at low temperatures.

1.40. Basalt Fibers

Basalt fibers is a recently developed reinforcements that is commercially used now a days. They are made by melting and extruding basalt, a volcanic rock. These fibers represent high fire resistance due to high melting point temperature. It has similar mechanical property alike HS glass.

1.41. Extended Chain Polyethylene Fibers

Extended chain polyethylene fiber is a newly developed fiber which is being widely used to manufacturing a composites material. One of the most common & very famous Extended chain polyethylene fibers, commercially available that is known as "Spectra". Spectra fibers are manufacturing from high-molecular-weight polyethylene with the help of gel spinning technique. Spectra polyethylene fibers composites have promising characteristics that is attributed as high impact resistance, high strength-to-weight ratio, low moisture absorption {1% as compared to Kevlar 49 (5%–6%)} and high abrasion resistance. Furthermore, it become melt at about 147°C therefore we have maximum safe operating or working temperature that is about 90°C. If temperature rises above this range then composites subjected to high level creep, reduction in strength and also cause of occurrence of thermal shrinkage. Moreover, low moisture absorption and high abrasion resistance characteristics of spectra fibers used to manufacturing marine composites. Marine composites work as structural composites and widely employed for construction of boat, ships, hulls and water skis, in addition to it, now a days a laminate composite of spectra fibers has been also apply for ballistic application since long period of time due to their extra-ordinary impact resistance for instance armours and helmets etc.

1.42. Hybrid Composites

In order to enhancing properties of composite material, material science researchers & material engineers develop a new advance material that is known as a hybrid composites in which two or more than two different types of fillers are reinforced in a single matrix so that we can attain improvement in property at as possible as minimum cost with respect to conventional composites. This composite called hybrid composites & sometimes also referred as a multi-component composite system. In such a case reinforcement is perform in similar manner like other composites material, means reinforcement is performed either by fibers, particulate fillers or both. Following type of hybrid composites are important from context point of view, these are:

1. **Sandwich Type:-** In sandwich type hybrids composites, one material is sandwiched between layers of another material.

2. **Interplay Hybrid Composites:-** In interplay hybrid composites, alternate layers of either two or more ceramic materials are stacked in regular pattern which overlap to each other. For example - Car bumpers are considered as hybrid composites that is made up of glass / epoxy layers to provide torsional rigidity and graphite/epoxy to give stiffness.

3. **Intra-ply Hybrid Composites:-** Intra-ply hybrid composites consist of two or more different composite fibers that is used in same ply. For examples - Golf clubs that uses graphite and aramid fibers. Graphite fibers provide the torsional rigidity and the aramid fibers provide tensile strength and toughness.

4. **Intimately Mixed Hybrid Composites:-** In intimately mixed type hybrid composites, constituents are mixed in such a manner that they mixed properly to each other but no concentration of either type may exist in the composite material.

5. **Resin Hybrid Composites:-** Resin hybrid laminates combine two or more resins instead of combining two or more fibers in a laminate. In this composite one resin is flexible and another has rigid nature. These resin hybrid laminates known for increased shear and fracture strength. Consider following table in which different kind of hybrid composites are given with their properties.

Table 1.25: Hybrid Composites & their Property

S.no	Hybrid composites	Properties
1	Sisal and glass fibers hybrid composites	Superior mechanical properties, mechanical properties enhance with increasing of glass fibers volume
2	Sisal / saw dust hybrid fiber composites with phenol formaldehyde resin	Tensile and flexural strength increased due to sisal fiber content. Because sisal fiber possesses higher strength and modulus than that of saw dust.
3	Sisal and pineapple hybrid composites	Enhance mechanical properties of hybrid composites. Because small of glass fibers is enough to develop strong mechanical property.
4	Glass fiber reinforced polyester hybrid composites	Enhance mechanical properties of hybrid composites. Because small of sisal fiber is enough to develop desired level of mechanical property in sisal fiber reinforced polyester matrix
5	Polypropylene / oil palm / glass fiber hybrid composites	Improved tensile and flexural strength
6	Hybrid composites that is made from Glass fiber mat and coir fiber mat in polyester matrix	Enhancement in flexural strength and reduced water absorption.
7	Hybrid composites made from phenol formaldehyde composite reinforced with oil palm and glass fibers.	Improved flexural modulus and impact strength.
8	Hybrid composites that is made from natural rubber composite reinforced with sisal/oil palm, sisal/coir hybrid fibers	Mechanical properties of the natural rubber hybrid composite is improved.

Further, except the above we have also some other hybrid composites, these are:

- Epoxy polyurethane-jute and epoxy polyurethane-jute rice/wheat husk hybrid composites.
- Wood flour / kenaf fiber polypropylene hybrid composites.
- PP/SEBS-MA/SGF hybrid composites.
- Nitrile rubber hybrid composites with hybrid filler.

1.43. Green Composites

Green composites is newly developed composites that is known for their outstanding characteristics like eco-friendly nature, fully degradability and sustainability It is generally manufactured by the composition of fibers with biodegradable resins. In this cadre for manufacturing green composites following bio-degradable matrices are used in conjunction with fibers. These are, polyamides, polyvinyl alcohol, polyvinyl acetate, polyglycolic acid, and polylactic acid, which are synthetic as well as polysaccharides, starch, chitin, cellulose, proteins, collagens/gelatine, lignin, and so on. Furthermore, natural fiber composites can be used instead of timber because it may be transformed easily into sheets, boards, gratings, pallets, frames, structural sections. They can not only work as substitute of wood, metal, or masonry for partitions, barricades, fences, railings, flooring, roofing, wall tiles but also used for pre-fabricating of housing, cubicles, kiosks, awnings, and sheds/shelters. Green composites may be used effectively in many applications such as mass-produced consumer products with short life cycles or products intended for one time or short time use before disposal.

1.44. Bio - Composite Materials

Bio-composite materials, draw the attention of our researchers due to their appealing property like low density, high specific mechanical strengths, availability, renewability, degradability and environmental-friendly nature. Usually bio composite material has natural cellulose fibers as reinforcements with polymer matrix phase. For example – Consider the following composition.

Composition

Fibers used for reinforcing (Jute, Sisal, Coconut, Areca and Banana)

+

Unsaturated polyester matrix

This composition has light weight and superior mechanical properties thereby it is used to manufacturing helmet. Now a days, Bio-degradable polymer materials or bio-composites has been expanded because polymers composites showing eco-friendly nature. A number of biological materials may be incorporated into bio-degradable polymer materials among those starch and fiber extracted from various types of plants are important. Further bio-degradable polymer materials can be used instead of synthetic polymer and also beneficial in perspective to environmentally and economically aspect.

Note

- Bio-degradable material also used in packaging, agriculture, medicine, and other areas.

1.45. Laminate Composites

Laminate composites are developed for augmenting mechanical property, physical property and thermal property of composite material. Laminate is defined as **"A process of stacking number of laminae that arranged alternatively in the direction of thickness or arrange in overlap position & then cemented together all the laminae for consolidation so that a complete accumulation can be transformed into single unit of enhance property and such a composite is known as laminate composite"** usually laminate composites consisting multiple layers of laminae as a whole.

"Whenever there is a single ply or lay-up in which all the layers stacking in same direction or orientation then such a lay-up is known as lamina on the other hand if layers are stacked at the orientation of different angle then lay-up is called laminate" Alike other composite material, laminate layer can be reinforced in different orientation by using particulates, fibers and flakes. These reinforcements may be performed either in unidirectional or bi-directional orientation of the constituent reinforcement. Furthermore, the final properties of laminate composites are always coinciding to the orientation of individual fibers throughout the layers of laminate. Layer of lamina may consisting continuous long fiber or discontinuous short fibers.

1.46. Lamina with Continuous Long Fiber

In this case, a continuous long fiber are reinforced with matrix phase so as to form a lamina. The thickness of such a lamina ranges between 0.1 to 1 mm. Throughout the alumina structure long fiber are arranged either in unidirectional orientation means all the continuous fibers has parallel alignment in one direction or in a bidirectional orientation means all the continuous fibers are arranged in two directions which are usually normal to each other or in a multidirectional orientation means all fibers are arranged in more than two directions, consider following diagram regarding to orientation of continuous fiber throughout the lamina.

Table 1.26: Arrangement of Continuous Fibers in Single Layer or Lamina

Fiber orientation throughout the lamina	Building block	Characteristics
Uni-directional orientation of continuous fibers		Continuous fibers align parallel to each other in one dimension
Bi- directional orientation of continuous fibers		Continuous fibers arranged perpendicular to each other in two dimensions
Multi - directional orientation of continuous fibers		Continuous fibers arranged in multi direction

A laminate composite that containing unidirectional fibers in lamina then composite material showing remarkable characteristics like highest strength and modulus in the longitudinal direction of the fibers on the contrary in transverse direction its strength and modulus found low. A composite in which lamina containing bi-directional fibers, the strength and modulus depending upon the counting of fibers in the longitudinal aswellas transverse directions i.e. we can say property variation coincide to fibers counting. Further if composites have balanced fiber in lamina then properties are same in both directions.

1.47. Lamina with Discontinuous Fiber

In this case, if dis-continuous fibers are reinforced with matrix phase so as to form a lamina. These discontinuous fibers can be arranged either in unidirectional orientation (say one direction) or in random orientation (say arrange in random manner). Discontinuous fiber-reinforced composites have lower strength and modulus than continuous fiber composites. However, with random orientation of fibers, we can obtain equal mechanical and physical properties in all directions in the plane of the lamina. Consider following diagram regarding to orientation of continuous fiber throughout the lamina.

Table 1.27: Arrangement of Dis-continuous fibers in Single Layer or Lamina

Fiber orientation throughout the lamina	Building block	Characteristics
Discontinuous uni-directional orientation of fibers		Dis- continuous fibers align parallel to each other in one dimension
Discontinuous random orientation of fibers		Dis-continuous fibers arranged randomly in the lamina

The thickness required to bear load or to maintain a given deflection in a fiber-reinforced laminate composite that is prepared by stacking several laminas in a specified sequence and then consolidating them to form a laminate. Consider following diagram which demonstrate the laminate composites.

Table 1.28: Arrangement of Lamina in Laminate Composites with their Fibers Orientation in Lamina

A consolidated laminate composite (Orientation of lamina)	Building block	Remark
Consolidated laminate composites with various layer of lamina of continuous fibers		➢ Top & bottom layer have unidirectional orientation of continuous fibers. ➢ 2nd layer has continuous fiber perpendicular to top lamina ➢ The successive 3rd, 4th, 5th and 6th layer of lamina has continuous fibers that are arranged at an angle of $30^{\circ}C$ from 2nd last or from top 2nd lamina. ➢ 2nd last lamina has continuous fiber which is perpendicular to the bottom layer of composites
Discontinuous & continuous fibers orientation in laminate composites		➢ Top & bottom layer have unidirectional orientation of continuous fibers. ➢ The successive 2nd, 3rd, 4th, 5th, 6th and 7th layer of lamina has uni-directional discontinuous fibers in lamina ➢ The last or bottom layer has Uni-directional continuous fibers in lamina.
Laminate composites with different lamina & fiber orientation		➢ Top & bottom layer have unidirectional orientation of continuous fibers. ➢ The successive 2nd, 3rd layer of lamina has inclined continuous fibers in lamina at an angle of $45^{\circ}C$ from top to toward left & right direction. ➢ 4th lamina has perpendicular fibers with respect to top layer ➢ 5th, 6th and 7th layer of lamina has further inclined fibers in lamina at an angle of 45° from top to toward left & right direction.

Moreover, we can also have fiber-reinforced polymer laminas which combined with thin aluminium or other metallic sheets to form metal–composite hybrids, commonly known as fiber metal laminates (FML). For example - Metal–composite hybrids are ARALL and GLARE. In ARALL laminate composites we have alternate layers of aluminium sheets and unidirectional aramid fiber–epoxy laminate.

A consolidated ARALL - laminate composite	Building block	Remark
Consolidated ARALL laminate composites with unidirectional continuous layer of aramid fiber/ epoxy in individuals		$\succ$ Thickness of each layer is 0.3 mm. $\succ$ 1st, 3rd and 5th layer is made up of aluminium alloys sheets. $\succ$ 2nd & 4th layer has continuous aramid fiber/ epoxy in unidirectional position.

On the other hand, in case of GLARE we have alternate layers of aluminium sheets and either unidirectional or bidirectional S-glass fiber–epoxy laminates. These laminate metal-hybrids composites are widely used in aircraft structures for manufacturing wing panels and fuselage sections.

Note

- Continuous-fiber laminated composites individual layers of laminae are arranged or oriented in such directions so that it will enhance the strength & improve load carrying capacity.

- Laminated composites fiber exhibits high tensile - strength that is about 3500 Mpa. A typical polymeric matrix normally has a tensile strength that ranges between 35 to 70 Mpa.

- In actual practise longitudinal tension and compression loads are carried by the fibers, while the function of matrix phase is to distributes the loads between the fibers in tension and further maintaining the load to inhibit buckling during compression. The matrix phase is also cooperating to handle shears that exists among the layers when laminate composites subjected to tension.

- Usually three layers are arranged alternatively for better bonding between reinforcement of polymer matrix phase. For instance, plywood and paper.

- A hybrid laminate can also be fabricated by the use of different constituent materials or of the same material with different reinforcing pattern.

- If a fiber reinforced composite consists of several layers with different fiber orientations, it is known as multilayer composite.

- In fabrication of laminate composites, lamina may be made of either same material or different material and fiber may be arranged in diverse manners throughout the structure.

- It is also possible to combine different kinds of fibers to form either an inter-ply or an intra-ply hybrid laminate. An inter-ply hybrid laminate consists of different kinds of fibers in different laminas, whereas an intra-ply hybrid laminate consists of two or more different kinds of fibers interspersed in the same lamina.

- Unidirectional (0°) laminae are extremely strong and stiff in the 0° direction. However, they are very weak in the 90° direction because the load must be carried by the much weaker polymeric matrix.

1.48. Nano-composites

Nano composite is as older as human birth or human civilization. It is based on the fact, a study that has been performed by the doctors & medical practitioner and found that nano-composite exists in nature in the form of **"Bones"**. The composition of, bone nano-composites is given below.

> **Composition = 30% (Matrix material phase) + 70% (Nanosized mineral particle phase)**

In bones matrix material phase is collagen fibres (Polymer) & Nanosized mineral particle phase is hydroxyapatite crystals.

Now a day's further evolvement in composites materials are identified as a nano-composites & it is defined as **"A composite material that has matrix phase in which the dispersed phase / reinforcement must have at least one of its dimensions in the nanometre range"** Nano means dispersed phase should have range of = 1 (NM) = 10^{-3} mm (micron) = 10^{-9} m.

For example – In this cadre polymer nano-composites are very popular in which reinforcement of polymer matrix phase is perform by nano size dispersed phase. It is always desirable & recommended for manufacturing of any nano-composite is that one of the constituent's phase must be less than 100 nm. Lighter weight due to low filler loading & Low cost due to fewer amount of filler are two distinct features of nano-composites. Even though, these nano-composites are also known for superior mechanical properties, superior thermal, electrical, optical, and other properties. The potential cause of improvement of such a property may be either because of higher interfacial interaction with the matrix phase (i.e. very high ratio of surface area to volume), in addition to it nano-reinforcements has superior promising property rather than fibers reinforcement. However, in some cases it has been identified that toughness and impact strength decreases. A newly developed nano-composite elements are known as nano-composites film that is apply for packaging applications due to enhance elastic modulus, better transmission rates for water vapor and heat dissipation etc. Nano-composites coincide to dimension of dispersed phase that is belonging to nano-meter range, but these nano - particles may also be differentiate on the basis of their dimensions, like,

1. The reinforcing phases may one dimension on a nano level say in the shape of platelets for example - Clays and layered silicates.

2. The reinforcing phases or fillers may have two dimensions in the nano-meter scale. For example - Carbon nanotube and cellulose whiskers.

3. The reinforcing phases may have three dimensions in the nano-meter range. Say reinforcement is performed by iso-dimensional nanofillers. For example - Spherical silica nanoparticles.

1.48.1. Polymer Nano-composites

Polymer Nano-composites is very famous in cadre of nano-composites and it is defined as "Polymer nano-composites are polymer matrix composites in which the reinforcement is done by filling a particle that must have at least one of its dispersed particles dimensions in the nano-meter range". These polymer nano-composites may be further classified on the basis nano reinforcements. For example, Nano-clay, Carbon nanofibers, and Carbon nanotubes. A brief description of those three are given below.

1.49. Nano Clay Composites

The reinforcement used in nano-clay composites is a layered silicate clay mineral, such as smectite clay which is a member of phyllosilicates family. In the natural form, the layered smectite clay particles has thickness that ranges between 6 to 10 mm thick and contain about 3000 planar layers. These smectite clay can be exfoliated or delaminated and dispersed as individual layers, each ~1 nm thick. The crystal structure of each layer of smectite clays contains two outer tetrahedral sheets between Si adopt a position on the other hand central octahedral sheet may be either of alumina or magnesia. Each & every layer has thickness of about 1 nm which is separated by a very small gap known as interlayer or the gallery. One of the common smectite clays used for nano-composite applications is called montmorillonite clay. Furthermore, whenever modified smectite clay is used with a polymer for manufacturing nanocomposite, following three types of dispersion may be possible. These are,

1.   Intercalated dispersion, in which one or more polymer molecules are intercalated between the silicate layers. The resulting material has a well-ordered multi-layered morphology of alternating polymer and silicate layers. The spacing between the silicate layers is between 2 and 3 nm.

2.   Exfoliated dispersion, in which the silicate layers are completely delaminated and are uniformly dispersed in the polymer matrix. The spacing between the silicate layers is between 8 and 10 nm. This is the most desirable dispersion for improved properties.

3.   Phase-separated dispersion, in which the polymer is unable to intercalate the silicate sheets and the silicate particles are dispersed as phase separated domains, called tactoids.

1.50. Carbon Nano Fibers

Carbon nanofibers are produced either in vapor-grown form or by electrospinning. They are typically 20–200 nm in diameter and 30–100 mm in length. These can be manufactured in continuous long length fiber that has diameter of 5 to10 mm. Carbon fibers are also made in vapor-grown form, but their diameter is in the range of 3–20 mm.

1.50.1. Manufacturing of Carbon Nano-fibers

Vapour Grown Carbon Nano-fibers are produced in vapor phase by decomposing carbon-containing gases, such as methane (CH_4), ethane (C_2H_6), acetylene (C_2H_2), carbon monoxide (CO), benzene, or coal gas in presence of floating metal catalyst particles inside a high-temperature reactor. Ultrafine particles of the catalyst are either carried by the flowing gas into the reactor or produced directly in the reactor by the decomposition of a catalyst precursor. The most common catalyst is iron, which is produced by the decomposition of ferrocene, $Fe(CO)_5$. A variety of other catalysts, containing nickel, cobalt, nickel–iron, and nickel–cobalt compounds can also be used. Depending on the carbon-containing gas, the decomposition temperature can range up to 1200°C. The reaction is conducted in presence of other gases, such as hydrogen sulfide and ammonia, which act as growth promoters. Cylindrical carbon nanofibers grow on the catalyst particles and are collected at the bottom of the reactor. The diameter of carbon nanofibers and the orientation of graphite layers in carbon nanofibers with respect to their axis depend on the carbon-containing gas, the catalyst type, and the processing conditions, such as gas flow rate and temperature. Carbon nano-fibers are available in platelet form, in which the graphite layers are stacked normal to the fiber axis or hollow tubular construction, in which the graphite layers are parallel to the fiber axis and fishbone or herringbone in which graphite layers inclined at an angle of either 10^0 or 45^0 with the fiber axis. Moreover, Single-wall and double-wall graphite layers exists in heat treated carbon nanofibers, in some cases folded graphite layers also obtained in both single-wall and double-wall. Carbon nanofibers have been incorporated into several different thermoplastic and thermoset polymers. In general, incorporation of carbon nanofibers in thermoplastics has shown modest to high improvement in modulus and strength, whereas their incorporation in thermosets has shown relatively smaller improvements.

1.51. Carbon Nano-tubes

Carbon nano-tubes are known for fascinating outstanding properties like mechanical, thermal, electrical, and optical properties. It has elastic modulus which is closer diamond and 3 to 4 times higher with respect to carbon fibers. Its maximum operating temperature has been found in vacuum, that is about 2800°C and their thermal conductivity is about twice as that of diamond and their electric conductivity is 1000 times higher than that of copper. This fibers can be reinforced with various matrices like polymers matrix, metals matrix and ceramics matrix for manufacturing composites material. Structural composites which is produced from carbon nano-particles & diverse matrix phase, now **a days used in** energy storage devices, electronic systems, biosensors, and drug delivery systems etc. Due to its unique characteristics carbon nano-tubes (CNT) may be of two types.

1. Single-walled carbon nanotube (SWCNT)
2. Multi-walled carbon nanotube (MWCNT)

1.51.1. Single-walled Carbon Nanotube (SWCNT)

SWCNT is a seamless hollow cylindrical tube that is formed by rolling a sheet of graphite layer. SWCNT are available in diameter, that ranges between 1 to 1.4 nm and its length falls between range of 50 and 100 mm. The structure of an SWCNT can be understand from how the graphite sheets is rolled up with desired chirality or helicity. Chirality is defined with the help of chiral angle and the chiral vector.

1.51.2. Multi-walled Carbon Nanotube (MWCNT)

MWCNT consists of a number of concentric SWNT. Both SWNT and MWNT are closed at the ends by dome-shaped caps. The concentric SWNTs inside an MWNT are also end-capped. The outer diameter of an MWNT ranges between 1.4 and 100 nm. The separation between the concentric SWNT cylinders in an MWNT is about 3.45 nm which is slightly greater than the distance between the graphite layers in a graphite crystal. The specific surface area of an MWNT depends on the number of walls. Furthermore, the young modulus of 1 to 2 nm diameter sized, SWNT is about1 TPa (1TPa = 10^3GPa) and that of MWNT it falls between 1.1 and 1.3 TPa. Consider the following table which shows some property of carbon nano-tubes.

Table 1.29: Mechanical Property of Various Carbon Nano Composites Tube

Type of nano-tubes	Young modulus (GPa)	Tensile strength (GPa)	Density (g/cm³)
SWCNT	1054	75	1.3
SWCNT bundle	563		1.3
MWCNT	1200		2.6
Graphite in plane	350	2.5	2.6
P – 100 carbon fiber	758	2.41	2.15

Where 1 TPa = 10^3 GPa

From the above table it has been concluded that young modulus of MWNT is higher than that of SWNT due to presence of Vander Waals forces between the concentric SWNTs in MWNT.

1.51.3. Properties of Carbon Nanotube–polymer Composites

Carbon nanotubes in a polymer matrix will create composites with very high modulus and strength further if proper dispersion of nanotubes is to be perform in the polymer matrix then mechanical properties can be enhanced as compare to the neat polymer. Consider the following example.

a. If MWNTs dispersed in a toluene solution of polystyrene then elastic modulus and tensile strength improve further if MWNT weight fraction increases then both characteristics would be increased. The addition of 5 % wt of MWNT perform then the elastic modulus of the composite was 120% higher and the tensile strength was 57% higher than that of polystyrene.

b. 1% weight MWNTs reinforced polyamide-6 composites then both tensile modulus and strength of the composite were significantly higher compared with the tensile modulus and strength of polyamide-6.

Note

- Now a days, polymer based organic / inorganic nanocomposites are very famous.
- Whenever carbon nanofiber added to epoxy. The epoxy resin was diluted by using acetone as the solvent. The diluted epoxy was then infused into the carbon nanofiber mat. After removing the solvent, the epoxy-soaked mat was cured at 120°C and then post cured. It has been observed that flexural modulus and strength were enhanced.

- Carbon nanofiber-reinforced polypropylene composites exhibits greater tensile modulus and strength as compared to polypropylene itself. If carbon nanofiber-reinforced polypropylene subjected to surface oxidation in a presence of CO_2 atmosphere then tensile modulus and strength of the composite increases that is about three times with respect to polypropylene.

- The Young's modulus of multi-walled carbon nanotube (MWCNT) ranges between 0.27 to 1.8 TPa and that of SWNT ranges from 0.32 to 1.47 TPa. Similarly, the strength of MWNT ranges from 11 to 63 GPa and that of SWNT from 10 to 52 GPa.

- The carbon nanotube exhibits high tensile strain (up to 15%) before fracture and also observed that carbon nanotubes indicate nonlinear elastic deformation under tensile, bending, as well as twisting loads. At high strains, they tend to buckle of the wall.

- Young's modulus and strength of nanotubes depending upon the tube length, diameter, number of walls, chirality, existence of impurities and even atomic structure.

- Carbon nano-tubes – reinforced alumina (CNT/Alumina) represent higher toughness in three cases these are, Crack deflection at the CNT/Alumina interface, Crack bridging by CNT and CNT pull-out on the crack plane.

1.52. Natural Fiber Composites

We have been studied about synthetic fiber & their respective composites. we have not only studied about their remarkable properties & structural and non-structural applications but also assimilate diverse form of the composite's material. However, they also indicate some adverse effect like resistant to biodegradation which further leads to environmental problems. In order to overcome the environmental issues or saving & protecting the environment, natural fiber can be used to manufacturing a composites material. Natural fibers like plant fiber exhibits very high strength, available in huge amount which can be extracted from renewable resource and also biodegradable but all natural fibers are hydrophilic in nature and have high moisture content, which is responsible for creating poor interface between fiber and hydrophobic matrix. Natural fiber composites may be defined as **"A composite material whose reinforcement is performed by using natural fibers, particles or platelets or we can say by renewable resources, known as natural fiber composites"**. Natural fibers are those which is extracted from plant (say, jute, bamboo, coir, sisal, and pineapple), Animal and other mineral sources. Natural fibers can be classified in following manner.

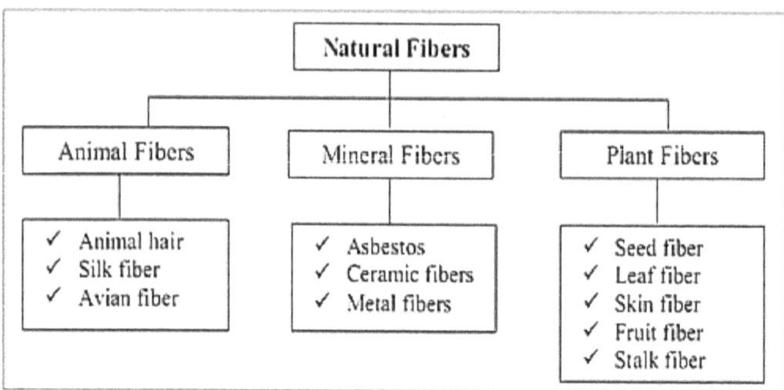

Diagram 1.17: Classification of Natural Fibers

A brief description of animal fiber, mineral fiber & plant fiber are given below.

Animal Fiber:- Animal fiber consisting proteins. For example, mohair, wool, silk, alpaca, angora. Animal hair (wool or hair) are the fibers taken from animals or hairy mammals like Sheep's wool, goat hair (cashmere, mohair), alpaca hair, horse hair etc. Silk fiber are the fibers collected from dried saliva of bugs or insects during the preparation of cocoons, For example silk from silk worms. Avian fiber are the fibers that obtained from birds for example feathers and feather fiber.

Mineral Fiber:- Mineral fibers are natural fiber which is extracted from minerals. For example, Asbestos, wood, serpentine, amphiboles, anthophyllite. Ceramic fibers includes glass fibers (Say Glass and Quartz), aluminium oxide, silicon carbide, and boron carbide. Metal fibers includes aluminium fibers etc.

Plant Fiber:- Plant fibers consisting cellulose, for example cotton, jute, flax, ramie, sisal and hemp. Cellulose fibers used to manufactured paper and cloth.

- Seed fiber are the fibers which collects from the seed and seed case e.g. cotton and kapok.

- Leaf fiber are the fibers collected from the leaves for example sisal and agave.

- Skin fiber are the fibers are collected from the skin or bast surrounding the stem of their respective plant. These fibers have higher tensile strength than other fibers. Therefore, these fibers are used for durable yarn, fabric, packaging, and paper. Some examples are flax, jute, banana, hemp, and soybean.

- Fruit fiber are the fibers which is collected from the fruit of the plant. For example, coconut (coir) fiber.

- Stalk fiber are the fibers that is actually the stalks of the plant. For example, straws of wheat, rice, barley and other crops including bamboo and grass. Tree wood is also such a fiber.

Consider the following table which shows some important property of natural fiber.

Table 1.30: Property of Natural Fibers

Plant Fibers	Density (Kg/m^3)	Tensile Strength (MPa)	Young'sModulus (GPa)
Jute Fiber	1300-1500	200-450	20-55
Sisal Fiber	1300-1500	80-840	9-22
Coconut Coir	1150-1250	106-175	6-8
Areca Nut	1050-1150	300-530	30-60

1.53. Advantages & Disadvantages of Composites Material

Due to their outstanding & remarkable characteristics & property of composites material these are used in diverse field of application. Its positive use always attracts our mind toward advantages on the contrary its detrimental effects coincide to unpleasant behavior of the composite material in perspective of application. However, we know that each & everything in nature showing both advantage & disadvantage in similar manner composites material also represent both advantage & disadvantage. We have plenty of advantage of composites material which expands their filed of application like, lighter weight, strength and stiffness, improved fatigue life, corrosion resistance, with good design aspect, low costs, specific strength, low density and high / low modulus (modulus/density), high strength fibers (like carbon), high weight savings, greater payloads, toughness, in some cases fatigue resistance (high durability), reduced assembly cost, long life and low manufacturing cost, high stiffness & strength (In one direction orientation), low stiffness & strength (In other direction orientation), low stiffness (In one multi direction orientation), better load transfer, high ductility & toughness (high surface interaction between fibers & matrix), dimensional stability and cost effective fabrication. In case of metal matrix composites material, high specific strength and modulus, corrosion resistance, low coefficient of thermal expansion (In case of metal matrix reinforced with graphite) and high strength stability at high temperature, superior mechanical property, flexibility for design, easily fabrication, impact resistance, excellent fatigue and many more. Ceramic matrix composites included high strength, hardness, high service temperature, chemical inertness, low density and many more. Further in case of carbon – carbon composites, gradual failure, high temperature stability, low creep at high temperature, low density, high thermal conductivity, low coefficient of thermal expansion, {High specific strength, High specific stiffness, High fracture resistance, Good abrasion resistance, Good impact resistance, Good corrosion resistance, Good fatigue resistance, low cost (Polymer matrix composites)}, high strength to weight ratio, high modulus, electrical conductivity, good thermal shock resistance, abrasion resistance, fracture resistance, excellent high temperature stability (in inert & vacuum environment) and superior corrosion resistance, one of the potential advantages of composites is the ability to cure or bond a number of detail parts together to reduce assembly costs and the number of required fasteners, are most common. Except the above advantages, composites material also have **certain disadvantages** which includes, high raw material costs, high fabrication & assembly costs, detrimental effects of temperature and moisture, complex mechanical characterization, uneasy repairing of composites structure, poor strength in the out of plane direction where the matrix carries the primary load, susceptibility to impact damage and delamination's or ply separations, greater difficulty in repairing them compared to metallic structures, Temperature affects mechanical properties of composites (Matrix-dominated mechanical properties decrease with increasing temperature). Cold temperatures affect to mechanical property, sensitive to moisture and humidity, moisture absorption rises if elevated temperatures increased, {Low operating temperature, {Low thermal resistance and High coefficient of thermal expansion (Polymer matrix composites)}, High CTE & CMEs, Low elastic property (Polymer matrix composites)}, brittle fracture mechanism, high material cost, high manufacturing cost, limited by temperature circumstances, mechanical property varies and so many other.

1.54. Properties and Applications of Composite Materials

Composites not a new material but it had been exists in nature in various form that is known as a natural composites material. like wood. We know that composites manufactured from two form that includes **"Reinforced phase + Matrix phase"**. The natural composites may be reinforced by different type of fibers like reinforcement of animal fiber (Animal hair, Silk fober, Avian fiber) or Mineral fiber (Asbestos fiber, Ceramic fibers, Metal fibers) and Plant fibers (Seed fiber, Leaf fiber, Skin fibe, Fruit fibers, Stalk fiber) with matrix phase. All those composites have its own characteristics & application. But in present scenario most of the application required enhanced characteristics & property therefore natural fiber replaced by synthetic fiber. Further we know that, composites material has four types like (Polymer matrix composites) PMCs, (Metal matrix composites) MMCs, (Carbon matrix composites) CMCs, and (Carbon – Carbon composites) CCCs. Composites material is referred as isotropic, heterogeneous and complex in nature and their properties are generally affected by many variables like reinforcement form, volume fraction, and geometry, properties of the interphase, interface (reinforcement and matrix are joined) and void content. It has been observed that properties of any composites influenced by reinforcement form, volume fraction, and geometry. For example, consider that case of E-glass fiber-reinforcing to polyester composites. The reinforcement can be performed by discontinuous fibers, woven roving or heavy fabric and straight i.e. parallel continuous fibers. Discontinuous reinforcement is not as efficient as continuous. However, discontinuous fibers does not impart enough strength as that of continuous fiber but their reinforcement enhanced some property. Consider following table in which we have to showing mechanical properties of E-glass fiber-reinforcing to polyester composites.

Table 1.31: Mechanical Property of E-glass Reinforced Fiber Polyester which Shows % Value of Various Composites

S.no	Glass content in % weight	Tensile strength (MPa)	Tensile modulus (GPa)
1	20% (Bulk molding compound)	45	9
2	30% (Sheet molding compound)	85	13
3	30% (Chopped strand mat)	95	7.7
4	50% (Woving roven)	250	16
5	70% (Unidirectional axial)	750	42
6	70% (Unidirectional transverse)	50	12

The above composites can be prepared by reinforcing by both particulate & fibrous reinforcement and hence it showing dimensional stability, smooth surface, reduce cost etc. Another composite is known as chopped strand mat (CSM), which is prepared by discontinuous fibers in random orientation that is about 25 mm long. Sheet molding compound (SMC) contains chopped fibers (Randomly oriented in two dimensions) whose length is ranges between 25–50 mm. Like BMC, SMC also contains particulate mineral fillers, such as calcium carbonate and clay. Composites reinforced with randomly oriented fibers tend to have lower volume fractions than those which made with aligned fibers or fabrics. Some composites with discontinuous-fiber reinforcement are made by chopping up composites reinforced with aligned continuous fibers or fabrics that have high fiber contents. It has been observed that, the strength and stiffness of polyester and most polymer matrices is lower than that of E-glass, carbon, and other reinforcing fibers. Due to the effect of reinforcement of particulates or improper orientation of fiber throughout the matrix phase modulus of SMC is always greater than that of CSM however both have the same fiber content. Particulate reinforcement in polymers responsible for augmenting the modulus but it does not increase strength. However, mostly it has been observed that particles reinforcement offers improvement of strengths & moduli of metal matrix composites. The moduli and strengths of the composites reinforced with fabrics and aligned fibers is greater than that of discontinuous fibers for example, the tensile strength of woven roving is more than twice than that of CSM. This property is measured parallel to the warp direction of the fabric. On the contrary if we measured the property perpendicular to the warp then elastic and strength properties become lowers slightly regarding to warp direction. Further, the elastic modulus, tensile strength, and compressive strength at 45^0 to the warp and fill directions of a fabric are much lower than the corresponding values in the warp and fill directions and axial modulus and tensile strength of the unidirectional composite are much greater than those of the fabric. However, the modulus and strength of the unidirectional composite in the transverse direction is lower. In case of transverse strength, it is lower than that of SMC and CSM. Now we will study under here properties of PMCs (Polymer matrix composites), MMCs (Metal matrix composites), CMCs (Ceramic matrix composites), and CCCs (Carbon – Carbon composites) and also describe application of composites in brief.

1.55. Properties of Polymer Matrix Composites

Polymer matrix composites also referred as a resin composites. These composites commercially availiable as polyester, vinyl ester, epoxy, phenolic, polyimide, polyamide, polypropylene, polyether ether ketone (PEEK), and others. In general, polymer is weak and showing low stiffness characteristics. For ensuring the development of their structural application, it is very imperative to enhance mechanical properties of PMCs which can be obtained by reinforcing the polymer matrix phase either by continuous or discontinuous fibres. To improve properties of polymer composites polymer matrix phase should be reinforced either by particulates / particle that may be either ceramic or metallic, fibers, flakes or arranged in laminates. particulates / particle reinforcement is performed either by aluminium, silver or permanent magnet particles. Aluminium and silver reinforcement polymer matrix showing increased electrical and thermal conductivity on the contrary ferrous and permanents magnet polymer composites are used to manufacturing composites which is employed as magnetic tape that is generally used for recording audio and video. Furthermore, if polymer matrix phase is reinforced by ceramic fibers such as aluminium oxides, aluminium nitrides, boron nitrides or diamonds then it has been found property like - physical property, thermal conductivity and thermal conductivity were improved. These properties are always desirable for electronic packaging application.

PMCs are known for low cost and easily fabrication, further low mechanical property like strength, modulus and impact resistance restricted its application as a structural composite's material. But whenever PMCs reinforced by strong fiber then we have improved in various property say, High specific strength, High specific stiffness, High fracture resistance, Good abrasion resistance, Good impact resistance, Good corrosion resistance, Good fatigue resistance, Low cost on the contrary it has showing adverse effects like, low thermal resistance and high coefficient of thermal expansion.

1.55.1. Elastic and Strength Properties of Polymer Matrix Composites

Polymer matrix composites that reinforced with aligned continuous fibers can be used as structural materials. Consider the following table in which elastic properties of reinforced unidirectional fiber polymer matrix composites are presented at room temperature. It is applied for fiber volume fraction that is about 60%. This table show axial modulus, transverse modulus, shear modulus & poison ratio of E-glass, aramid, boron, SM PAN-based carbon, UHS PAN-based carbon, UHM PAN-based carbon, and UHM pitch-based carbon.

Table 1.32: Elastic Property of Unidirectional Polymer Matrix Composites & 7075 – T6 Aluminium

Fiber	Poisson ratio	Axial modulus in GPa	Transverse modulus in GPa	Shear modulus in GPa
7075 – T6 aluminium	0.33	70	70	27
E- Glass	0.28	45	12	5.5
Aramid fiber	0.34	76	5.5	2.1
Boron fiber	0.25	210	19	4.8
SM Carbon (PAN)	0.25	145	10	4.1
UHS (IM) Carbon (PAN)		180	9	4.0
HM Carbon (PAN)		225	7	3.9
UHM Carbon (PAN)	0.20	310	9	4.1
UHM Carbon (PITCH)	0.25	480	9	4.1

These polymer matrices are widely manufactured from epoxies which has good mechanical properties. The strength, properties of polymer matrix composites always depending upon the temperature which is responsible for variation in mechanical properties. Some polymers for instance, polyimides, have good properties at elevated-temperature. Polyimide matrices are used to manufacturing aircraft gas turbine engine components which has operating temperature about 290°C. Another property of composites is poison ratio and defined as the ratio between transverse strain to axial strain when the composite material is subjected to axial loading direction. It has been observed that composite transverse modulus and shear moduli are much lower than that of axial values. Furthermore, in case of carbon fiber composites, moduli increase if tensile stress increases and decreasing with increasing the compressive stress. Due to low transverse strengths of unidirectional laminates, they are used in structural applications however laminates, layers in various directions can be improve strength, stiffness and buckling like properties. The elastic properties of quasi-isotropic laminates composites and 7075-T6 aluminium is given below in following table and it has been observed that moduli and strengths are much lower than the axial values of unidirectional laminates made of the same material. In overall sense we can say that laminate geometry plays a vital role in most of the applications.

Table 1.33: Elastic Property of Quasi-isotropic Matrix Composites & 7075 – T6 Aluminium

Composites Fiber	Poisson ratio	In – plane shear MPa	In plane extensional modulus GPa
7075 – T6 aluminium	0.33	27	70
E- Glass	0.28	9	23
Aramid fiber	0.32	10	28
Boron fiber	0.33	29	80
SM Carbon (PAN)	0.31	20	54
UHS (IM)Carbon (PAN)	0.31	20	63
UHM Carbon (PAN)	0.32	41	110
UHM Carbon (PITCH)	0.32	62	165

The specific elastic moduli and densities of quasi-isotropic polymer matrix composites and 7075-T6 aluminium is given below in following table, It was concluded that specific moduli of composites reinforced with E-glass and aramid fibers are much lower than that of aluminium on the other hand the specific moduli of SM carbon fiber-reinforced epoxy are similar to aluminium. While the values of composites using UHS (IM) and UHM carbon fibers are greater than that of aluminium.

Table 1.34: Showing Specific Gravity & Specific Modulus

Composites Fiber	Specific gravity { g/cm³}	Specific modulus (GPa)
7075 – T6 aluminium	2.8	26
E- Glass	2.0	11
Aramid fiber	1.4	21
Boron fiber	2.1	40
SM Carbon (PAN)	1.6	24
UHS(IM) Carbon (PAN)	1.6	39
UHM Carbon (PAN)	1.7	66
UHM Carbon (PITCH)	1.8	92

Consider the following table, which shows the strength properties of unidirectional polymer matrix composites and 7075-T6 aluminium & concluded that composite strength in transverse and shear direction is lower than that of axial direction. Further, compression strength of carbon fiber-reinforced composites decreases with increasing fiber modulus.

Table 1.35: Showing, Strength in Axial, Transverse, Axial Compression, Transverse Compression & Shear

Composites Fiber	In – plane shear MPa	Axial tension in MPa	Transverse tension in MPa	Axial compression in MPa	Transverse compression in MPa
7075 – T6 aluminium	331	551	551	517	514
E- Glass	70	1020	40	620	140
Aramid fiber	60	1240	30	280	140
Boron fiber	90	1240	70	3310	278
SM Carbon (PAN)	80	1550	41	1380	169
HM Carbon (PAN)	80	3500	40	1380	168
UHM Carbon (PAN)	80	1380	41	760	170
UHM Carbon (PITCH)	41	900	20	280	99

Also consider the following table which shows, specific axial strengths of unidirectional composites and 7075-T6 aluminium, strength properties of quasi-isotropic polymer matrix composites and 7075-T6 aluminium and specific strength properties of quasi-isotropic polymer matrix composites and 7075-T6 aluminium.

Table 1.36: Showing Specific Axial Strengths of Unidirectional Composites and 7075-T6 Aluminium

Composites Fiber	Axial tension in MPa	Axial compression in MPa
7075 – T6 aluminium	205	190
E- Glass	485	300
Aramid fiber	688	200
Boron fiber	620	1620
SM Carbon (PAN)	960	850
UHS(IM) Carbon (PAN)	2165	860
UHM Carbon (PAN)	831	450
UHM Carbon (PITCH)	500	265

Table 1.37: Showing Strength Properties of Quasi-isotropic Polymer Matrix Composites

and 7075-T6 Aluminium

Composites Fiber	Tension in MPa	Compression in MPa	Shear in MPa
7075 – T6 aluminium	551	517	330
E- Glass	550	330	250
Aramid fiber	450	190	65
Boron fiber	480	1100	350
SM Carbon (PAN)	580	580	400
UHS(IM) Carbon (PAN)	1350	580	405
UHM Carbon (PAN)	450	250	205
UHM Carbon (PITCH)	300	95	73

Table 1.38: Showing Specific Strength Properties of Quasi-isotropic Polymer Matrix

Composites and 7075-T6 Aluminium

Composites Fiber	Tension in MPa	Compression in MPa
7075 – T6 aluminium	280	190
E- Glass	250	158
Aramid fiber	330	140
Boron fiber	240	550
SM Carbon (PAN)	350	370
UHS(IM) Carbon (PAN)	2200	360
UHM Carbon (PAN)	300	160
UHM Carbon (PITCH)	169	50

1.56. Physical Properties of Polymer Matrix Composites

Physical property of polymer composites are attributed by so many characteristics like density, coefficient of thermal expansion and thermal conductivity. It has been found that composites reinforced with pitch-based UHK carbon fibers showing same mechanical properties than that of UHM carbon fibers that made from pitch however they have higher thermal conductivity. Some composite material showing negative coefficient of thermal expansion means that when they experience heat contraction would be achieved. Consider following table which shows physical properties of unidirectional and quasi-isotropic polymer matrix composites that includes density, thermal conductivity, and CTE.

Table 1.39: Showing Physical Properties of Unidirectional Polymer Matrix Composites &
7075 – T 6 Aluminium

Composites Fiber	Tension in MPa	Compression in MPa
7075 – T6 aluminium	280	190
E- Glass	250	158
Aramid fiber	330	140
Boron fiber	240	550
SM Carbon (PAN)	350	370
UHS(IM) Carbon (PAN)	2200	360
UHM Carbon (PAN)	300	160
UHM Carbon (PITCH)	169	50

Further consider another table which shows physical properties of unidirectional and quasi-isotropic polymer matrix composites & 7075 – T 6 aluminium.

Table 1.40: Showing Physical Properties of Quasi-isotropic Polymer Matrix Composites &
7075 – T6 Aluminium

Composites Fiber	Density {g/cm^3}	In plane coefficient of thermal expansion 10^{-6}/k	In plane thermal conductivity W/M-k	Thermal conductivity W/M-k (Along thickness)
7075 – T6 aluminium	0.33	24	130	130
E- Glass	0.28	10	0.9	0.6
Aramid fiber	0.32	1.4	0.9	0.1
Boron fiber	0.33	6.5	1.4	0.7
SM Carbon (PAN)	1.6	3.0	2.8	0.5
HM Carbon (PAN)	1.7	2.3	6	0.5
UHM Carbon (PAN)	1.8	-0.4	23	0.5
UHM Carbon (PITCH)	1.8	-0.4	194	10

It has been concluded that throughout the thickness, thermal conductivities of unidirectional and quasi-isotropic composites are similar to the transverse thermal conductivity of unidirectional composites.

1.57. Fatigue Properties of Polymer Matrix Composites

Fatigue is nothing but it is a phenomenon of failure of any material when subjected to cyclic or fluctuating loading. It has been observed that fatigue behaviour of unidirectional composites showing favourable result in condition of tension. But in case of polymer matrix composites that was reinforced with carbon, boron, and aramid fibres, SN-curve are relatively flat. Glass fiber-reinforced composites shows a greater reduction in strength with increasing number of cycles. Still, polymer matrix composites reinforced with HS glass are shows fatigue resistance and hence used in manufacturing helicopter rotors where fatigue resistance does matter. Furthermore, mostly metals are failed due to fatigue by crack propagation when subjected to fluctuating tensile rather than compressive load. But composites showing high susceptibility to fatigue failure when it experiencing compression rather than tension. Consider the following curve which shows the cycles to failure as a function of maximum stress for carbon fiber-reinforced epoxy laminates when it is subjected to tension–tension and compression–compression fatigue. The laminates have 60% of their layers oriented at 0^0, 20% at $+45^0$, and 20% at -45^0. They are subjected to a fluctuating load in the 0^0 direction.

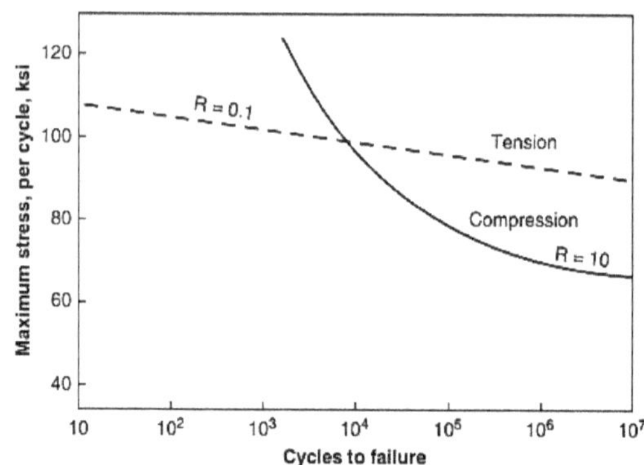

Curve – Cycle required to failure at maximum stress for carbon -reinforced epoxy laminates (loading condition = tension – tension = R = 0.1 & compression – compression = R = 10).

Where R = {Minimum stress / Maximum stress}, For tensile R = 0.1 & for compression R = 10. It has been concluded high strength reduction were achieved in case of compression–compression fatigue instead of tension – tension.

1.58. Creep and Creep Rupture of Polymer Matrix Composites

Polymer matrix composites that reinforced with carbon and boron fiber indicates high resistant to deformation and failure under sustained static load if they are loaded in a fiber-direction. This phenomenon known as creep and creep rupture. In case of aramid fiber-reinforced polymer matrix then creep and creep rupture behaviour not favourable. Glass fibers reinforced composites exhibits superior creep, and creep rupture resistance. But polymers showing creep if & only if, it does not constrain with fibers and creep can only be considered in polymer matrix composites when stress is loaded in matrix dominated direction rather than fiber. For example - laminate through-thickness direction.

Note

- The orientation of included Particles, Flakes, Fibers in matrix phase also a responsible factor for properties improvements.

1.59. Application of PMCs Composites Material

Polymer Matrix Composites falls under the category of synthetic polymer which is further classified as a thermoplastic & thermosets, the basic purpose of synthetization is to enhance or develop the desired level of property in the PMCs thereby we can take advantages in diverse field of application. It has plenty of fascinating characteristics that is attributed as high mechanical properties, design flexibility, easy fabrication, light weight, low cost, erosion resistance, corrosion resistance, impact resistance, superior fatigue strength, high strength & stiffness. If it is reinforced - continuous fibers with matrix phase then we have got improved productivity, durability, high performance, weight saving, low cost and many more. Now a days these appealing characteristics are required & desirable in aerospace application, transportation, construction application, marine application, sporting goods, and recently used in infrastructure, construction and transportation, space shuttle application, fighter jet, corvette leaf springs, snow skis, I – Beam, pressure vessels and sports (tennis). Basically, PMCs found application in, Aerospace structures, Marine application, Automotive, Sports, Bulletproof, Chemical application, Biomedical applications, Bridges manufacturing and Electrical field etc. The most important & widely used polymer composites are fibre-reinforced polymer (laminar structure) which is formed by accumulating or stacking a number of layer and thereafter bond together to achieve desired thickness of polymer. The property of such a polymer composites can be altered by changing the fibre orientation among layers in the laminate structures. Further except the laminar structure, a prominent work has been performed by the polymer matrix that is reinforced by E-glass, carbon, aramid and boron, Glass fiber-reinforced polymers (GFRPs), Glass fibers reinforced with thermosetting polyester and vinyl ester resins, Carbon fiber-reinforced polymers (CFRPs) {high performance application}, graphite / epoxy polymer composites (Used as ultra-light leg prostheses-for athletes) Boron fiber-reinforced polymer (BFRP) composites and Aramid fiber-reinforced composite (AFRP) etc. We will be discussed under here some application of polymer matrix composites.

1.59.1. Aircraft and Military Applications

Now a days polymer matrix composites (PMCs) are widely used as structural composites in aircraft manufacturing and military applications where weight saving and high speed does matter. In order to full fill the desired purpose fiber-reinforced polymer composites are employed. The basic motive of application of fiber-reinforced polymers in aircraft and helicopter applications is not only weight saving but also increase fuel saving and payload. Following advantages were noticed with polymer matrix composites.

1. We have reduced number of components and fasteners & consequently fabrication and assembly costs reduces. For example, if components & fasteners are made from carbon fiber-reinforced epoxy of Lockheed L-1011 assembly then 25% weight saving can be achieved.

2. Fins of helicopter usually made by carbon fiber-reinforced epoxy because it has higher fatigue resistance and corrosion resistance as a result of which maintenance & repair costs, replacement and rebuilding cost is low rather than metal fins.

3. The laminated construction used with fiber-reinforced polymers allows the possibility of aero-elastically tailoring the stiffness of the airframe structure. For example, the air-foil shape of an aircraft wing can be controlled by appropriately adjusting the fiber orientation angle in each lamina and the stacking sequence to resist the varying lift and drag loads along its span.

In this boron fiber-reinforced epoxy, carbon fiber-reinforced epoxy (Used to manufacturing wings, fuse Lage, empennage components that indicate superior structural integrity & durability and consequently mostly use in military aircrafts) are important. For example - Airframe of aircraft "AV-8B" that is a 'VSTOL' showing overall weight saving approximately 25% due to application of carbon fiber reinforced epoxy, the fighter aircraft F-22 provide benefit of 25% by weight by using carbon fiber-reinforced polymer. Consider the following table which shows the application of various polymer composites material that is used in airframe manufacturing for military application & also indicate overall weight saving.

Table 1.41: Fiber – Reinforced Polymer Composites Used in Military Air Craft

S. no	Type of aircraft	Fiber -reinforced Polymer composites Material	Component manufactured	Over all weight saving in %
1	F – 14	Boron fiber – epoxy	Skin on the horizontal stabilizers box	19
2	F – 11	Caron fiber – epoxy	Under the wing's fairings	25
3	F – 15	Boron fiber – epoxy	Fin, rudder and stabilizer skin	25
4	F – 16	Caron fiber – epoxy	Skin on vertical fin box, fin leading edge	23
5	F / A-18	Caron fiber – epoxy	Wing skin, horizontal and vertical tail box, wing & tail control surface	35
6	AV – 8B	Caron fiber – epoxy	Wing skin and substructure, forward fuge Lage , horizontal stabilizers , flaps , ailerons	25

Now a day's polymer matrix composite material like high strength carbon fiber-reinforced epoxy are also utilizing to manufacturing a structural component of commercial aircrafts so that we can save the weight of commercial aircraft & flight. Consider the following table which shows the % weight saving of commercial aircraft by using fiber reinforced polymer composites.

Table 1.42: Fiber – Reinforced Polymer Composites Used in Commercial Air Craft

Name of aircraft	Type of aircraft	Component manufactured	Over all weight saving in %
Boeing	727	Elevator face sheets	25
	737	Horizontal stabilizer	22
	737	Wing spoilers	37
	756	Ailerons, rudders, elevators, fairings etc.	31
McDonnell – Douglas	DC - 10	Upper rudders	26
	DC - 10	Vertical stabilizer	17
Lock-heed	L-1011	Aileron	23
	L-1011	Vertical stabilizer	25

Let be example of commercial aircraft that is Boeing – 777 that is made up of two structural material.

1.   Aluminium alloys structure whose proportion is about 50%.
2.   Polymer matrix composites structure has 50% portion that is made up of carbon fiber-reinforced epoxy.

Boeing components like, horizontal stabilizer, vertical stabilizer, elevator, rudder, control surfaces, engine cowlings, and fuselage floor beams are manufactured from polymer matrix composites (Carbon fiber-reinforced epoxy). This composites only cooperate in % weight saving furthermore it indicates only 10% weight. Consider the following diagram in which fiber reinforced composites are used to manufacturing the components.

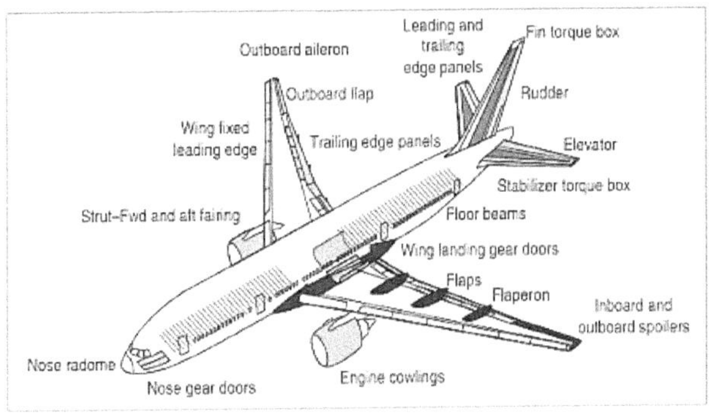

Diagram 1.18: Fiber Reinforced Components of Boeing – 777

Airplanes like Boeing - 787, in which carbon fiber-reinforced epoxy consisting large proportion in construction and rest of the constructional materials including aluminium alloys (20%), titanium alloys (15%) and steel (10%). Further polymer composites material also plays a prominent role in small aircraft which is used to manufacturing aircraft's airframe. The carbon fiber–epoxy and Kevlar- 49 are generally used to manufacturing wing skins, main spar, fuselage, empennage, and various control surfaces etc.

The initial applications of carbon reinforced composites come on the front line in, aerospace / defence, military aircraft, sports equipment, commercial aircraft structures, unmanned aerial vehicle (UAV) structures, aircraft gas turbine fan blades, spacecraft structures, satellite antennas, optomechanical systems, ships etc. But now a days it can also be used in industrial applications that includes compressed natural gas and hydrogen tanks, wind turbine blades, x-ray equipment, robots and other machine components such as rollers, energy storage flywheels, optomechanical equipment, antennas, uranium centrifuges, fuel cells, fire fighter compressed air tanks, prosthetics & orthotics, oil and natural gas exploration and production, transmission cable cores, train structures and interiors, civil engineering structure repair and reinforcement etc. Moreover, it also has a consumer application, that including sports equipment, like golf clubs, snowboards, skis, bicycles, softball bats, hockey sticks, bows and arrows, fishing rods, and tennis rackets, musical instruments, head phones, furniture, notebook computer cases. Automobile components, such as structural components and drive shafts, and boats.

Moreover, fiber reinforced polymer composites that was a Glass fiber-reinforced polymers (GFRPs) has great significance in cadre of polymer matrix composite. These fiber are manufactured by reinforcing either continuous unidirectional fibers, discontinuous fiber or fabrics in polymer matrix phase. Continuous unidirectional fibers reinforcement in polymer matrix composites exhibits high structural capacity rather than discontinuous fiber. GFRPs consisting number of applications in the commercial, consumer and industrial field. In commercial & consumer, aerospace/defence structural applications are important. Aerospace/defense applications including, solid fuel rocket motor cases, rocket launchers, aircraft fairings flooring, interiors, and cargo containers, helicopter rotor blades, patrol boats, mine hunter ships, and ballistic armour. In case of industrial applications, truck cabs and trailers, compressed natural gas tanks, wind turbine blades, ships & boats, chemical industry equipment like tanks and pipes, water treatment equipment are important. In case of civil engineering application, it includes, building components, bridges, concrete rebars, concrete forms, transmission line poles. In addition to it we have also application in endoscopic surgery equipment, electrical insulation products, oil and gas production, fire fighter compressed air tank and glass fiber-reinforced parts consumer products. It has also contribution in sporting application & automobile application like boats, sports equipment, gliders, shower enclosures and bathtubs, automobile components such as bodies, engine parts, and interior components, and recreational vehicles. Furthermore, fiber-reinforced polymers are used in many military and commercial helicopters for making baggage doors, fairings, vertical fins, tail rotor spars, and so on. One key helicopter application of composite materials is the rotor blades. Carbon or glass fiber-reinforced epoxy is used in this application. In addition to significant weight reduction over aluminium, they provide a better control over the vibration characteristics of the blades. Fiber-reinforced polymers can control critical flopping & twisting frequency by varying the type, concentration, distribution, as well as orientation of fibers along the blade's chord length. Another advantage of fiber-reinforced polymers is blade manufacturing flexibility. These composite blades can be easily transformed into filament-wound form & also molded into complex air-foil shapes with somewhat additional manufacturing costs. But carbon fiber-reinforced epoxy application also imposes some limitation in aircraft structures like high cost, low impact damage tolerance and sensitivity to light, in addition to it they also exhibit galvanic corrosion when interaction takes place with metal like aluminium or titanium such a corrosion can inhibit by coating the contacting surfaces by paints. Eventually, composite materials are widely used to manufacturing components of Airbus and Boeing like bulkheads, fuselages, wings, frame structure and many more but except the aircraft application it is also used to manufacturing other components for instance. Air-foil surfaces, antenna structures, compressor blades, engine bay doors, fan blades, flywheels, helicopter transmission structures, jet engines, radar, rocket engines, solar reflectors, satellite structures, turbine blades, turbine shafts, rotor shafts in helicopters, wing box structures, etc.

Note

- Airbus was the first commercial aircraft manufacturer, which extensively use of composites in their A-310 aircraft. These components show 10% of the aircraft's weight only. Such a components are lower access panels and top panels of the wing leading edge, outer deflector doors, nose wheel doors, main wheel leg fairing doors, engine cowling panels, elevators and fin box, leading and trailing edges of fins, flap track fairings, flap access doors, vertical stabilizer, rear and forward wing–body fairings, pylon fairings, nose radome, cooling air inlet fairings, tail leading edges, upper surface skin panels above the main wheel bay, glide slope antenna cover, and rudder.

- The Airbus A-320 that is a commercial aircraft to use an all-composite at tail portion, which includes the tail cone, vertical stabilizer, and horizontal stabilizer. Its 25% of weight is made from composites.

- In air bus A-380, central torsion box, rear-pressure bulkhead, the tail, and flight control surfaces, such as the flaps, spoilers, and ailerons are made from polymer matrix composites.

- The major structural applications for fiber-reinforced composites are in the field of military and commercial aircrafts, for which weight reduction is critical for higher speeds and increased payloads.

- Carbon fiber-reinforced epoxy become a primary material in many wings, fuselage, and empennage components.

- Now a days E-glass, carbon, aramid and boron fibers reinforced polymer matrix composite has plenty of application. Thermosets are dominant matrix materials for structural applications on the other hand thermoplastic matrices often reinforced either by discontinuous fibers or continuous fibers so as to achieve desired level of property throughout the composite's material.

- Glass fiber-reinforced polymers (GFRPs) is very popular & most widely used polymer matrix composites. In this polymer glass fibers is reinforced in thermosetting polyester or vinyl ester resins. It is mostly used as structural composite that can be manufactured at low cost.

- Carbon fiber-reinforced polymers (CFRPs) play a vital role in high-performance applications although it is more expensive than that of Glass fiber-reinforced polymers (GFRPs).

- Boron fiber-reinforced polymer (BFRP) composites falls under the category of advanced composite that had been apply in aerospace & defence applications. For example, flight application that has boron fiber-reinforced epoxy F-14 horizontal stabilizer, fighter aircraft and UAVs.

- Aramid fiber-reinforced composite (AFRP) applications include ballistic armour, such as helmets, commercial aircraft pressure bottles & freight containers, aircraft engine containment rings, brakes and other friction parts, patrol and service boats, leisure boats, canoes, kayaks, and luxury yachts.

1.59.2. Space Applications

We have been study the application of polymer matrix composites for aircraft & military application and their respective importance. In similar manner, space application required weight saving which always in prime consideration so that we can achieve fuel saving of space vehicles. In order to full fill the desired purpose fiber-reinforced composite is preferred in many space vehicle applications as a structural member. For example - Structure of space shuttle that has mid-fuselage truss structure are made from boron fiber-reinforced aluminium tubes, payload bay door made from sandwich laminate of carbon fiber-reinforced epoxy face sheets and aluminium honeycomb core, remote manipulator arm is made from ultrahigh-modulus carbon fiber-reinforced epoxy tube, and pressure vessels is made from Kevlar-49 fiber-reinforced epoxy. In case whenever large structural components is required then, fiber-reinforced polymers are used for support structures & manufacturing of smaller components like - solar arrays, antennas, optical platforms. The characteristics & property of these ceramic components of supporting structure like dimensional stability always influenced by temperature variation unless thermal expansion or distortion would be affected the efficiency of focusing characteristics of telescope so therefore selection of composites material for any particular or specific space application does matter. For example - Carbon fiber-reinforced epoxy laminates composites are used in artificial satellites and space telescope due to outstanding characteristics like - lower density, higher strength, higher stiffness–weight ratio, superior mechanical property and approximate zero coefficient of thermal expansion. If temperature variation experienced then large changes in the relative positions of mirrors or lenses takes place due to either thermal expansion or distortion, that may create serious problems in focusing the telescope. To solve such a complicated problem low weight & low coefficient of thermal expansion, carbon fiber-reinforced epoxy tubes are used in building truss structures for low earth orbit (LEO) satellites and interplanetary satellites. This truss structures provide necessary support to optical benches, solar array panels, antenna reflectors, and other modules. But we also have another major drawback with epoxy-based composites in LEO satellites which is susceptible to degradation, when it interacts to atomic oxygen that absorb from the atmosphere. This complication can be solved by wrapping them with thin aluminium foils. Furthermore, fiber-reinforced polymers in the space environment also exhibits two issues these are outgassing of the polymer matrix when interact to vacuum in space and embrittlement due to particle radiation as a result of which outgassing has accountability to dimensional changes and existence of microcrack, appears due to embrittlement. Outgassing results, lowering the efficiency or may lead to improper functioning of installed sensors or solar cells.

1.59.3. Automotive and Transportation

The basic purpose of introducing a composites material in an automotive & transportation field is to achieve weight saving thereby we can attain better mileage with minimum fuel consumption. In addition to enabling ground breaking vehicle designs, composites also cooperate to manufacturing light weight of vehicle. These composite materials are used to manufacturing bearing materials, bodies, connecting rod, crankshafts, cylinder, engines, piston, etc. Fiber-reinforced composites play a dominating role in automotive industry due to their extraordinary characteristics. The fiber-reinforced composites can be used to manufacturing following components of vehicles these are:

1. Manufacturing of body components by composites material
2. Chassis components manufactured by composites material
3. Engine components manufactured by composites material

1.59.3.1. Body Components

Body means, the exterior – structure of the vehicle, in exterior body components high stiffness, dent resistance in conjunction with "Class A" surface finishing for good appearance are always imperative. In order to full fill the desired purpose composite material is used for that components which made from E-glass fiber-reinforced sheet molding compound (SMC) composites. It is manufacturing by reinforcing randomly oriented discontinuous glass fibers into a matrix phase of polyester or vinyl ester resin. Usually fibre length required for SMC that is approximately 25 mm in length. The basic advantage of E-glass fiber is low cost rather than carbon fiber. Exterior body components include hood or door panels. SMC can also be used to manufacturing underbody or under hood components that includes radiator supports, bumper beams, roof frames, door frames, engine valve covers, timing chain covers, oil pans. Furthermore, small and large components of automobiles, light trucks, and heavy trucks is also manufactured by SMC thereby we can gain low tooling cost and weight reduction. These small & large parts are hoods, pickup boxes, deck lids, doors, fenders, spoilers etc.

1.59.3.2. Chassis Components

In case of chassis components, Corvette rear leaf spring, Uni-leaf spring are manufactured from fiber-reinforced composites. E-glass fiber-reinforced epoxy composites are used to manufacturing uni-leaf spring. The replacement of multi-leaf steel springs by Uni-leaf spring not only facilitate to weight saving but also enhance fuel saving of vehicles. Moreover, other structural chassis components like drive shafts and road wheels are also manufactured by the composite's material. But the major drawback of such a structural part is high manufacturing cost regarding to steel.

1.59.3.3. Engine Components

In case of engine components, it is imperative that material must not only have high temperature stability but also should have superior fatigue strength. In order to full fill the desired purpose, engine components are manufactured either from high temperature polymers, metal matrix or ceramic matrix composites. The polymer matrix used for this application may either be a polyester, vinyl ester or polyurethane. In addition to it, fiber-reinforced composites like glass fiber-reinforced polyester has been used in sports racing car since long period of time, instead of aluminium body panels that will impart weight saving, higher speed and also fuel saving which is absolutely necessary for racing car but now a days carbon fiber-reinforced epoxy composites adopt a position of GFRC in race car. Body, chassis, interior, suspension components, gear box and survival cell, which protects & prevent the driver in during crash are made from CAFR- EPOXY. A recent application of carbon fiber-reinforced polymers that comes on the front line in high cost vehicle like BMW M6 in which roof panel is made from GFRP and consequently we have significant weight saving of the roof panel.

Note

- The manufacturing process used for making SMC parts is called compression molding.

- Discontinuous randomly oriented, E-glass fiber-reinforced vinyl ester composites is used to support composite radiator.

- SMC composites employed for part integration of station wagon tailgate assembly for that load-bearing is desirable. Usually tailgate consisting two pieces i.e. outer SMC shell and inner reinforcing SMC piece further both are bonded together by means of urethane adhesive. This composite tailgate assembly can be manufacturing at low cost and also has an advantage of weight saving. These outer & inner portions are made from discontinuous randomly oriented E-glass fiber-reinforced with polyester matrix.

- Vehicle panel of composites is manufactured by applying structural reaction injection molding (SRIM) technique. This, panel are usually made from E-glass fibers that is discontinuous in nature aswellas randomly oriented in the structure of matrix phase. Matrix phase may be either polyurethane or polyurea.

- Carbon fiber-reinforced polymers may also apply in automobile application instead of glass fiber-reinforced polymers composite because they can fulfil the same purpose alike a GFRC say weight saving, fuel saving but the major drawback is highly expensive & indicate higher strength–weight and modulus–weight ratios.

1.59.4. Contribution of Polymer Matrix Composites in Sport Applications

Fascinating characteristics of fiber-reinforced polymers composites material which can be attributed as, easy moulding or design flexibility, high elastic modulus, high strength, light weight, Stiffness, vibration resistance, durability and good corrosion resistance etc. are imperative for manufacturing sporting good or sport equipment's. or in other word we can say that it is a promising material for manufacturing a sport equipment's. Polymer composites has diverse form among those carbon fiber-reinforced epoxies exhibit valuable contribution in sport application.

- The basic purpose of utilizing a carbon fiber is to weight saving. These weights saving mostly required in bicycle race & canoe races where higher speeds and quick maneuvering does matter. Bicycle frames for racing bikes today are mostly made from carbon fiber reinforced epoxy tubes, fitted together by titanium fittings and inserts.

- Sandwich type constructions of composites that is manufactured by composition of two things, one is a carbon or boron fiber-reinforced epoxies called (Skin material) and second is a soft & light weight urethane foam called (Core material). We have final product that has stiffness aswellas high weight saving, and now a days these properties are desirable for production of tennis rackets or snow skis. Finally, we can say that this composites material applies for manufacturing tennis rackets or snow skis.

- The advantage of light weight is also taken in golf clubs. In golf club, golf shaft is made from carbon fiber-reinforced epoxy composites. The basic reason behind is that composite golf shaft has reduced weight rather than steel golf shafts as a result of which club head can strike golf ball with desired force and hence faster swing & long drive can be achieved.

- Glass and carbon fiber-reinforced epoxy thin stick has been used as a fishing rods due to their higher modulus & light weight. For fly-fishing rods, carbon fiber-reinforced epoxy is preferred because it create a small tip deflection and "wobble-free" action. It also dampens the vibrations more rapidly and reduces the transmission of vibration waves along the fly line. Thus, the casting can be longer, quieter, and more accurate, and the angler has a better "feel" for the catch. Furthermore, carbon fiber-reinforced epoxy rods recover their original shape much faster than the other rods.

- In order to reduce the material cost of carbon fibers reinforced composites sometimes carbon fibers may either be mixed with glass or Kevlar 49 fibers.

- Glass fiber-reinforced epoxy is preferred in pole-vault poles due to its high strain energy storage capacity. A good pole must have a high stiffness and high elastic limit stress so that the strain energy of the bent pole can be recovered to propel the athlete above the horizontal bar. As the pole is bent to store the energy, it prevents plastic deformation and fracture.

- Faster damping of vibrations provided by fiber-reinforced polymers reduces the shock transmitted to the player's arm in tennis or racket ball games and provides a better "feel" for the ball.

- In archery bows and pole-vault poles, the high stiffness–weight ratio of fiber-reinforced composites is used to store high elastic energy per unit weight, which helps in propelling the arrow over a longer distance or the pole-vaulter to jump a greater height.

Consider the following table which shows some most common application of fibre reinforced polymer composites.

Table 1.43: Application of Fiber Reinforced Polymer Matrix Composites in Sports

Application of polymer matrix composites in sports equipments (Fiber reinforced polymer composites)	
Tennis rackets	Racket ball rackets
Golf club shaft	Finishing rods
Bicycle frames	Snow water skis
Ski poles & pole vault poles	Hockey sticks
Base bats	Sail boats & kayaks
Oars & paddles	Canoe hulls
Surfboards & snow boards	Arrows
Archery bows	Javelins
Helmets	Exercise equipments
Atheletic shoes	Atheletic ssoles & heels
Bobsleds fishing poles	Props
Football helmets	Horizontal bars
Jumping board	Parallel bars
To rowing	

Eventually, we can say that the fibre-reinforced composite materials indicate remarkable property like, easy moulding, high elastic modulus, high strength, light weight, good corrosion resistance etc. so therefore, fibre-reinforced composite materials usually employed for manufacturing of sports equipment. For instance, bicycle frames, bobsleds fishing poles, football helmets, hockey sticks, horizontal bars, jumping board, kayaks, parallel bars, props, tennis rackets, to rowing furthermore carbon fibres and fibre glass composite materials provide moral support for athletes to perform own self with highest confidence.

1.59.5. Marine Application of Polymer Composites Material

The application of composites material in naval ships comes on the front line around 1950s and then further evolvement, expands their field of application. Alike other diverse field of application, polymer composites material also has a number of potential applications in the marine industries. In this application we can say that a composites material which has been used to manufacturing a components or parts of marine vessels known as marine composite. Furthermore, marine application always required some specific property say light weight, damage resistance, corrosion resistance etc. These composite materials mostly used to manufacture, boat hulls, bulkheads, deck, mast, propeller, and other components for military, commercial, and recreational boats and ships. Let be consider the application of some polymer composites like glass fiber-reinforced polyesters, Kevlar 49-fiber reinforced composites and carbon fiber-reinforced epoxy which is being extensively used in marine application now a days.

1.59.5.1. Glass Fiber-reinforced Polyesters

In present scenario, all most maximum components are manufactured either from glass fiber-reinforced polyester or glass fiber-reinforced vinyl ester resin. These components usually manufactured with the help of contact moulding technique. In this, different type of boats says sail boats, fishing boats, dinghies, life boats, yachts, hulls, decks, and various interior components are important. It has low manufacturing cost and hence can be handled by small companies for boats creations.

1.59.5.2. Kevlar 49-fiber Reinforced Composites

In some application, whenever weight saving in conjunction with higher cruising speed, acceleration, maneuverability, fuel saving, higher tensile strength to weight ratio and modulus to weight ratio is imperative then Kevlar 49-fiber reinforced composites is used. This composite material is widely used to manufacturing boat hulls, decks, bulkheads, frames, masts, and spars etc. Hence we can say that Kevlar 49-fiber reinforced composites is a best substitute of glass fiber-reinforced composites for such a applications.

1.59.5.3. Carbon Fiber-reinforced Epoxy

Carbon fiber-reinforced epoxy is used to manufacturing racing boats in which weight reduction does matter. In these boats, the complete hull, deck, mast, keel, boom, and many other structural components are manufactured from the carbon fiber-reinforced epoxy. In order to fulfil the desired purpose carbon fiber-reinforced epoxy laminates and sandwich laminates of carbon fiber-reinforced epoxy skins with either honeycomb core or plastic foam core are used. If we want to further enhance the impact resistance and weight saving of boats then carbon fibers may be hybridized with lower density and higher strain-to-failure fibers like - high-modulus polyethylene fibers etc. Vividly we can say that composites materials are used in hulls, decks, bulkheads, masts, propulsion shafts, rudders, and others of mine hunters, frigates, destroyers, and aircraft carriers. A largest composite ship of the world recognizes as **"Royal Swedish Navy"** that is made from fiber-reinforced polymer. Another is a long combat ship known as **"Stiletto"** whose primary construction material is carbon fiber reinforced epoxy.

Note

- Selection of composites material for any marine application always coincide to some specific factors like design specification, light weight, high strength, sometimes speed. payload capability of ship, maneuverability and stealth characteristics.

1.60. Application of Polymer in Construction and Infrastructure

Now a days wide range of application & larger contribution of composites material comes on the front line in the field of construction, architecture and infrastructure. Its application includes, doors, fixtures, moulding, roofing, shower stalls, swimming pools, vanity sinks, wall panels, window frames and decoration of home and offices as well as repairing of various infrastructure like buildings, bridges, roads, railways, and pilings. Usually reinforced – steel & concrete composition play a vital role in construction of bridges, buildings and other civil engineering infrastructures since long period of time. But reinforced concrete construction may become deteriorate after certain length of time due to corrosion of steel-reinforcing bars & expansion of steel bar during hot weather as a result of which crack propagation takes place around the steel bars & finally construction either may fails or load-carrying capacity aswellas sustainability below down. To solve a problem of such a construction failure, reinforced concrete and steel construction may be replaced by fiber-reinforced polymers. This replacement imparts superior corrosion resistance, long life of construction as well las low maintenance and repair costs. Fiber-reinforced polymer not only apply for new bridge construction or replacement of steel-reinforced concrete composition but also used for upgrading, retrofitting, strengthening damaged, deteriorating, or substandard concrete or steel structures, for upgrading purpose, strips and plates that is made from composite is fixed over the crack or damaged areas of the concrete structure by using adhesive, wet layup or resin infusion. Retrofitting is usually performed of steel components like steel girder and heavy steel construction components etc.

For that composite plates are attached with the help of flanges and consequently stiffness and strength of flange enhanced. The strengthening of reinforced concrete columns is required at that regions where high earthquake occurrences does matter. In order to protect the structure from earth-quick we will wrap the reinforced concrete columns with fiber-reinforced composite jackets in which the fibers are primarily in the hoop direction. The basic advantage of such a wrapping is that it will be enhanced longitudinal stiffness, easy installation and showing superior corrosion resistance rather than steel construction.

Note

- Fiber-reinforced polymers do not corrode and another advantage of using fiber-reinforced polymers for large bridge structures due to their lightweight, easy transportation, easier hauling & installation and less injuries in case of earthquake. Light weight facilitates to construct a bridge with longer span between the supports.

- Composite traffic bridge had been constructed by "Lockheed Martin Research Laboratories" in which E-glass fiber-reinforced polyester components, like composite deck (Sandwich laminate), face sheets, core (In the form of tubes) were used. The composites dock was supported on U-shaped beams of E-glass fabric-reinforced polyester. It was observed that design was modular, components were stackable and have easy transportation and assembly.

- Now a day's number of composite bridge decks is being constructed from glass fabric-reinforced polyester matrix that consists of either full-depth hexagonal or half-depth trapezoidal profiles. The profiles are supported on steel beams and road surface is a polymer-modified concrete.

- Composite materials offer various advantages like corrosion resistance, design flexibility, durability, light weight, and strength. These advantages not only move our journey toward innovation & application in the field of construction and infrastructure but also indicate equal contribution in other field for instance - medical applications, oil and gas, transportation, sports, aerospace, renewable energy (wind turbine blades) and many more. Some applications, such as rocket ships, does not have any ground without the composite materials.

- In construction field, pultruded fiberglass re-bar is used to strengthen concrete, and glass fibers are used in some shining materials.

1.61. Application of Polymer Composites in Corrosive Environments

Whenever corrosion comes on the front line in any application then this problem enforcing us for using composite material instead of traditional materials in such a condition fibre-reinforced polymer composite become a solution. The basic advantage of polymer composite is long-term resistance to severe chemical conditions and temperature environments. The application of these composites includes, chemical handling applications, corrosive environments, outdoor exposure, and other severe environments such as chemical processing plants, oil and gas refineries, pulp and paper industries, and water treatment. Some other very most common applications including, cabinets, ducts, oil & gas applications, fans, sewer application, grating, hoods, pumps, and tanks etc.

1.62. Application of Polymer Composites in Electrical & Electronics Field

In the field of electrical & electronics, components of polymer composites play a valuable role because there is a requirement of high dielectric properties, arc & trac resistance. Mostly electrical & electronics components are manufactured from thermoset polymers. Some most common applications include, arc chutes, arc shields, bus supports, lighting components, circuit breakers, control system components, metering devices, microwave antennas, motor controls, standoff insulators, standoffs and pole line hardware and printed wiring boards, substation equipment, switchgear, terminal blocks, and terminal boards etc.

1.63. Application of Polymer Composites in Energy Field

Composite material are also used to manufacturing a structural components of energy generation plant like renewable energy (Wind and solar power) say wind turbine blade. The application also coincides to same property such as superior strength, light weight of components, flexibility in processing etc. The application of polymer reinforced composites in wind and solar power sector not only augmenting the safety & efficiency of power plant but also take part in improving the power output. At the end of this section, it is explicit that the evolvement of polymer composites has diverse field of application due to their appealing characteristics. Further a short brief application of PMCs is given below.

- **Aerospace Structures:-** The military aircraft industry has mainly led the use of polymer composites. In commercial airlines, the use of composites is gradually increasing. Space shuttle and satellite systems use graphite/ epoxy for many structural parts.

- **Marine Application:-** Boat bodies, canoes, kayaks etc are manufactured from composites material.

- **Automotive & Transportation Application:-** Body panels, leaf springs, driving shaft, bumpers, doors, racing car bodies etc are manufactured from composites material.

- **Sports Goods Application:-** Golf clubs, skis, fishing rods, tennis rackets and so on, manufactured from composites material.

- **Bulletproof** vests and other armor parts.

- **Chemical** storage tanks, pressure vessels, piping, pump body, valves, and so on.

- **Biomedical Applications:-** Medical implants, ortho-pedic devices, X-ray tables etc. are made from composites material.

- **Constructional Application:- Bridges made from polymer composite** materials are gaining wide acceptance due to their lower weight, corrosion resistance, longer life cycle, and limited earthquake damage.

- **Electrical & Electronic Application:-** Panels, housing, switchgear, insulators, and connectors. And many more components are manufactured from composites material.

1.64. Property of Metal Matrix Composites

We know that metal matrix composites consisting two phases i.e.

Composition = Reinforced phase / Dispersed phase + Metal matrix phase

Reinforcement may either be performed by fiber (Which may be either continuous or discontinuous fiber), Particulates / Particles with the metal matrix phase. Whenever fiber, particulates / particles reinforced to metal matrix phase & then composite known as fiber – reinforced metal matrix composites and particulates / particles – reinforced metal matrix composites respectively. For example - Aluminum fiber reinforced – Aluminum metal, Silicon carbide fiber reinforced – Aluminum metal, Graphite fiber reinforced – Aluminum metal, Tungsten carbide particles reinforced – silver metal, Titanium carbide particles reinforced – steel, Diamond particles reinforced – steel, Tungsten carbide particles reinforced – Cobalt matrix and so many others. But in actual practice characteristics & behavior of any metal matrix composites depending upon the type of reinforcement means which kind of reinforcement is being perform i.e. fiber or particulates. However High specific strength, good modulus, lower coefficient of thermal expansion, high strength at high temperature, high temperature stability, toughness and corrosion resistance are some common property.

We are very familiar with metals & metallic alloy and their respective application in diverse field of mechanical engineering due to fascinating property like strength, hardness, ductility, ultimate tensile strength, toughness, brittleness and so many others. But whenever we want to enhance mechanical property a new material has been developed which is known as metal matrix composites. Such a metal matrix composites are manufactured by reinforcing, continuous fibers, discontinuous fibers, whiskers and particulates with the metal matrix phase and consequently we have got improvement in strength, stiffness, wear resistance on the contrary low coefficient of thermal expansion. Higher strength and modulus were achieved with continuous fibers but it has extremely high cost which limit the circumstances of the application of continuous fibers reinforced – metal matrix composites. Instead of continuous fibers reinforced – metal matrix composites, now a days discontinuous fibers or particles reinforced metal matrix composites are widely used due to low cost & high productivity. Furthermore, we will study about physical & mechanical property of some common metal matrix composites.

1.65. Mechanical Properties of Continuous Fiber-Reinforced Metal Matrix Composites

Whenever continuous fiber-reinforced with metal matrix composites in which fiber orientation always remain unidirectional exhibits higher transverse strengths. If we consider the case of elastic properties then it has been observed that axial moduli of metal matrix composite are higher than that of metal matrix phase when metal matrix phase of Al, Al, Ti are reinforced with continuous fiber of boron, alumina, and silicon carbide (SiC). Consider the following table which shows elastic property and density of unidirectional metal matrix composites.

Table 1.44: Elastic Property and Density of Unidirectional Metal Matrix Composites

Fiber used for reinforcement	Metal matrix phase	Density in g/cm^3	Axial modulus in GPa	Transverse modulus in GPa
Boron fiber	Aluminum	2.6	210	140
Alumina fiber	Aluminum	3.2	240	130
Silicon carbide fiber	Titanium	3.6	260	170

Consider another table which represents specific axial moduli of unidirectional metal matrix composites in which boron, aluminium & SiC fiber - reinforced with metal matrix phase of aluminium and titanium respectively. Further it has been observed that boron-fiber reinforced has high specific axial modulus rather than aluminium fiber - reinforced & SiC-Fiber reinforced metal matrix composites.

Table 1.45: Specific Axial Modulus of Unidirectional Metal Matrix Composites

Fiber used for reinforcement	Metal matrix phase	Specific Axial modulus in GPa
Boron fiber	Aluminum	81
Alumina fiber	Aluminum	75
Silicon carbide fiber	Titanium	72

Consider next table which shows strength of metal matrix composites in terms of axial tensile strength, transverse tensile strength, density and axial compressive strength of unidirectional metal matrix composites in which boron, aluminium & SiC fiber - reinforced with metal matrix phase of aluminium and titanium respectively. Further it has been observed that SiC-fiber reinforced - Ti metal has higher axial tensile strength, transverse tensile strength and axial compressive strength in comparison to boron fiber reinforced – Al metal & aluminium fiber reinforced – Al metal.

Table 1.46: Axial Tensile Strength, Axial Compressive Strength and Transverse Tensile Strength of Unidirectional Metal Matrix Composites

Fiber used for reinforcement	Metal matrix phase	Density in g/cm^3	Axial tension in MPa	Transverse tension in MPa	Axial compression in MPa
Boron fiber	Aluminum	2.6	1240	140	1720
Alumina fiber	Aluminum	3.2	1700	120	1800
SiC - fiber	Titanium	3.6	1700	340	2760

Consider next table which shows specific strength of metal matrix composites in terms of axial tensile strength, and axial compressive strength of unidirectional metal matrix composites in which boron, aluminium & SiC fiber - reinforced with metal matrix phase of aluminium and titanium respectively. Further it has been observed that Al-fiber reinforced - Al metal has higher axial tensile strength than those of boron fiber reinforced - Al metal & SiC-fiber reinforced - Ti metal on the other hand axial compressive strength of SiC-fiber reinforced - Ti metal is higher in comparison to boron fiber reinforced – Al metal & Al-fiber reinforced - Al metal.

Table 1.47: Axial Tensile Strength, Axial Compressive Strength of Unidirectional Metal Matrix Composites

Fiber used for reinforcement	Metal matrix phase	Axial tension in MPa	Axial compression in MPa
Boron fiber	Aluminum	480	660
Alumina fiber	Aluminum	530	560
SiC - fiber	Titanium	470	780

1.66. Physical Properties of Unidirectional Metal Matrix Composites

In case of unidirectional metal matrix composites which has been reinforced with continuous fibers, we have to consider so many physical properties among those density and thermal conductivity are demonstrated in following table in which boron fiber, aluminium fiber and SiC fiber are reinforced in aluminium and titanium metal matrix phase. It has been observed that silicon carbide fiber reinforced metal matrix composite exhibits higher density rather than boron & aluminium fiber reinforced metal matrix composites. Further boron fiber, aluminium fiber and silicon carbide fiber reinforced metal matrix composites showing axial coefficient of thermal expansion of 6, 7 & 5.9 respectively while alumina fiber reinforced aluminium metal matrix phase indicate transverse coefficient of thermal expansion, axial thermal conductivity & transverse thermal conductivity that is about 16, 80, 45 w/mk respectively. Consider following table in which all the data are tabulated below.

Table 1.48: Physical Property of Unidirectional Metal Matrix Composites

Fiber used for reinforcement	Metal matrix phase	Density in g/cm^3	Axial CTE $10^{-6}/K$	Transverse CTE $10^{-6}/K$	Axial thermal conductivity in W/m-k	Transvese thermal conductivity in W/m-k
Boron fiber	Aluminum	2.6	6	--	--	--
Alumina fiber	Aluminum	3.4	7	16	80	45
SiC - fiber	Titanium	3.9	5.9	--	--	--

1.67. Property of Discontinuous Fiber-reinforced Metal Matrix Composites

Discontinuous fiber-reinforced metal matrix composites has number of remarkable property which leads to stimulation for manufacturing various type of mechanical engineering components. Now a days, a most common and widely used application of discontinuous fiber-reinforced metal matrix composites is in manufacturing of internal engine components because for any kind of internal engine application, wear resistance, strength at elevated-temperature, higher thermal conductivities, low thermal stresses and high fatigue strength are always imperative. In order to fulfil the desired purpose internal engine components are usually manufactured either by discontinuous alumina fiber reinforced with aluminium metal matrix phase, discontinuous silica fibers reinforced with aluminium metal matrix phase or sometimes discontinuous carbon fibers reinforced with alumina metal matrix phase. The basic purpose of fibers reinforcement with aluminium metal matrix phase is to augmenting wear resistance, high temperature strength stability and fatigue strength, furthermore due to the improvement in wear resistance there is no requirement of other components like cast iron sleeves in engine blocks and cast-iron inserts in pistons. In addition to it, fiber-reinforced aluminium composites also exhibit higher thermal conductivities than that of cast iron. If fiber volume fractions is low in composites then their CTEs is approximately equal to unreinforced aluminium as a result of which thermal stresses would be reduced. In case of carbon fibers reinforced alumina metal matrix, wear resistance enhanced and coefficient of friction of cylinder walls also reduces.

1.68. Properties of Particle-Reinforced Metal Matrix Composites

There are plenty of particle-reinforced metal matrix composites which falls under the special category of metal matrix composites. Among those particles reinforced metal matrix composites, tungsten carbide particles embedded in a cobalt matrix phase (Sometimes referred as cermet, cemented carbide), Diamond particles embedded in a cobalt matrix (Referred as polycrystalline diamond) and tungsten carbide particle-reinforced silver are very common & most widely used in industrial application or engineering applications since long period of time. Tungsten carbide particles embedded in a cobalt matrix phase showing superior fracture toughness rather than monolithic tungsten carbide and now a days it is widely used to manufacturing of cutting tools and dies etc. Diamond particles reinforced cobalt matrix composites that is known as polycrystalline diamond are widely used in rock drill bits and cutting tools. But now a day's particle-reinforced metals are also widely used in electronic application, photonic thermal management and packaging applications. A metal matrix composite that is tungsten carbide particle-reinforced silver indicate better electrical conductivity, high hardness and wear resistance than that of monolithic silver and consequently best suitable for electrical application like circuit breaker contact pad material. Titanium carbide particles reinforced with Ferrous alloys is also considered in the class of particulate reinforced metal matrix composite showing higher wear resistance, stiffness and low density and hence it has been used in industrial applications since long period of time.

1.69. Mechanical Properties of Titanium Carbide Particle-reinforced Steel

Titanium carbide particle-reinforced steel also consider in the cadre of metal matrix composites in which titanium carbide particle-reinforced either with steel or steel alloys. It is usually available & recognized as **"Ferro-Tic."**. They are widely used to manufacturing cutting blades, tools and dies, and wear parts of machines since long period of time. In order to understand the mechanical behaviour of particle reinforced metal matrix composites, consider the example of titanium carbide particles reinforced austenitic stainless steel in which titanium carbide particles have fraction of 45% volume. It has been observed that modulus is very high as compare to monolithic base metal, low density in comparison to monolithic matrix and consequently, we have higher specific stiffness of metal matrix composite regarding to unreinforced metal and low density on the other hand we obtain reduced wear but composite indicate higher inherent wear resistance due to the reinforcement of TiC particles.

1.69.1. Mechanical Properties of Silicon Carbide Particle-reinforced Aluminium

Silicon carbide particles reinforced aluminium metal matrix, are belongs to category of special material. The property development of such a metal matrix composites always based upon the size of particle, volume fraction of particles in composites, matrix phase and type of manufacturing process. Consider the following table which shows mechanical properties of aluminium, titanium, and steel alloys and also indicate mechanical property when silicon carbide particles that has volume proportion of 25, 55, and 70% in aluminium, titanium, and steel alloys matrix phase.

Table 1.49: Mechanical Property of Silicon Carbide Particles Reinfoced with Aluminium, Titanium & Steel

Mechanical property						
	Before reinforcing			After reinforcing silicon carbide particles , with aluminium , titanium & steel		
	Aluminium 6061 – T6	Titanium 6Al – 4V	Steel 4340	Volume fraction in %		
				25%	75%	70%
Modulus (GPa)	69	113	200	114	186	265
Tensile yield strength (MPa)	275	1000	1480	400	495	225
Ultimate tensile strength (MPa)	310	1100	1790	485	530	225
Elongation (%)	15	5	10	3.8	0.6	0.1
Density g/cm^3	2.8	4.5	7.8	2.69	3	3
Specific Modulus (GPa)	25	26	26	40	63	88

Usually it has been found that if particle volume fraction increases, modulus, yield strength increase and fracture toughness and tensile ultimate strain decrease. Particle reinforcement also improves short-term elevated-temperature strength properties and fatigue resistance Furthermore, it has been also observed that modulus and specific modulus of the composite is higher than that of aluminium, titanium, and steel.

1.70. Physical Properties of Silicon Carbide Particle-reinforced Aluminium

Physical properties of silicon carbide particle-reinforced aluminium is depending upon the variation of proportion of silicon carbide particle in metal matrix phase. Consider the following table which shows physical properties of metal matrix composite in which silicon carbide particle is embedded in aluminium, titanium, and steel alloys metal matrix phase. The % volume proportion of silicon carbide particle is taken, 25, 55 & 70% respectively.

Table 1.50: Physical Property of Silicon Carbide Particle-reinforced Aluminium

Physical property						
	Before reinforcing			After reinforcing silicon carbide particles , with aluminium , titanium & steel		
	Aluminium 6061 – T6	Titanium 6Al – 4V	Steel 4340	Volume fraction in %		
				25%	75%	70%
Cofficient of thermal expansion 10^{-6} / k	23	9.5	12	16.5	10.5	6.2
Thermal conductivity W/ m-k	218	16	17	160 to 220	16 to 220	16 to 220
Density g/cm^3	2.8	4.5	7.8	2.9	3	3

It has been observed that if volume fraction increases, CTE decreases and density increases slightly and thermal conductivities remain in the range of aluminium alloys. The low CTEs are desirable in electronic application, photonic thermal management and packaging.

1.71. Mechanical Properties of Alumina Particle-Reinforced Aluminium

Alumina particles are used to reinforcing aluminium composite which can be used instead of silicon carbide particles reinforced composites because it does not react with metal matrix even though at high temperatures. Consequently, alumina-reinforced composites can be used in a wide range of processes and applications. However, the stiffness and thermal conductivity of alumina are lower than the corresponding properties of silicon carbide, and these characteristics are reflected in somewhat lower values for composite properties. Metal matrix composites (MMCs) is advanced composites material that contains metals or metallic alloy, that used as matrix phase (Primary phase) in which a second phase (Dispersed phase) is embedded or fixed firmly & evenly distributed throughout the composites structure so that we can augmenting the various property. The mechanical properties MMCs composites depending upon the microstructure. Stiffness, yield strength and ultimate tensile strength may improve by the weight fraction of the reinforcement phase in the matrix. Furthermore, we know that physical and mechanical properties of MMCs is enhanced due to the combined effects of metallic and ceramic properties. These enhanced mechanical property of MMCs are tensile and compressive strength, creep, notch resistance, and tribology while improved, physical properties are intermediate density, thermal expansion, and thermal conductivity etc. But they inhibit application due to lake of thermal fatigue, thermo-chemical compatibility and lower transverse creep resistance.

1.72. Application of Metal Matrix Composite

We have variety of metal matrix composites material which is either fiber (Continuous or Discontinuous fiber) reinforced or particulate reinforced. For example Alumina fiber-reinforced aluminium, Titanium boride particle-reinforced titanium, Silicon carbide particle-reinforced aluminium, Boron carbide particle-reinforced aluminium, Carbon fiber-reinforced aluminium, Silicon–aluminium composites, Carbon fibers reinforced aluminium, Carbon fibers reinforced copper, Diamond particle-reinforced aluminium, Diamond particle-reinforced copper, Diamond particle-reinforced silver, titanium carbide particle-reinforced ferrous alloys, tungsten carbide particle-reinforced cobalt and diamond particle-reinforced cobalt, and so many others. All those MMCs has different mechanical, Thermal & physical properties and hence those can be applied in various field of application. One factors which influence the application is type of fiber used i.e. metal matrix is reinforced either by continuous fiber or discontinuous fiber. Whenever metal matrix composites reinforced with continuous fibers then manufacturing cost become high which limit its application while discontinuous fiber reinforced metal matrix composites can be manufactured at low cost. Let be see some application of metal matrix composites.

- Boron fiber-reinforced aluminium tubes is used in space shuttle orbiter mid-fuselage section that has rib trusses structure.

- Carbon fiber-reinforced aluminium exhibits high axial stiffness and low axial coefficient of thermal expansion & now a days it is apply for manufacturing of Hubble Space Telescope high-gain antenna mast / waveguides whose function is to maintain antenna pointing accuracy.

- Silicon carbide fiber-reinforced titanium tubes represent extremely high axial moduli and strength and hence used as aircraft actuators.

- A metal matrix composite that consisting continuous fibers i.e. alumina fiber-reinforced aluminium is used in transmission lines. Furthermore, high specific stiffness (stiffness-to-density) and specific strength (strength-to-density) property, used for construction of support towers.

- Silicon particle-reinforced aluminium is used in automobile engine cylinder liners for wear resistance.

- Boron carbide particle-reinforced aluminium is being used for spent nuclear fuel containers.

- Cast iron inserts is replaced by discontinuous ceramic fibers composites that is widely used in automobile and truck engines to enhance cylinder bore and piston wear resistance and high-temperature strength.

- Titanium boride particle-reinforced titanium is used to manufacturing engine valves.

- Silicon carbide particle-reinforced aluminium has been used in aerospace, commercial, optical structural applications, electronic and photonic applications, optical systems, high-performance automobile and manufacturing of race car components etc. Aerospace/defence applications have included F-16 replacement fuel access doors and ventral fins, gas turbine engine fins and helicopter rotor blade sleeves. Sports equipment applications have included bicycle structural and mechanical parts, golf clubs, and baseball bats. Moreover, characteristics of high specific stiffness is being used in integrated circuit robot end effectors and high-speed machine parts.

- Silicon carbide particle-reinforced aluminium exhibits low Coefficient of thermal expansion and low density and hence it is widely used in electronic and photonic thermal management and packaging applications. In the electronic/photonic packaging industry, these materials are commonly known as Al-SiC.

Metal Matrix Composites also manufactured by reinforcing ceramic with metal matrix phase which not only augmenting specific strength, specific stiffness, wear resistance, corrosion resistance but also improve elastic modulus. In this case composites material have combine effects of both the phases i.e. primary phase & secondary as well. Means final composites consisting property of both metal matrix say ductility and toughness as well as property of reinforcement phase say high strength and high modulus and consequently. Eventually we have got high strength in shear & compression and higher temperature stability. Due to remarkable property, these ceramics are extensively used for high performance applications like aircraft engines & in automotive sector. Furthermore, aluminium oxide (Al_2O_3) and silicon carbide (SiC) powders in the form of fibers and particulates are used to reinforce metal matrix composite. Such type of composites has application in the automotive and aircraft industries for manufacturing of engine pistons and cylinder heads. Due to its high specific strength and corrosion resistance properties, MMCs can also be used in various engineering applications. Metal matrix composites extensively used in diverse field of engineering application because of their metallic characteristics like Electrical conductivity, thermal conductivity and noninflammability, matrix shear strength, high strength, high stiffness, decreased weight, higher service temperature, improved wear resistance, higher elastic module, ductility, higher toughness, abrasion resistance, facility of coating, joining, forming and heat treatment etc. But it limits the application due to higher density and complication in part processing. The low weight of MMCs are used in aerospace industry. If discontinuously reinforced aluminium composites that composed of high strength aluminium alloys which reinforced with silicon carbide particles or whiskers not only enhanced stiffness, strength and low weight but also improve fabricability and consequently MMCs behave as a promising candidate for structural components. Some other most important applications of MMCs are used to manufacturing, flat panels, cylinders, I–beam, channels, complex shape like engine shaft of gas engine, gas turbine engine components (Due to high strength, low weight and high temperature stability), structural components of aerospace planes (Due to weight saving & high temperature stability),In advanced fighter plane for manufacturing air fuselage structure, landing gears, steering gears, drag braces (Due to long life cycles, cost savings and weight savings), In gas turbine engine components manufacturing like exhaust nozzles, link vanes, blade cases and shaft rings (Due to light weight & high operating temperature).

1.73. Property of Ceramic Matrix Composites

We know that composites material consisting two phases i.e. Primary phase (Dispersed phase) and Secondary phase (Matrix phase).

$$Composition = \{(Reinforced / Dispersed\ Phase) + (Matrix\ Phase)\}$$

In case of ceramic composites, we have same provision alike other composites that is fiber may either be continuous or discontinuous fiber and particulates/particles, works as dispersed phase/reinforced phase which dispersed with ceramic matrix phase {Like – Alumina, Calcium alumino-silicate (CAS), Lithium alumino silicates (LAS)}. Furthermore, ceramic composites have remarkable property. However, they are manufactured for some specific application due to their extremely high manufacturing cost. Among those ceramic composites, some important & most common are Silicon carbide fiber-reinforced calcium alumino-silicates(CAS), Carbon fiber-reinforced lithium alumino-silicates (LAS), Carbon particles-reinforced lithium alumino-silicates (LAS), Silicon carbide particle-reinforced alumina, Silicon carbide whisker-reinforced alumina (Aluminium Oxide), Silicon carbide based fabric fiber reinforced with silicon carbide matrix, Silicon carbide fibers reinforced-silicon carbide matrix, Zirconia particles reinforced – alumina and Silicon carbide continuous fiber reinforced -silicon carbide are most important.

Behaviour, Property & characteristics of any ceramic composites depending upon the inclusion of % volume or % weight of dispersed phase / Reinforced phase with matrix phase. It has been observed that accumulated characteristics / property mostly coincides to amount of dispersed phase only. In addition to it, types of fiber used, particle size, final micro-structure of composites material and type of reinforcement material also responsible. Further I will be naming here some accumulated property without any specific bifurcation of ceramic composite material. These are, high strength, hardness, high temperature stability, chemical inertness, low density, corrosion resistance, high stiffness, wear resistance, oxidation resistance, fracture toughness improved (By using continuous fibers), low modulus, excellent thermal shock (By using continuous fibers), thermal conductivity, weight saving, fuel saving, high specific strength, improved thermal & mechanical behaviour when after reinforcing high specific stiffness and so many others. Except the above positive characteristics ceramic composites also has certain disadvantages like – low tensile strength, low fracture toughness, low resistance to thermal & mechanical shock, low thermal fatigue, low thermo-mechanical compatibility and low transverse creep resistance. These disadvantages potentially can be transformed in to advantages by modification the composition & processing. Usually ceramics material falls under the class of high stiffness, hardness, wear resistance, corrosion resistance, oxidation resistance and high-temperature application material on the other hand they also indicated certain non-favourable characteristics like low fracture toughness (Due to existence of flaws), low strength, low thermal and mechanical shock, but the lower value of these ceramics can be improved by transforming the ceramic material into ceramic composites and consequently we can use it in various valuable engineering applications. Further we will try to describe the mechanical & physical properties of ceramic composites material.

1.74. Mechanical Properties of Ceramic Matrix Composites

We have number of mechanical properties of ceramic matrix composites, among those some of them are tabulated under here,

Table 1.51: Fracture Toughness & Other Mechanical Property of Ceramic Composites Material

Materix material	Reinforcing material	Specific gravity	Young modulus	Ultimate tensile strength	Cofficient of Thermal expansion	Fracture toughness in MPa-m$^{\frac{1}{2}}$
Aluminimum		26	100	340	128	30 to 45
Steel		78	300	940	65	40 to 65
Alumina						3 to 5
Silicon carbide						3 to 4
Aluminium alloys						35
SiC	Al_2O_3				3	27
SiC	SiC					30
Alumina	Zirconia particles					6 to 15
Alumina	Silicon carbide whishkers					5 to 10
SiC	Continious SiC Fibers					30
Litium aluminium silicste (LAS)	SiC	19	13	69	2	43
Calcium aluminium silicste (CAS)	SiC	23	15	55	25	49

The basic problem with ceramic material is that it has low fracture toughness which can be augmented by reinforcing fibers, whiskers, and particles with ceramic matrix phase. Whenever such a ceramic composite is manufactured then we have not only found enhancement in fracture toughness but also notice improvement in strength aswellas thermal & mechanical shock resistance. Further it has been observed that fracture toughness highly improved when continuous fiber were reinforced with in the ceramic matrix phase. Consider following table which shows fracture toughness of structural metallic alloys, monolithic ceramics and ceramic matrix composites reinforced which reinforced with whiskers and with continuous fibers.

Table 1.52: Fracture Toughness of Some Common Ceramic Composites Material

Materix material	Reinforcing material	Fracture toughness in MPa-m$^{\frac{1}{2}}$
Aluminimum	None	30 to 45
Steel	None	40 to 65
Alumina	None	3 to 5
Silicon carbide	None	3 to 4
Silicon carbide (SiC)	Aluminia (Al_2O_3)	27
Silicon carbide (SiC)	Si-C	30
Alumina	Zirconia particles (ZrO_2)	6 to 15
Alumina	Silicon carbide whishkers	5 to 10
Silicon carbide (SiC)	Continious Si-C Fibers	30
Litium aluminium silicste (LAS)	Silicon carbide (SiC)	43
Calcium aluminium silicate (CAS)	Silicon carbide (SiC)	49

It has been observed that when continuous fibers are reinforced then fracture toughness of ceramic composites increases for example – If silicon carbide fibers reinforcement is performed with silicon carbide matrix phase then fracture toughness increases that become equivalent to aluminium alloys. Such a reinforcement is also greatly reducing the chance of failure. Furthermore, fracture toughness of monolithic ceramics increases somewhat due to presence of flaws. Monolithic ceramics that contains flaw size that ranges between 20 to 80 µ-m may fail at the stress value of 700 MPa. Silicon carbide matrix reinforced with fabric made from silicon carbide-based fibers referred as SiC / SiC. It has density of 2.86 g/cm3. If fiber volume fraction is taken 26% then composite can sustain strength up to 1315°C. It has also been found that continuous fiber reinforcement, SiC / SiC represent excellent thermal shock resistance. Further consider another table which shows, some important mechanical properties of SiC / SiC at temperature of 816°C, 1038°C and 1315°C.

Table 1.53: Mechanical Property of Silicon Carbide Fiber Reinforced – Silicon Carbide Matrix

Mechanical properties	At temperture of 816°C	At temperture of 1038°C	At temperture of 1315°C
Wrap tensile modulud in GPa	208	209	158
Tensile strength in MPa	362	325	295
Tensile proportional limit in MPa	117	168	163
Plane shear strength in MPa	47.2	None	None

Moreover, at room temperature, density & modulus of SiC / SiC ceramic composites is about 2.86 g/cm³ & 208 GPa respectively. Properties of such a range is required in aerospace application and industrial applications on the other hand ultimate tensile strength (325 MPa) of such a ceramic composite material that is observed at 1038°C used to manufacturing aircraft gas turbine engines components.

1.75. Physical Properties of Ceramic Matrix Composites

Physical property involves density, thermal conductivity and coefficient of thermal conductivity. Since we have number of ceramic matrix composites among those let be consider the case of silicon carbide fiber-reinforced silicon carbide (SiC / SiC) in which fiber volume fraction is about 40% then we have got density of 2.5 g/cm³, coefficient of thermal expansion and thermal conductivity is about 19 W/m·K and throughout-thickness its value is 9.5 W/m·K.

1.75.1. Application of Ceramic Matrix Composite

We have limited application of ceramic based components due high manufacturing cost. But some applications are given below.

- Continuous fibers reinforced - ceramic composites are widely used as structural materials.

- Silicon carbide matrices reinforced by carbon (SiC/SC) or silicon carbide fibers reinforced with carbon matrix (SiC/C) exhibits weight saving / weight reduction in addition to fuel saving and hence it is widely used in military aircraft engine flaps. Further C/SiC ceramic composites are also used in spacecraft optical systems.

- Silicon carbide whisker-reinforced alumina (Aluminium oxide) represents superior fracture toughness due to that it is extensively used in cutting tools.

- Silicon carbide particle-reinforced alumina has been widely used to make parts for abrasive slurry pumps. C/SiC brakes are being used in high-end automobiles.

- Ceramic composites also apply to manufacturing a components of heat exchangers which is used in corrosive environment and high-temperature environments.

Except the above application, some other field of application of ceramic based composites may also be possible.

1.76. Carbon – Carbon Composites

Carbon – carbon composites is also a new invention in the field of material science & technology, in which carbon fibers – reinforced with carbon matrix phase. In such a composite, carbon fiber may be presence as a continuous fiber or in the form discontinuous fiber or in the form of particulates / particles form with in the carbon matrix phase. Carbon-carbon composites known for excellent temperature stability, resistance to erosion at high temperature, low creep at high temperature, high fatigue resistance, low density, high thermal conductivity, low coefficient of thermal expansion, high strength to weight ratio, high modulus, low density, high coefficient of friction, electrical conductivity, good thermal shock resistance, good tensile and compressive strengths, abrasion resistance, good fracture toughness, superior high temperature stability in presence of vacuum & inert atmosphere, good corrosion resistance. But we know that each & every material has certain drawback or negative aspects, in similar manner carbon – carbon composites showing gradual failure, high cost, low shear strength, low strength, low stiffness and low oxidation resistance (up to $370^{o}C$). Although these carbon – carbon composites now a days extensively used in space application (Space shuttle nose cone), high temperature application and air brakes (Due to weight saving, high durability and high specific height). Furthermore, property variation of carbon–carbon composites depending upon the reinforcement i.e. it may either fiber (Continuous or Discontinuous), particulates / particles reinforcement, % volume fraction of reinforcement in composites and also coincide to characteristics of matrix phase. There are a wide range of composite materials available that fall under this category, among those carbon matrix composites consist of carbon or ceramic fibers embedded in carbon matrices are most common. Another very important composite known as silicon carbide fiber-reinforced carbon showing coefficient of thermal expansion equivalent to ceramic coatings which will not only reducing coating cracking but also prevent spallation and consequently used to manufacturing of aircraft gas turbine engine flaps. One of the most common drawbacks which exists due to oxidation characteristics that is achieved between 370°C – 500°C, further oxidation characteristics may be controlled by including oxidation inhibitors in the matrix phase as a result of which ceramic coatings may increase the oxidation temperature. It has been observed that in presence of vacuum or inserts atmosphere carbon–carbon composites sustain their properties up to 1380°C. Now we will discuss about the mechanical & physical property of such a carbon - carbon composites.

1.77. Mechanical Properties of Carbon Matrix Composites

We know that carbon matrix phase is a low strength, brittle and low-stiffness materials. Hence elastic modulus and strength both are low in transverse direction for unidirectional carbon–carbon composites throughout the thickness. Furthermore, fabric-reinforced materials indicate poor strength through out - thickness direction. It has been observed that in direction of fibrous reinforcement, we can obtain modulus as high as 340 GPa, tensile strengths as high as 700 MPa, and compressive strengths as high as 800 MPa. On the contrary directions perpendicular to fiber directions, elastic modulus may obtain about 10 MPa, tensile strengths 14 MPa and compressive strengths 34 MPa. Consider following table which shows, mechanical property of some common carbon–carbon composites.

Table 1.54: Mechanical Property of Carbon - Carbon Composites

Material	Reinforced carbon – carbon	Advanced carbon – carbon
Fiber	WCA (Rayon)	T 300 (PAN)
Fabric	Plain weave	Eight harness strain
Density g/cm³	1.4 to 1.7	1.8
In – plane		
Tensile modulus (GPa)	14	110
Tensile strength (MPa)	120	330
Compressive strength (MPa)	88	180
Shera strength (MPa)		6.9
compression modulus (GPa)	17	110
Through out thickness		
Tensile strength (MPa)	5.5	3.4

In the above table WCA (Rayon) that reinforced with carbon fibers made from a rayon precursor are used in space shuttle orbiter application. This material is known as reinforced carbon–carbon (RCC) or reusable carbon–carbon. Another material is reinforced by PAN-based standard modulus fiber with T300.

1.78. Physical Properties of Carbon Matrix Composites

We have number of carbon–carbon composites which showing variety of physical properties. We know that any kind of properties depending up on the type of fibers used, precursor and manufacturing processes that apply to produce the matrix. Densities generally fall between the range of 1.4 to 1.8 g/cm^3. In-plane, thermal conductivities range from less than 10 to about 350 W/m·K. Throughout the thickness conductivities observed very low. Further coefficient of thermal expansions is also very low and usually falls between -1 to +1 ppm/K. Carbon fiber may also be reinforced with polymer matrix phase, consider following table which shows some important mechanical & other properties of carbon-based polymer composites.

Table 1.55: Mechanical Property & Other Property of Carbon Fiber Reinforced – Polymer Composites

Composite material	Density in g/cm^3	Modulus in (GPa)	Tensile strength in (MPa)	Ratio of modulus to weight	Ratio of tensile strength to weight	Elongation in %
High strength carbon fiber reinforced – Epoxy matrix (Unidirectional)	1.55	138	1550	9.06	101.9	---
High Modulus carbon fiber reinforced – Epoxy matrix (Unidirectional)	1.63	215	1240	13.44	77.5	---
Carbon fiber reinforced – epoxy matrix (quasi-isotropic)	1.55	45.5	579	299	38	---
30% carbon fiber reinforced – nylon 66 matrix	1.20 to 1.57	16 to 33	193 to 260	---	---	0.9 to 4.0
62% carbon fiber reinforced – PEEK	1.6	50	750	---	---	---
40 to 60% carbon fiber reinforced – epoxy matrix	1.15 to 2.25	26 to 520	4.6 to 3220	---	---	0.4 to 11
5 to 40% carbon fiber reinforced – polycarbonates	1.15 To 1.43	2.1 to 25.5	46 to 186	---	---	0.9 to 11.8
20 to 30% carbon fiber reinforced – polyimide	1.38 to 1.68	4.5 to 29	36.5 to 241	---	---	0.8 to 5.5
30% carbon fiber reinforced – PP	1.07	17	117	---	---	1
Carbon fiber reinforced – vinyl ester	1.5 to 1.65	136	900 to 1200	---	---	1.4

Furthermore, we know that carbon matrix composites consisting carbon fibers in carbon matrix. These composites exhibit maximum temperature stability up to 3000°C in a presence of inert or vacuum environment. But whenever interaction takes place with oxygen the degradation starts due to oxidation at the temperature of 500°C. In order to protect C–C composites against oxidation at high temperatures we will provide external coating. Coating may be of either oxide coatings (SiO_2 and Al_2O_3) or non-oxide coatings (SiC, Si_3N_4, and HFB_2). C – C composites can have fiber either in continuous or discontinuous form. Continuous fibers are usually applied in structural applications.

1.79. Application of Carbon Matrix Composite

Following are some important application of carbon matrix composite.

- **Space Shuttle Nose Cones:-** Carbon–carbon composite is used to manufacturing nose cone for space shuttle due to weight saving, high thermal conductivity (Prevent surface cracking), high specific heat (Absorb large heat flux) and high thermal shock resistance. All those properties cooperate for re-entering the shuttles into earth atmosphere because during re-entry, shuttle experiencing about temperature of 1700°C and also has capability to maintain space temperature. Further, the basic advantage of carbon–carbon composites is that nose remains undamaged & provide long service life. Resistance to erosion at high temperature is a basic objective of application of carbon–carbon composites in aerospace application. It is usually used to manufacturing re-entry vehicle nose tips, rocket nozzles, and exit cones. In case of space shuttle orbiter, carbon–carbon components play a vital role and used to manufacturing leading edges and nose cap.

- **Aircraft Brakes:-** Carbon–carbon composites are extensively used to manufacturing an air brakes which apply in all kind of aircraft / plan due to their fascinating property like, weight saving & fuel savings, high durability, high specific heat, Effective braking capacity and less braking times. These brakes are used in aircraft or commercial aircraft flight such as Airbus. A comparative study has been performed between performance of steel & carbon–carbon composites brakes then it is explicit that carbon–carbon composites air – brakes performance & efficiency is superior with respect to steel. Furthermore carbon–carbon composite material also used to manufacture brakes and clutches of racing car and motorcycle.

- **Mechanical Fasteners:-** Carbon–carbon composites are also used to manufacturing a mechanical fastener whose function is to join two assembly or parts, non-permanently. Such kind of fasteners can be applying at high temperature application where strength does matter.

Except the above application, now a day's carbon–carbon composite also extensively used in industrial field for instance glass-making equipment, heat treatment racks, and water-heating elements. Carbon–carbon composites represent thermal characteristics like thermal conductivities. These composites available in various range of thermal conductivities. An application of high-conductivity carbon–carbon composites is used to manufacturing radiator panel of earth Observing Satellite (NASA EO1), Optical bench of GOCE spacecraft made from carbon–carbon composites, silicon carbide fiber-reinforced carbon composites used to manufacture military aircraft engine flaps and many more.

Note

- Carbon–carbon composites use carbon fibers in a carbon matrix. These composites are employed for high-temperature application that is about 3000°C.

- C–C composites can perform satisfactorily in oxidizing condition with coatings up to temperature of 1700°C.

- High tensile & compressive strength, low shear strength and low modulus and low failure strain of C – C composites matrix limit the circumstances of application.

- Over large number of advantages & wide span of application it also includes certain drawbacks like high cost, low shear strength, and susceptibility to oxidations at high temperatures.

- Carbon–carbon composites are widely used in aerospace, defence and commercial applications.

- Carbon–carbon aircraft brakes showing high weight savings over steel.

1.80. Applications of Bio-composites

Bio composites are also considered in the class of composites materials. Its application includes fabrication of the interiors of the aircraft like - floors, ceilings, sidewalls and stowage bins. In addition to, now a days it is being used to manufacturing, Sporting goods (Tennis rackets, Golf clubs), Aircraft (Door panels, Dashboard), Automobile (Bumpers, Engine components) and shipbuilding etc. Some other application includes, Paints and coatings, Electrical systems and electronics, Aircraft industry (Doors and elevators), Consumer & marine applications, Aerospace applications, Chemical industry (Tanks, Pipes, Pressure vessels) and Automotive body frames, engine components.

1.81. Applications of Advanced Composites

Now a days advanced composite is widely used to manufacturing prototype structure of military aircraft systems, such as wing and fuselage section components, empennage, helicopter rotors, etc. Other possible example includes manufacturing of gear wheels for lubricating and coolant pumps, gear case housing, golf club shafts, rackets, and fishing rods. In the case of gear wheels, carbon reinforced plastic is a promising candidate due to light weight, self-lubricating gears. Following are advantages of fiber reinforced plastic gear wheels such as self-lubricating, simplifying design, rust free, chemically inert, Low maintenance, low weight, high strengths & stiffness, low coefficient of thermal expansion, low coefficient of friction and low wear are most common.

1.82. Bonding in Composites Material

Bonding is nothing but it is a process of creation, a strong force of attraction among individual particles throughout the structure so that after consolidation we can have improvement in mechanical property, physical property, chemical property, thermal property and corrosion property and others. We know that composites material consisting two phase primary phase & secondary phase. In primary phase fiber (Continuous or Discontinuous fiber) and particulates / particles, are reinforced with metal matrix phase. Here we have same purpose i.e. to develop strong force of attraction either of fibers or particulates / particles with in the matrix phase so that after agglomeration or consolidation of composites we can attain improvement in either mechanical property, physical property, chemical property, thermal property or corrosion property. Bonding characteristics always depending upon the, size of fibers & type of fibers used, size of particulates i.e. fine grains, small grains & coarse grains and orientation of fibers throughout the matrix phase. Sometimes characteristics or property also depending upon the micro-structure of composites. Basically, following type of bonding exists in composites materials, these are Mechanical bond, Dis-solution & wetting bonding, Oxide bonding, Reaction bonding, Exchange bond, Diffusion Bonding, Plasma Spray Bonding, Hot Rolling Bonding and Mixed bonds. A possible brief description of all those are given below.

1.82.1. Mechanical Bonding

Mechanical bond is generated by contraction of the matrix phase on the filament or reinforcement / dispersed phase. Whenever contraction takes place some kind of frictional effects exists between interface of reinforcement & matrix phase and consequently mechanical interlocking starts eventually mechanical bonding consolidate the composites material. One of the most important things is that it does not have any kind of chemical source. Since no chemical source included therefore composite represent weakness under transverse loads.

1.82.2. Dissolution and Wetting Bond

A contact angle of less than 90° occurs in wetting and is also characteristic of dissolution. If wetting is assumed to be accompanied by some dissolution, however small, then this bonding characteristic covers both extremes of mutual solubility. Elimination of adsorbed gases and of contaminant films must be achieved before element-to-element contact can occur and result in wetting and dissolution.

1.82.3. Reaction Bond

The reaction bond formed if new chemical compound is formed at the interface of matrix phase & dispersed phase or sometimes any kind of other chemical agent is included throughout the manufacturing of composites which not only work as catalyses but also secure bonding of desired level in the composite material so that we can achieve specific properties. For example - formation of titanium diboride at the interface between boron and titanium. Where boron fiber is a dispersed phase & titanium is metal matrix phase. Sometimes we have more than two components in the composites then reaction bond consisting two or more reactions at the interface. Consider following example.

$$\text{Ti (Al) Solid Solution} + \text{B} \longrightarrow (\text{Ti , Al}) \, \text{B}_2$$
$$\text{Ti} + \text{Ti (Al) B}_2 \longrightarrow \text{Ti B}_2 + \text{Ti (Al) Solid Solution}$$

In the above example, we have two metallic phase & single boron ceramic dispersed phase, for manufacturing composites material overall reaction is being accomplished in two steps i.e. reaction between Titanium-Aluminium Ti (Al) solid solution and boron.

1.82.4. Oxide Bond

Whenever composites material is manufactured, presence of oxygen is responsible for formation of oxides bonds. These oxides bonds are existing in the form of oxide film at the interface of reinforced phase & matrix phase. For example - bond between silver & alumina whisker, where alumina whisker works as dispersed phase & silver recognize as a matrix phase. This oxide bond appears as an oxide films at the interface of both. Another example is nickel-alumina bonding, in which oxide bonds exists between interface of nickel metal phase & alumina dispersed phase, furthermore bond formed between the oxide-coated surfaces of aluminium and boron that takes place due to reaction of two oxides. This composite consisting oxide film at the interface of both the phases i.e. primary phase & secondary phase.

1.82.5. Diffusion Bonding

Diffusion bonding exists during the consolidation of metal matrix composites i.e. consolidation of metal matrix phase & dispersed phase. Such a bonding formed due to application of heat and pressure during the manufacturing process. This bonding is known as diffusion bonding. Diffusion bonding are generally found in (Boron/Al), (Borsic /Al), (Ti / SiC) and (Ti / Borsic) composites.

1.82.6. Plasma Spray Bonding

Generally, monolayer tapes is consolidated into thick form by diffusion and such a process can only be accomplished by plasma spray bonding. The rate of powder feeding influenced the spray deposit and also quality of matrix filament bond formation.

1.82.7. Hot Rolling Bonding

Manufacturing or fabrication of composite material always required application of heat and pressure for certain length of time to complete the reaction, so that composites material phases may be consolidated with diffusion bond. Further to minimise such a reaction length of time, fabrication or manufacturing of composites is performed by hot rolling process and a bonding process is called hot rolling process & bond is called hot rolling bonds. The hot rolling process has short period of reaction time between the interface of filaments (reinforced phase) and matrix phase, at that bonding temperature. But this method does not give satisfactory result for tapes of either monolayers or a few layers thick composites. Furthermore, in addition to, it is also inhibiting high volume loading of the filaments, cross - and angle-ply configurations and consolidation of thin tapes. For example – boron reinforced – titanium.

1.82.8. Mixed Bonding

Composites material also have a mixed bonding.

1.83. Failure Criteria of Composites Materials

We know that the composite material composed of two components i.e. primary phase which is known as dispersed phase or reinforcement and secondary phase which is known as a secondary phase hence we can say that,

$$
\begin{array}{ccccc}
\text{Composite} & = & \textbf{Primary phase} & + & \textbf{Secondary phase} \\
\text{composition} & & \text{(Dispersed phase or} & & \text{(Matrix phase)} \\
& & \text{Reinforcement)} & &
\end{array}
$$

Composite material failure may occur either due to fiber failure, matrix failure, delamination, interfacial failure or buckling. Actually, basic objective of failure criteria is to identify, how much load or stress can be carried out by the composite material before failure or fracture. A value of load or stress before the failure can be considered as a maximum strength while a load at which material breaks or fail or fractured then it is called fracture strength of the composites, both type of strength depending upon the position of fibers i.e. continuous or discontinuous fiber and their orientation further more sometimes it is also depending upon the arrangement of layer i.e. it may either be in unidirectional or multidirectional. In actual practice we will only observe maximum strength, for specific value of stress.

- Maximum longitudinal tensile strength
- Maximum longitudinal compressive strength
- Maximum transverse tensile strength
- Maximum transverse compressive strength
- Maximum longitudinal shear strength
- Maximum impact strength
- Maximum tensile strength at elevated temperature

Another failure criterion can also be identified i.e. failure of composites due to residual stress.

1.83.1. Fundamental of Failure

In case of iso-tropic material or conventional material, we have been evaluated either highest value of tensile, compressive or shear stress regarding to failure or failure mechanism. one of the most important inherent characteristics of iso-tropic material or conventional material is that it has constant strength. During that we will be estimating the value of applied stress & a time at which failure occurs. A maximum value of stress before failure of material is known as ultimate strength. But in case of composite, we will have to calculate strength for unidirectional or multidirectional composites material.

1.83.2. Failure Criteria for Unidirectional Composite Material

There are two method by which we can evaluate the failure criteria of unidirectional composite material. Such a criteria coincide to axial stress and respective value of strain.

1. **Maximum Stress & Strain Criteria:-** In case of maximum stress criteria, we will have to calculate the value of applied stress either in x – Direction loading (Longitudinal loading), Y – Direction loading (Transverse loading) and Z – Direction loading (Shear loading), Consider following conditions.

$$\sigma_x \leq P_x \quad \text{................................. condition (i)}$$

$$\sigma_y \leq P_y \quad \text{............................... condition (ii)}$$

$$\sigma_z \leq P_z \quad \text{............................... condition (iii)}$$

Where,

P_x = Load applied in X–Direction (Longitudinal loading)

P_y = Load applied in Y–Direction (Transverse loading)

P_z = Shear load applied

And, σ_x, σ_y, σ_z are the value of stress in Longitudinal loading, Transverse loading and shear stress.

For obtaining failure criteria of composites any one of the conditions must be satisfied by the composite material which represent failure criteria due to maximum stress.

Furthermore, the above failure condition can also be transformed into maximum strain criteria, Consider following equation.

$$e_x = P_x / E_x \quad \text{................................Condition (i)}$$

$$e_y = P_y / E_y \quad \text{................................. Condition (ii)}$$

$$e_z = S_z \quad \text{.......................................Condition (iii)}$$

Where,

E_X & E_Y are young modulus of composite material, the value of E_X & E_Y would be equal for same material but when composite is made from various layers then it becomes change.

and

P_x = Load applied in X–Direction (Longitudinal loading)

P_y = Load applied in Y–Direction (Transverse loading)

S_z = Applied shear load

For failure of composites any one of the conditions must be fulfilled. Sometimes we have two same value of the criteria that is possible if & only if when Poisson ratio of unidirectional composite material is same.

1.83.3. Quadratic Interaction Failure Criteria

Quadratic interaction failure criteria can also be used for unidirectional & multidirectional composites. Consider following equation in context of quadratic interaction failure criteria.

$$P_{ij}\ \sigma_i\ \sigma_j\ +\ Pi\ \sigma_i\ =1 \tag{1}$$

The above equation only deals for maximum stress value.

Where,

i, j are vector direction of stress & applied load and

P_{ij}, Pi & $\sigma_i\ \sigma_j$ are the value of load applied & stresses in i & j vector direction respectively.

For failure criteria, the above equation must be satisfied by the composite material at the maximum stress value.

This maximum force criteria can also be transformed in strain criteria.

$$P_{ij}\ e_i\ e_j\ +\ Pi\ e_i\ =1 \tag{2}$$

The above condition must be satisfied by composite material which is based on the strain failure criteria.

Where, P_{ij}, Pi & e_i, e_j are the value of load applied & strain value in i & j vector direction respectively

Note: The analyses of failure criteria play a vital role in designing and sizing of composites laminate.

In case of two-dimensional stress condition, maximum stress equation may be written as,

$$P_{xx}.\ \sigma_y^2\ +2\ P_x\ \sigma_x\ \sigma_y\ +\ P_{yy}\ \sigma_y^2+\ P_{zz}\ \sigma_z^2+\ P_y\ \sigma_y\ =1 \tag{3}$$

If, X = Longitudinal tensile strength & X^l = Longitudinal compressive strength

Then the value of strength,

$$P_{xx}=\{1\ /\ X\ X^l\},\ P_x=\{(1\ /\ X)+(1/X^l)\} \tag{4}$$

Similarly, if Y = Transverse tensile strength & Y^l = Transverse compressive strength

Then the value of strength,

$$P_{xx}=\{1\ /\ Y\ Y^l\},\ P_x=\{(1\ /\ Y)-(1/\ Y^l)\} \tag{5}$$

Similarly, for longitudinal shear strength can be calculated by the following relationship,

$$P_{zz}=\{1\ /\ S^2\} \tag{6}$$

Where,

S = Longitudinal shear strength

Moreover, Unidirectional composites possess excellent strength and stiffness in the longitudinal direction because major load is shared by fibers while in other loading conditions, the load sharing takes place by both i.e. fibers and matrix. In addition to it, another parameter is also a matter of concern which plays a vital role in the strength of composites that is interface region of fiber phase and matrix phase. Furthermore, whenever composites material subjected to loading condition, failure initiation takes place at the weak point or mostly at the weak interfaces. Load sharing by constituent phases depending up on the type of loading. Hence strength of unidirectional composites can be evaluated in five type of loading condition these are, longitudinal tension and compression, transverse tension and compression, and shear. Let be consider the case of unidirectional composites when it is subjected to unidirectional tensile load along the fiber. In such a condition we have two assumption, (i) Fiber is perfect linear & elastic up to fracture and (ii) Matrix is linear initially, then nonlinear as strain increases.

We know that composites has number of parallel fiber of equal length & diameter, whenever tensile load is applied then all the fibers does not equally elongated because some of them may be fail as a result of which fiber breaking also generate failure of the composites material and total failure is depending upon the characteristics of the interface that exists between matrix phase and dispersed phase. If the matrix is brittle and the interface strong, then cracks will be created due to fiber breaking which will further propagate through the matrix across the neighboring fibers, leading to the composite failure. If the interface is weak, then interfacial failure can be initiated at the fiber breaks and the fiber matrix debonding will grow along the broken fibers. A longitudinal damage growth may also generate due to yielding of matrix phase & fibers. Hence, we observed failure mode under a longitudinal tension thereby we have got transverse crack propagation mode and longitudinal damage growth mode. The transverse crack propagation mode is in fact what is observed in brittle, homogeneous materials. In this failure mode, the strength of stronger fibers cannot be fully utilized, and hence the composite strength is not optimum. But in case of complete longitudinal damage growth mode, broken fibers are simply separated from intact ones as far as load sharing is concerned, and the composite behaves like a dry bundle of fibers.

1.84. Tensile Properties of Composites Material

In order to understand the tensile property of composite, consider the example of unidirectional boron-reinforced aluminum and titanium composites which has been tested in parallel direction of boron fiber (at 0°) & perpendicular direction of fiber or enforcement (at 90°). We have got following conclusions.

At the initial application of load, both phases deform elastically & rule-of-mixture help in evaluation of modulus. Furthermore, as load application increases then yield strength of the matrix phase improves & the matrix phase flow plastically and hence stiffness of composite become lowers. Composite stiffness can be measured by the weighted average of the modulus of the reinforcement and instantaneous strain hardening rate of the matrix. In the case of Boron / Aluminum, the young modulus for aluminum is somewhat lower as compare to modulus of boron filament. Further, the stress-strain behavior in the second region is neither elastic nor linear. During second stage, strain hardening rate of matrix vary and permanent deformation takes place due to matrix flow and breaking of several weakened filaments and filaments breaking is notice until up to the slope of the stress-strain diagram does not decrease and third stage initiation not achieved. Further at the end of second stage & initiation of 3rd stage composite suddenly fractured. It gives a failure criterion at that specific load & also at that strain condition. Another thought is also observed that yield strain of the matrix also enhanced fracture strain of the brittle filament. However, the transition from stage-I to stage-II depending up on the yield strength of the matrix and the magnitude of residual stresses which exists during the consolidation process. If ductile metal wire is being reinforcement with matrix phase then composite both phases deform plastically which will continue up to stage-III and during third stage failure of composites may occurs. Furthermore, if we want to evaluate composite tensile strength by rule-of-mixture then matrix grain size, impurity distribution throughout the composites, consolidation induced defects during consolidation, inter-diffusion of components, presence of reaction, filament strength and strain-to-failure ratio would be a matter of great concern which will not only affects the chemical properties but also responsible for mechanical property reduction. Tensile strength of filaments also responsible for load transfer & failure of composites material. Strength of composites mostly affected by space between filaments, stiffness of filaments, strength of filaments, stress flow through matrix and chemical change that occurs due to reaction of filaments with the matrix phase should be considered.

1.85. Failure of Composites Due to Residual Stress

Residual stress comes on the front line during the consolidation of composition due to variation of thermal expansion of reinforcement phase & matrix phase, which may also be a cause of composite failure. Such a differences in thermal expansion coefficient will result in the formation of longitudinal and radial residual compressive stresses on the filaments and corresponding tensile stresses in the matrix. Sometimes it has been found that if matrix yielding is performed during cooling by applying tensile load then elastic deformation does not obtain but on the other hand if filament matrix bond is weak then during cooling slippage takes place instead of matrix strain hardening as a result of which we have got reduction in residual stress. Furthermore, low composite strain rate is achieved in cross & angle piled material. If such a composite subjected to cooling then thermal expansion variation exists between reinforcement phase & matrix phase but on heating it will increase the composite strain failure. Residual strain mostly observed in boron filaments – reinforced aluminum, boron filaments – reinforced titanium and titanium – 6 Al – 4V.

1.86. Transverse Tensile Property of Composites Material

In unidirectional reinforced composites material, tensile modulus, strength & ductility is always lower in transverse test rather than longitudinal test, due to lower modulus & strength, we can get lower strain criteria and hence we are mostly unable to evaluate stiffness & strength of composites on the contrary if filaments are strongly bonded with matrix phase, composite is free from any kind of impurities/defects then it has been noticed that transverse strength, greater than that of the strength of bulk matrix alloys. When load is applied then it has been observed that after fracture, volume of filaments damage is less than that of % volume of total filaments in composites. For example – In case of composites which is reinforced by boron filaments is heated then composites exhibit high strength without presence of majority of boron filaments in the composites. Furthermore, we can say that transverse strength always coincides to amount of boron filaments and failure can be controlled by inhibiting the filaments breaking. Since relative area of split filaments on the fracture is larger than the volume % of boron, it is clear that filament offers least resistance to crack propagation & consequently we can control the composite failure. Higher transfer strength of heat-treated reinforced composites always affects the strength & stress concentration ability of the matrix phase of the composites. For example – boron filaments reinforced – aluminum matrix, transverse properties can be enhanced either by addition of 3rd phase or by heat treatment of matrix phase. Another alternate for improving transverse properties is to use cross or angle – plied boron/aluminum. Moreover, transverse strength can also be improve by increasing number of filaments, by improving the diameter of filaments or by using another filaments which is made from either SiC, Al_2O_3 or graphite aswellas transverse strength is also enhanced by addition of third phase, for example addition of stainless steel wire with the layer of titanium foil. In such an addition transverse strength increases without affecting longitudinal properties.

1.87. Micro-mechanics of Composites

Micro-mechanics of composites deals with the micro-scopic analyses of structural composites material that includes study of fiber, matrix & voids. We know that composites consisting two phases i.e. primary phase and secondary phase. The overall volume & mass of composites can be represented by the following relationship.

M_T = Total mass of the composites

V_T = Total volume of the composites

Consider following formula say,

Composite composition = Fibers reinforced phase / Primary phase + Matrix phase or

Secondary phase

Total mass is a summation of mass of fibers & mass of matrix.

$$M_T = M_F + M_M \tag{1}$$

And total volume of composites is a summation of followings.

$$V_T = V_F + V_M + V_V \tag{2}$$

Where,

M_T = Total mass of composites

M_F = Total mass of fibers

M_M = Total mass of matrix

V_T = Total volume of composites

V_F = Total volume of fiber

V_M = Total volume of matrix

V_V = Total volume of voids

Calculation of Mass Fraction & Volume Fraction:- Divide the equation 1 & 2 by M_T & V_T respectively, then we have got,

$$\text{Mass fraction} = m_f + m_m = 1 \tag{3}$$

$$\text{Volume fraction} = v_f + v_m + v_v = 1 \tag{4}$$

Where subscript f, m & v represent fiber, matrix & voids respectively.

Calculation of Composite Density:- Divide equation 1 by 2 we have,

$$\rho = (M_T / V_T) = \{(\rho_f V_f + \rho_m V_m) / V_T \} = ((\rho_f V_f + \rho_m V_m) \tag{5}$$

In term of mass fraction density becomes,

$$\rho = 1 / \{(m_f / \rho_f) + (m_m / \rho_m) + (V_v / \rho) \tag{6}$$

Voids fraction can be calculated as,

$$V_v = 1 - \rho / \{(m_f / \rho_f) + (m_m / \rho_m) \tag{7}$$

Furthermore, in structural composites, matrix surrounds the fibers. fiber has stiffness & strong nature and known as a load bearing member of composites while matrix is soft & weak in nature. The basic function of matrix is to protect fiber from splitting & other environment effect. In order to study the overall composites material, we will consider volume element known as representative volume which is an accumulated form of fiber reinforcement with in a matrix phase. In micro-mechanics, volume fraction, mass fraction and load bearing capacity are evaluated but in addition to it we can also analyze stress & strain behavior under the micro-mechanics. Moreover, rules of mixture are also another technique by which we can understand the micro-mechanics in which we studied about stress, Poisson & load transfer mechanics. In rules of mixture we will assume that fiber & matrix both are elongated equally & uniformly. In this condition following assumption are made.

- Elastic and plastic isotropy.

- Perfect mechanics continuum at interface.

- No chemical reaction between constituents.

- An absence of residual stress.

- An absence of rheological interaction at interfaces.

- Definability of in-situ properties of both fiber and matrix properties at the strains in question.

According to Kelly & Davies, only longitudinal modulus & Poisson ratio can be determined by simple test but on the other hand by using rules of mixtures longitudinal strength can calculate.

1.88. Reinforcement Material

Reinforcement / dispersed phase is a constituent of composite materials. We know that composites material requires two phases that is primary phase called reinforcement phase or dispersed phase and another is called matrix phase (work as binder) or secondary phase. Consider the following diagram to understand the consolidated composite material.

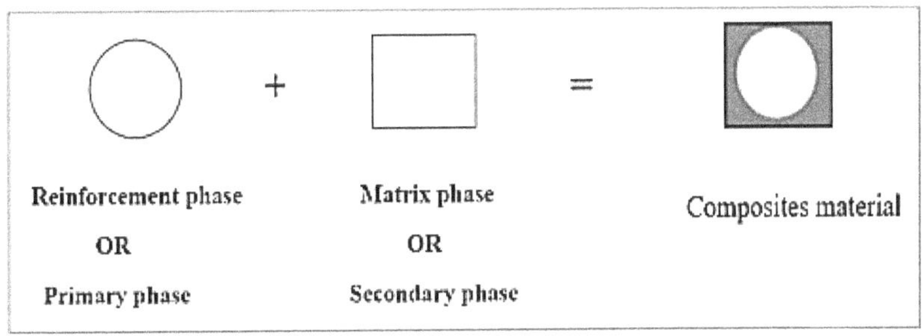

Diagram 1.19: Composite Material Combination

Reinforcement of composites can be performed by using natural fibers or synthetically manufactured diverse form of reinforcement. Consider following ray diagram which shows various types of natural fibers and synthetic form of reinforcement. But now a days we have wide range of application of synthetic form instead of natural fibers because it significantly enhancing the mechanical, physical, chemical, thermal and many other properties.

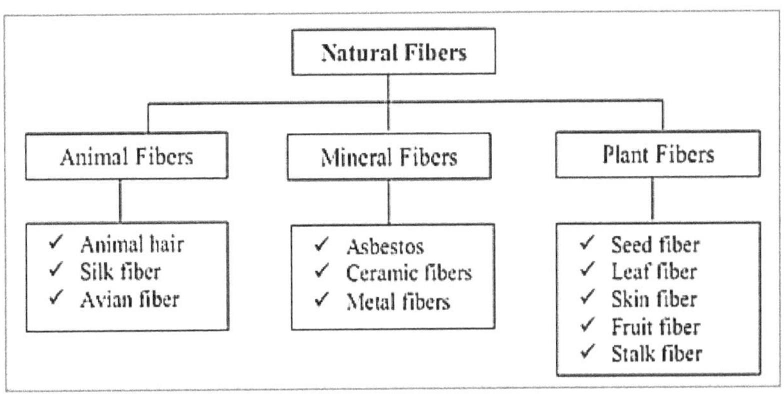

Diagram 1.20: Natural Fiber Used in Composites as a Reinforcement

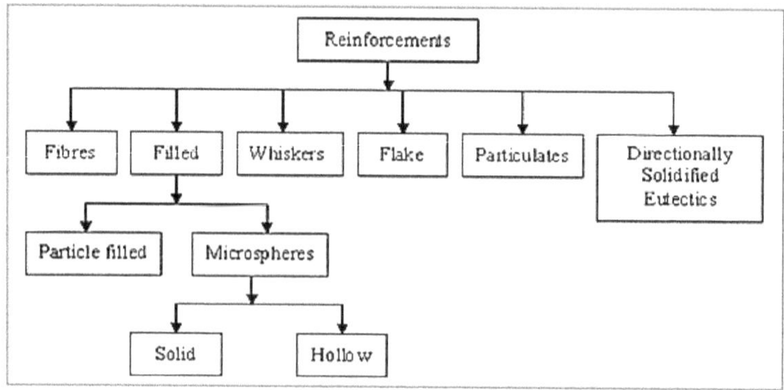

Diagram 1.21: Synthetic form Used in Composites as a Reinforcement

Reinforcement of the composites may be performed by using either fibers, particulates / particles, flakes, fabrics or whiskers. Every reinforcement form has its own characteristics that is depending upon the which kind of material is being used to manufacturing a reinforcement form. These reinforcement materials also cooperate to characterized the composites material & their field of application because reinforcement material always influence the basic inherent property of the composite's materials. Furthermore, for manufacturing a composites material, another phase is required in which reinforcement is done, known as matrix phase. Matrix phase may either be polymer matrix phase, Metal matrix phase, Ceramic matrix phase or Carbon matrix phase.

Among the various form of reinforcement, fiber forms are widely used to manufacturing composites materials because fibres have high aspect ratio (Ratio between length to diameter) thereby it can have enough capability to transferring shear stress between both the phases i.e. matrix phase and the fibres phase. Fibers can attribute by its long axis despite of other two axes either often circular or near circular. Particles have no preferred orientation and so does their shape. Whiskers have a preferred shape but has smaller diameter aswellas length as compared to fibres. Reinforcing constituents in composites not only play an inevitable role for augmenting the strength but also equally cooperate to improve heat resistance, corrosion resistance, provide necessary rigidity, impact resistance, stiffness and so many others. These fibres may be used as continuous fibers, discontinuous fibers, whiskers (Elongated single crystals) and in the form of particles. Continuous aligned fibers form, imparts remarkable property to the composite material and hence widely used in high-performance applications. In addition to it fibres phase also cooperate in manufacturing process as a result of which we can transformed the composites in various shapes & size. Glasses fiber, carbons (Graphite) fiber, ceramics fibers and high-modulus organics are widely used to manufacturing a composites material which is being mostly apply for mechanical engineering applications. Fibers can also be manufactured in the form bundle in which so many fibers are accumulated that is known as multifilament bundles or strands whose ends may either be in twisted form or untwisted forms, if ends are twisted then it is known as yarns. Fibers may also be manufactured in monofilaments that has larger diameters than that of strand filaments. It has been observed that strengths of monolithic ceramic decreases with increasing fiber material volume that is responsible for existence of flaws in composites material and consequently strength become reduces. Consider the following table which shows mechanical and thermal property of some most important reinforcing material.

Table 1.56: Property of Some Important Reinforcing Material (Fiber Material)

Reinforcing material (fiber)	Density in g/cm^3	Axial modulus in GPa	Tensile strength in MPa	Axial cofficient of thermal expansion ppm/k	Axial thermal conductivity W/m-k
E – glass	2.6	70	2000	5.0	0.9
HS – glass	2.5	83	4200	4.0	0.9
Aramid fiber	1.4	124	3200	-5.2	0.04
Boron fiber	2.6	400	3600	4.5	--
SM Carbon (PAN)	1.7	235	3200	-0.5	9.0
SM Carbon (PAN)	1.7	235	3200	-0.5	9.0
UHM Carbon (PAN)	1.9	590	3800	-1.0	18
UHS Carbon (PAN)	1.8	290	7000	-1.5	160
UHM Carbon (Pitch)	2.2	895	2200	-1.6	640
UHK Carbon (Pitch)	2.2	830	2200	-1.6	1100
SiC mono-filaments	3.0	400	3600	4.9	--
SiC multi-filaments	3.0	400	3100	--	--
Si -C -O	2.6	190	2900	3.9	1.4
Si -Ti - C- O	2.4	190	3300	3.1	--
Aluminium oxide	3.9	370	1900	7.9	--
High density polyethylene	0.97	172	3000	--	--
Basalt	2.7	100	2900	5.5	1.7

Let be discuss some important types of fibers material with their field of application.

1.88.1. Glass Fibers Reinforcing Material

Glass fiber are manufactured from silica by adding other oxides with in it. These glass fiber are used to reinforcing composite material and known as a reinforcement material. Glass fiber are best suitable for high performance application aswellas other mechanical engineering application due to their good characteristics and low manufacturing cost. E-glass and HS glass fiber composites both are applying in the field of mechanical engineering applications. E-glass fiber stand for electrical application and known for high electrical resistance. It is widely used to reinforced and manufacturing of structural composite which is apply in electrical insulation applications. E-glass fibers have low cost and indicate low elastic moduli & susceptibility to creep. 'S-glass fiber' material is also utilizing to manufacturing composites where strength does matter. Basically 'S-glass fiber' coincide to strength only. Another glass fiber material is known as a corrosion-resistant glass fibers which is abbreviate as "CR", the composites material which is reinforced by CR-fiber has been applied in chemical industry since long period of time. In the class of glass fiber, HS glass fiber is also used to reinforcing the composites. Such a glass fibers has high stiffness, strength, better resistance to fatigue and creep than that of E-glass. Glass fibers can be manufactured in the form of multifilament bundles in which filament diameters ranges between 3 to 20 μm. In addition to it glass fibers material also apply for reinforcing rubber, short fiber for NR and SBR, plastic and elastomers etc. whenever glass fiber reinforced with elastomer then composites exhibit enhancement in strength and wear resistance, and these outstanding properties utilized in auto tires, V-belts and conveyor belts manufacturing. Furthermore, it has been observed that the tensile strength of glass fiber reinforced - rubber composite is affected by humidity if store it, i.e. decrease with increasing atmospheric humidity. Consider the following which shows mechanical property of some common glass fiber.

Table 1.57: Mechanical Property of Some Most Common Glass Fiber Reinforcing Material

Reinforcing material (fiber)	Density in kg/m³	Young modulus in GPa	Virgin filament strength in MPa	Roving strength in MPa	% strain to failure
A – alkali glass fiber	2460	73	3100	2760	3.6
C – chemical A – alkali	2460	74	3100	2350	--
D- dielectric glass fiber	2140	55	2500	--	--
E – electrical glass fiber	2550	71	3400	2400	3.37
R – glass fiber	2550	86	4400	3100	5.2
S – glass fiber	2500	85	4580	3900	4.6
S$_2$ - glass fiber	2450	90	3600	--	--
S$_3$ - glass fiber	2830	99	3250	--	--

1.88.2. Carbon Fibers Reinforcing Material

Carbon fiber also refer as a graphite fiber that is known for high stiffness, high strength, low density, low weight, low coefficient of thermal expansion, excellent creep resistance, stress rupture, fatigue and corrosion resistance but it has major drawback that they indicate oxidation at high temperature. These fibers are available in wide range of strength & modulus. It has been observed that carbon fiber indicate higher tensile modulus & maximum tensile strength as high as 895 GPa & 7000 MPa respectively. Furthermore, carbon fibers are also available in multifilament bundles in which carbon fiber diameter ranges between 4 to 10 μm. These fibers usually applied for reinforcement of polymers matrix, metal matrix, ceramics matrix, and carbon matrix. Carbon fiber are manufactured from precursor materials for instance, PAN, petroleum pitch, coal tar pitch. Rayon-based fibers or sometimes lignin & polyethylene. PAN based carbon fibers are widely used to manufacturing composite material due to excellent tensile & compressive strength that has maximum tensile strength of 7 GPa and compressive strength depending upon the modulus, as modulus increases then compressive strength decreases moreover PAN based carbon fibers are available in three types these are, SM Carbon fibers, UHM carbon fibers and UHS carbon fibers. SM carbon fiber are mostly used to reinforcing the composites because of its low cost. UHS carbon fibers sometimes refer as intermediate modulus (IM) carbon fibers which is stronger & their axial modulus falls between SM & high modulus carbon. Another variety of carbon fiber is known as a pitch-based carbon fiber, in this category we have two type of pitch-based carbon fibers, these are UHM carbon fiber (Pitch based) and UHK carbon fiber (Pitch based). It has been found that pitch-based carbon fibers exhibits high axial modulus rather than PAN based. For example UHM pitch based carbon fiber showing maximum modulus value up to 895 GPa. Basically, strength of carbon fiber depending upon the modulus because modulus of carbon fibers is a function of degree of alignment, if imperfection occurs then voids exists along the axis as a result of which stress rises and it becomes a cause of weakness. Furthermore, if carbon fibers composition consisting 99 % carbon value, having three-dimensional order of the atoms and also have tensile modulus above 350 GPa and after that such a carbon fiber is heat treated above 1600°C then carbon fiber is known as high modulus carbon fibers. High modulus & high strength are represented by the carbon fibers that has diameter of range of 7 to 8 μm which consisting crystals of "turbostratic" graphite further for achieving excellent modulus & strength layer planes of graphite must be aligned & parallel to the fiber axis. UHK pitch based carbon fibers is known for extraordinary high axial thermal conductivity that is about double with respect to copper. But whenever matrix is reinforced with pitch-based carbon fiber then composite behave slightly poor in tension, compression & shear than that of PAN based reinforcement. Carbon fiber play a vital role in high performance application. If short fiber are reinforced with polymer matrix then we have short fiber polymer composites which has low weight & superior mechanical property which is imperative for aero-space application. Consider following table which shows some mechanical properties of pitch-based carbon fiber, PAN based carbon fiber and rayon-based carbon fibers.

Table 1.58: Some Important Property of Carbon Fiber Reinforced Material

Precursor	Pan	Pan	Pitch	Pitch	Rayon	Pitch (K13D2U)
Modulus	Low	High	Low	High	Low	Ultra-high
Tensile modulus in GPa	231	390	160	385	41	930
Tensile strength in GPa	3.4	2.5	1.4	1.8	1.1	3.7
% Strain for failure	1.4	0.6	0.9	0.4	2.5	0.4
Relative density	1.8	1.9	1.9	2.0	1.6	2.2
Carbon assay in %	94	100	97	99	99	Greater than 99

Consider another table which shows mechanical property of carbon fiber like HS graphite fiber reinforcing material & HM graphite fiber reinforcing material.

Table 1.59: Mechanical Property of Graphite Fibers

Reinforcing material (fiber)	Density in Mg/cm^3	Modulus in GPa	Breaking stress in MPa	% Elongation
HS – Graphite fiber	1.8	253	4.5	1.1
HM – Graphite fiber	1.85	520	2.4	0.6

Due to remarkable property like high stiffness, high tensile strength, low weight, high chemical resistance, and high temperature stability before oxidation. These carbon fibres are used to reinforcing polymers, metals, ceramics, and carbon matrix phase for manufacturing a composite material. Now a days these composites are widely used in the field of aero-space, automotive, sporting goods equipment's, consumer goods and many more application.

1.88.3. Boron Fibers Reinforcing Material

Boron fibers are generally used to reinforce polymers matrix and metals matrix. Boron fibers are produced in monofilaments or single filaments by depositing a layer of boron over the substrate of tungsten wire. These fibers are available in the diameters range of 100–140 μm. The properties of boron fibers is depending upon the ratio between of overall fiber diameter to the tungsten core. For example, fiber specific gravity is 2.57 for 100-μm diameter boron fibers and 2.49 for 140-μm diameter boron fibers. Consider the following table which show mechanical properties of boron fiber.

Table 1.60: Mechanical Property of Boron Fibers (Reinforcement Material)

Reinforcing material (fiber)	Density in g/cm^3	Axial modulus in GPa	Tensile strength in MPa	Axial cofficient of thermal expansion ppm/k	Axial thermal conductivity W/m-k
Boron fiber	2.6	400	3600	4.5	--

Note:

- Boron fiber are manufactured by chemical vapour deposition (CVD) technique in which boron layer is deposited over the substrate of tungsten wire.

1.88.4. Silicon Carbide Fiber Reinforcing Material

Silicon-carbide-based fibers are primarily used to reinforce metals and ceramics. There are two types of silicon carbides fibers are famous, these are monofilament silicon carbide fibers & multifilament's silicon carbide fibers. monofilaments carbide fiber are manufactured by depositing a layer of high-purity silicon carbide upon the substates of carbon monofilament fiber with the help of chemical deposition technique on the other hand we have plenty of multifilament silicon-carbide-based fibers which is made by pyrolysis method of preceramic polymers. These fibers form by pure silicon carbide (SiC) which contains few amounts of silicon, carbon, oxygen, titanium, nitrogen, zirconium, and hydrogen. Consider following table which show mechanical property of monofilament silicon carbide fibers & multifilament's silicon carbide fibers.

Table 1.61: Mechanical Property of Monofilament & Multifilament Silicon Carbide Fibers

Reinforcing material (fiber)	Density in g/cm^3	Axial modulus in GPa	Tensile strength in MPa	Axial cofficient of thermal expansion ppm/k	Axial thermal conductivity W/m-k
SiC mono-filaments	3.0	400	3600	4.9	--
SiC multi-filaments	3.0	400	3100	--	--

1.88.5. Alumina Fiber Reinforcement Material

Alumina-based fibers are primarily used to reinforce metals and ceramics, although they have been used to produce polymer matrix composite aircraft firewalls. The primary constituents, in addition to alumina, are boria, silica, and zirconia.

Table 1.62: Mechanical Property of Alumina Fibers

Reinforcing material (fiber)	Density in g/cm³	Axial modulus in GPa	Tensile strength in MPa	Axial cofficient of thermal expansion ppm/k	Axial thermal conductivity W/m-k
Aluminium oxide	3.9	370	1900	7.9	--

1.89. Aramid Fibers

Aramid, or aromatic polyamide fiber, is a high-modulus organic reinforcement which is used to reinforce polymers and cement which is used for ballistic protection. P-aramid fiber that is known as Kevlar fiber invented in year 1971 and then further expands as Kevlar fiber – 29, Kevlar fiber – 49 and Kevlar fiber – 149. Another aramid fiber also comes on the front line that is known as "Twaron". Consider following table which shows mechanical property of aramid fibers.

Table 1.63: Mechanical Property of Aramid Fiber

Reinforcing material (fiber)	Density in g/cm³	Axial modulus in GPa	Tensile strength in MPa	Axial cofficient of thermal expansion ppm/k	Axial thermal conductivity W/m-k
Aramid fiber	1.4	124	3200	-5.2	0.04

Consider another table which shows characteristics & mechanical property of aramid fiber like Kevlar fiber – 29, Kevlar fiber – 49 and Kevlar fiber – 149.

Table 1.64: Characteristics & Mechanical Property of Aramid Fibers (Kevlar Fiber-29, Kevlar Fiber-49 and Kevlar Fiber-149)

Reinforcing material (Aramid fiber)	Modulus in GPa	Tensile strength in MPa	% Elongation	Property
Kevlar fiber – 29	83	3.6	4.0	High toughness , high strength , intermediate modulus for tyre cord reinforcements
Kevlar fiber – 49	131	3.6	2.8	High modulus , high strength for composite reinforcement
Kevlar fiber – 149	186	3.4	2.0	Ultrs high modulus

Note:- Alike carbon fibers, Kevlar 49 fibers also suffer from weak interfacial adhesion with most matrix resins.

1.90. High-Density Polyethylene Fibers

High-density polyethylene fibers are primarily used to reinforce polymers which is apply for ballistic protection. The properties of high-density polyethylene depending upon the temperature, as temperature increases then its property decreases for example it creeps under load even at low temperatures.

Table 1.65: Mechanical Property of High Density Polyethylene Fiber

Reinforcing material (fiber)	Density in g/cm^3	Axial modulus in GPa	Tensile strength in MPa	Axial cofficient of thermal expansion ppm/k	Axial thermal conductivity W/m-k
High density polyethylene	0.97	172	3000	--	--

1.91. Basalt Fibers

Basalt fibers are recently developed reinforcements. They are made by melting and extruding basalt, a volcanic rock. The basic advantage of such a fibers is that it has high fire resistance & high melting point temperature. The mechanical properties of basalt fibers are similar to those of HS glass.

Table 1.66: Mechanical Property of Basalt Fiber

Reinforcing material (fiber)	Density in g/cm^3	Axial modulus in GPa	Tensile strength in MPa	Axial cofficient of thermal expansion ppm/k	Axial thermal conductivity W/m-k
Basalt	2.7	100	2900	5.5	1.7

1.92. Nylon Fiber Reinfocement Material

Nylon fiber are also used to reinforcing the matrix phase. nylon fiber are known for high strength, high elastic recovery, abrasion resistance, lustre, wash-ability, resistance to oil & chemicals, high resilience, high durability, superior physical property, light weight and high melting point temperature stability. Consider following table which shows some physical property of nylon fiber at 21ºC & 65 % relative humidity for continuous filaments & staple.

Table 1.67: Physical Property of Continious & Staple Fiber

Reinforcing material (fiber)	Property	For continious filaments	For staple
Nylon fibers	Hardnes at breaking	0.4 to 0.71	0.35 to 0.44
	Elongation at breaking	15 to 30	30 to 45
	Elastic modulus	3.5	3.5
	Moisture regrain	4 to 4.5	4 to 4.5
	Specific gravity	1.14	1.14

Physical property at $21^{\circ}C$ & 65 % relative humidity

Table 1.68: Mechanical Property of Nylon - 6 & Nylon - 66 Fiber

Reinforcing material (fiber)	Density in g/cm³	Modulus in GPa	Tensile strength in MPa	Cofficient of thermal expansion ppm/k	% Elongation up to breaking	Thermal conductivity W/m-k
Nylon -6	1.0	2.8	81.4	--	60	--
Nylon -66	1.14	1.4 to 2.8	60 to 75	90	40 to 80	0.2

Nylon are extracted from diamine & dicarboxylic acid. Commercially two type of nylon fiber are available in market, one is Nylon – 6 & another is Nylon-66. Nylon – 66 fiber are made from adipic acid and hexamethylene diamine on the other hand Nylon-6 fibers are produced from caprolactam. Nnylon fiber are ususally applied to reinforcing rubber, short nylon fiber used to reinforced NBR and CR. Whenever short nylon-6 fiber reinforced to NR, NBR, SBR then we have enhancement in mechanical property furthermore short nylon fiber reinforced SBR composites used for manufacturing v-belt.

Note

- Carbon fibres (AS4, IM7), Glass fibre (E-glass, S-glass), Aramid fibres (Kevlar and Twaron), and boron fibres are used to reinforcing polymer matrix composites.

- Strengths of monolithic ceramics decrease with increasing material volume, as material volume increases flaws comes on the front line and it is called size effects. Due to size effect, mean fiber strength become lower.

- High aspect ratio can be obtained with glass fibers but breakage of fibers comes on the front-line during processing due to brittleness.

- Glass fiber-reinforced PMCs are usually applied as thermal and electrical insulators due to their low coefficient of thermal expansion and low electrical conductivities of glass fibers.

- Coefficient of thermal expansion of glass fibers is low as compare to most of the metals.

- Oxidation at very high temperatures results excessive pitting on the carbon fiber surface as a result of which fiber strength reduces, In order to reducing the chance of oxidation, surface treatment is performed, in surface treatment non-oxidative coating is deposited over the carbon fiber surface like - Organic polymer and polymer coating such as styrene–maleic anhydride copolymers, methyl acrylate–acrylonitrile copolymer, and polyamides.

- It has been observed that some carbon fibers has extremely high thermal conductivities such a property is imperative for electronic packaging for that heat dissipation and thermal control does matter.

- In most of carbon fiber, stiffness, tensile strength and compression strength and thermal conductivity in axial direction always higher than that of radial direction but it has been noticed that carbon fibers usually have negative axial CTEs (which means that they get shorter when heated) and positive CTEs in radial direction.

1.93. Matrix Materials

Reinforce material & matrix phase material are essential constituents of composites, without which composite could not be manufactured. We have been already discussed about the reinforcement material, further we are discussing here in brief about the matrix material. Usually we have four category of matrix materials these are polymers matrix material, metal matrix material, ceramic matrix material and carbon matrix material. Furthermore, we have plenty of sub category of all those matrix materials, consider following table which shows some common type of matrix materials which is being used to manufacture a composite material.

Table 1.69: Mechanical & Thermal Property of Some Common Matrix Material

Matrix material	Class of matrix material	Density in g/cm³	Modulus in GPa	Tensile strength in MPa	Cofficient of thermal expansion ppm/k	% Elongation up to breaking	Thermal conductivity W/m-k
Epoxy	Polymer	1.8	3.5	70	60	3	0.1
Aluminium (6061)	Metal	2.7	69	300	23	10	180
Titanium (6Al − 4 V)	Metal	4.4	1.5	1100	23	10	26
Silicon carbide	Ceramic	2.9	520	--	4.9	< 0.1	81
Alumina	Ceramic	3.9	380	--	6.7	< 0.1	20
Glass (boro-silicate)	Ceramic	2.2	63	--	5	< 0.1	20
Caron	Carbon	1.8	20	--	2	< 0.1	5 to 90

1.93.1. Polymer Matrix Materials

Usually two types of polymers matrix are in composites materials thermosets matrix resins and thermoplastics matrix resins. Thermosets matrix resins are available in various types, these are epoxies, bismaleimides, thermosetting polyimides, cyanate esters, thermosetting polyesters, vinyl esters, benzoxazines, and phenolics. A brief description of all those are given below.

- **Epoxies:-** A composites material manufactured by epoxy are used in airframe structures and aerospace applications. Generally, epoxy is a brittle material but toughening can improve its impact resistance and after toughening structural properties is also enhanced. It indicates maximum service temperature for airframe that is about 120°C. But a major drawback of epoxies resins is moisture sensitivity.
- **Bismaleimide Matrix:-** A composites material manufactured by bismaleimide matrix resins are used for aerospace applications in which temperature resistance requires up to 200°C.
- **Thermosetting Polyimides Matrix:-** A composites material manufactured by thermosetting polyimide is being applied for high temperature application that ranges between 250– 290°C.
- **Cyanate Ester Resins Matrix:-** Cyanate ester resins matrix is less moisture sensitive than that of epoxies. A composites material manufactured by using cyanate ester resins matrix can be apply up to maximum temperature range of 205°C.
- **Thermosetting Polyesters Resins Matrix:-** Thermosetting polyesters resins matrix are widely used in commercial applications. They are relatively inexpensive, easy to process, and corrosion resistant.
- **Vinyl Esters Matrix:-** Vinyl esters matrix is also used in commercial applications. It represents superior corrosion resistance than that of polyesters but it is expensive.
- **Phenolic Resins Matrix:-** A composites material manufactured by phenolic resins matrix is most widely used where fire resistance does matter for instance – Manufacturing of Aircraft interiors and offshore oil platform structures. It produces less smoke and toxic products if burning takes place at very high temperature.

1.93.2. Thermoplastic Resins Matrix

Thermoplastics matrix resin has three main group, Amorphous thermoplastics, Crystalline thermoplastics and Liquid crystal thermoplastics. In Amorphous thermoplastic - Polycarbonate, acrylonitrile–butadiene–styrene (ABS), polystyrene, poly-sulfone, and polyetherimide are important. Crystalline thermoplastic consisting nylon, polyethylene, polyphenylene sulphide, polypropylene, acetal, polyether sulfone, and polyether ether ketone (PEEK). Amorphous thermoplastics indicate very low resistance to solvent. For example, thermoplastics like nylon are extensively used with chopped E-glass fiber reinforcements. Consider following table which shows mechanical property of some important polymer matrix material (Thermosetting & Thermoplastic matrix material).

Table 1.70: Mechanical Property of Some Important Polymer Matrix Material (Thermosetting & Thermoplastic Matrix Material)

Matrix material	Density in g/cm^3	Modulus in GPa	Tensile strength in MPa	Coffficient of thermal expansion ppm/k	% Elongation up to breaking	Thermal conductivity W/m-k
Epoxy	1.1 to 1.4	3 to 6	35 to 100	60	1 to 6	0.1
Thermosetting polester (Thermosetting matrix)	1.2 to 1.5	2 to 4.5	40 to 90	100 to 200	2	0.2
Polypropylene (Thermoplastic matrix)	4.4	1.5	1100	23	10	26
Nylon-66 (Thermoplastic matrix)	1.14	1.4 to 2.4	60 to 75	90	40 to 80	0.2
Polycarbonate (Thermoplastic matrix)	1.06 to 1.20	2.2 to 2.4	45 to 70	70	50 to 100	0.2
Polysulfone (Thermoplastic matrix)	1.25	2.2	76	56	50 to 100	--
Polyethermide (Thermoplastic matrix)	1.27	3.3	110	62	60	--
Polamideimide (Thermoplastic matrix)	1.4	4.8	190	63	17	--
Polyphenylene sulfide (Thermoplastic matrix)	1.36	3.5	65	--	4	54
Polyether ether ketone (Thermoplastic matrix)	1.26 to 1.32	3.6	93	--	50	47

Note

- Thermosets matrix resin are most widely used for structural applications.
- Polymer matrices are weak in nature has low-stiffness and indicate viscoelastic nature but its strength and stiffness can be improved by reinforcing the fibers with Polymer matrix.
- After curing process, thermosets materials become rigid and unable to recycle or reformed. Further thermosets exhibit resistant to solvents and corrosive environments than that of thermoplastics.
- Thermoplastics can be softened and reformed again by application of heat. Mostly it behaves as a wax.
- Selection of any matrix is depending upon the maximum service temperature of composite material. It has been observed that elastic and strength properties of polymers decrease with increasing temperature.
- Temperature resistance of polymers is detected by the glass transition temperature (T_g) which is a temperature at which transition of polymer takes place from rigid ness to a rubbery one.
- It has been observed that strength and stiffness of polymer is significantly reduce above the glass transition temperatures. But advancement, can improve this characteristic for example, carbon fiber-reinforced polyimides have replaced titanium in some aircraft gas turbine engine parts.
- The major adverse effect of the polymer matrices is sensitivity to moisture. For example – resins has characteristics of absorbing water as a result of which not only weight gain, & dimensional change occurs but also reduction in strength & stiffness is noticed and also variation in T_g. Furthermore, resins also de-absorb moisture when kept in dry atmosphere and capacity of absorption and desorption always depending upon the temperature.
- Vacuum resin exhibits outgassing when it interacting with organic and inorganic chemicals that further leads to affects the thermal control, absorptivity and emissivity.

1.94. Metal Matrix Material

Metallic material can also be used as a matrix phase. It has some remarkable property with respect to polymer matrix like high temperature stability, suitability for thermal and mechanical treatments (Plastic deformation is possible) and High yield strength and modulus than that of polymers which require to manufacturing composite of high transverse strength, high modulus as well as compressive strength. On the contrary it has some unwanted properties like high densities, high melting point temperature (High process temperatures) and sometimes corrosion at the fiber–matrix interface. Now a days we have plenty of metal matrix material like aluminium, aluminium alloys, titanium, iron, copper, lead, magnesium, cobalt, silver, and superalloys. Among those aluminium and titanium matrix are widely used due to low densities and existence in various alloy forms. A fascinating property of aluminium and its alloys enforcing for using as matrix material for manufacturing metal matrix composites. For example - Pure aluminium matrix material known for good corrosion resistance & Aluminium alloys matrix material like – Al-201, Al-6061 and Al -1100 are known for higher tensile strength–weight ratio. Including to titanium, Titanium alloys say α & β alloys are also used as a metal matrix material for example - Ti-6Al-9V and metastable β – alloys say Ti-10V-2Fe-3Al. Both the titanium alloys not only indicate high tensile strength to weight ratio but also showing better strength sustainability up to $400^{\circ}C$–$500^{\circ}C$ with respect to aluminium alloys matrix material. But we have major drawback i.e. titanium alloys has high reactivity with boron and Al_2O_3 fibers at normal fabrication temperatures on the other hand borsic (Boron fibers coated with silicon carbide) and silicon carbide (SiC) fibers show less reactivity with titanium further we have enhanced tensile strength sustainability due to the effects of coating of boron and SiC fibers with carbon-rich layers.

Note

- The coefficient of thermal expansion of titanium alloys is approximately closer to that of reinforcing fibers which reduces the chance of thermal mismatch between both reinforcing phase & matrix phase.
- Nickel-and cobalt-based superalloys have also been used as matrix but a major drawback is that it fiber oxidized at elevated temperatures.
- It has been observed that whenever carbon fiber is reinforced with aluminium alloys carbon reacts with aluminium & formation of aluminium carbide (Al_4C_3) takes place which affects the mechanical properties of the composite, in order to prevent it a protective layer either titanium boride (TiB_2) or Sodium coating has been generally provided over the carbon fibers which reduce the fiber degradation.
- Magnesium is a light metal but could not be used as metal matrix due to atmospheric corrosion. In similar manner beryllium is also a lighter metal which has a tensile modulus higher than that of steel but due to higher brittleness it cannot be used as matrix material.

1.95. Ceramic Matrix

Ceramic material has outstanding property like high temperature stability, high thermal shock resistance, high modulus, high hardness, high corrosion resistance, and low density on the other hand it exhibits brittle nature & low resistance to crack propagation and low fracture toughness. Ceramic material not only used as reinforcing material but also used as a matrix material, when it is used as a reinforcing material its fracture toughness increases. There are number of ceramic matrix material among those, Alumina (Al_2O_3), Calcium alumino silicate (CAS), Lithium alumino silicates (LAS), Mullite (Al_2O_3 – SiO_2) and cements are important. Structural ceramics used as matrix materials can be categorized as either oxides or non-oxides. Alumina (Al_2O_3) and mullite (Al_2O_3–SiO_2) are the two most commonly used oxide ceramics. They are known for their thermal and chemical stability. The common non-oxide ceramics are silicon carbide (SiC), Silicon nitride (Si_3N_4), Boron carbide (B_4C) and aluminium nitride (AlN). SiC is used where high modulus and high temperature stability does matter. Si_3N_4 is known for high strength while AlN is popular due to high thermal conductivity.

Note

- The fracture toughness of ceramics ranges between of 3 to 6 MPa·m$^{1/2}$.

- Ceramic material also used as reinforcement material with ceramic matrix for manufacturing ceramic composites for example - silicon carbide fiber reinforced calcium alumino silicate (CAS) and carbon fiber reinforced with lithium alumino silicate (LAS) & so many other. Furthermore, ceramic reinforcement materials are available in the form of SiC, Si_3N_4, AlN, and other ceramic. SiC fiber mostly used as reinforcement due to its thermal stability and compatibility for both oxide and non-oxide ceramic matrices. These reinforcing materials either used in the form of whiskers, platelets, particulates or in the form of monofilament and multifilament continuous fibers with the ceramic matrix material.

1.96. Carbon Matrix Materials

Carbon can also be used as a matrix material. Carbon is weak & has brittle nature and whose thermal conductivities ranges from very low to high which is depending upon the precursor materials and processes of manufacturing. In this category carbon fiber reinforced with carbon matrix is very popular.

1.97. Composite Manufacturing Process

Fabrication processes are also responsible factor which influenced & having significant impact for the developing a desired level of property with in the composite material or final finished product. Basically manufacturing process governs shaping of composites, kind of composite material that is being processed, design specification of composite parts, application of finished product, required performance & efficiency, orientation of fibers / reinforced phase throughout the composites material where alignment & uniformity of fibers does matter. In order to fullfil the desired purpose several manufacturing or fabrication process are availiable for composite material processing, further composite manufacturing process can be further broadly classify into three categeory i.e.

(A) Open mould process

(B) Closed mould process

(C) Liquid composite moulding process

Both the above process can further classify as **Open mould process** [Hand lay-up process, spray up process, Vacuum-bag auto clave process, Filament winding process] and **Closed mould process includes** Compression moulding, Injection moulding, Sheet moulding compound (SMC) process, Continuous pultrusion process. Moreover, some other process has also comes on the front line which take parts in fabrication or manufacturing of composites material among those resin infusion process, automated fiber placement process, additive manufacturing, preformed moulding, resin transfer moulding (Liquid composites moulding process), reaction injection moulding (Structural reaction injection moulding), reinforced reaction injection moulding, resin film infusion process, elastic reservoir moulding process, tube rolling process, matched die forming process, hydroforming process, thermoforming process are important. A possible brief description of all those processes are given below.

1.97.1. Hand Lay-up Technique

A name hand layup implies that the reinforcing phase is laid manually to manufacturing a composite. In this process, initially mould is prepared which may have a flat surface, a cavity or a positive-shaped mold that is either made from wood, metal, plastic, or a combination of all those materials & then either dry fibres, dry fabrics or fibres that has form of woven, knitted, stitched are fills inside a mould by hand and after that liquid resin is poured inside the mould. In subsequent process, wetting of fibers takes place and air releases that has been entrapped into the lay-ups can be performed by using rollers or brushes. Mostly nip-roller-type impregnators are widely used that create desired pressure on to resin for filling the fabrics which can be accomplished either by the action of rotating rollers or a bath of resin. Consider the following schematic diagram regarding to hand layup process.

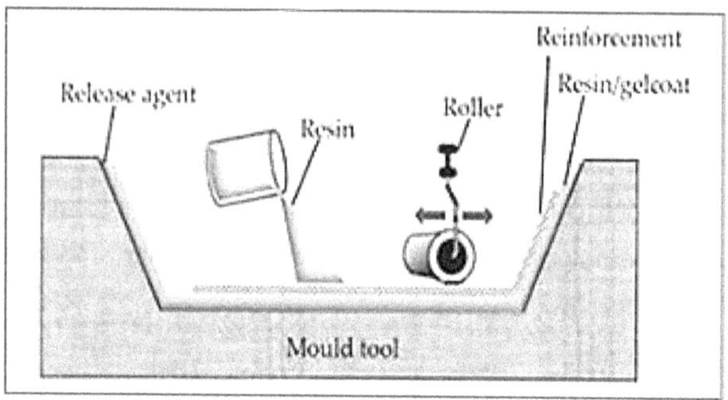

Diagram 1.22: Hand Layup Technique

In case of number of layered composites, fibers layer are wetted & arranged in laminate form and then cure under standard atmospheric conditions, after completing the curing process we can again added next layers & then cure to achieve finished product. This technique is applied for making glass fibre composite products such as bathtubs, boat hulls and decks, fenders, RV components, shower stalls, spas, truck cabs and simple shape articles. Furthermore, the operating maximum temperature for layup temperature depends upon following.

1. The maximum cure temperature
2. The heating rate in $^{\circ}$C/min
3. The composite layup thicknesses

The maximum cure temperature is generally identify & determine by the prepreg manufacturer for any particular resin–catalyst system and also evaluate time–temperature–viscosity characteristics of that resin–catalyst system. It has been observed that heating at low rates exhibits uniformity in temperature distribution within the layup on the contrary heating at high rates requires large thickness of layup, the heat generated by the curing reaction is faster than the heat transferred to the mold surfaces and a temperature "overshoot" occurs. Moreover, the maximum pressure requires for resin flow in the layup depending upon the following factors.

1. Layup thickness
2. Rate of heating
3. Rate of pressure application

It has been found that cure pressure is sufficient to squeeze out all excess resin from 16 to 32 layups but if the heating rate is very high at that cure pressure, the resin may start to gel before the excess resin is squeezed out from every ply in the layup. Eventually we can say that each & every composites material has different cure temperature, heating rate and cure time which is wholly depending upon the thickness of layup, consider the following temperature which shows cure temperature, heating rate and cure time for 32 – Ply carbon fiber epoxy laminates composites.

Table 1.71: Cure Time for 90% Degree of Cure in 32-ply Carbon Fiber – Epoxy Laminate

Cure temperature in ^{o}C	Heating rate ^{o}C/ min	Cure time in min
135	2.8	236
163	2.9	111
177	2.8	90
178	5.7	66
177	11.2	52

1.97.2. Spray-up Technique

Spray-up technique is a modified form of **hand layup technique** which also falls under the class of open-mould technique that is used for manufacturing composite. In this process, liquid resin matrix and chopped reinforcing fibers are sprayed by two separate sprays onto the mold surface. The fibers are chopped into fibers that has 25–50 mm length and then sprayed by an air jet simultaneously with a resin spray at a specific ratio between of reinforcing phase and matrix phase. Consider the following schematic diagram for spray-Up technique.

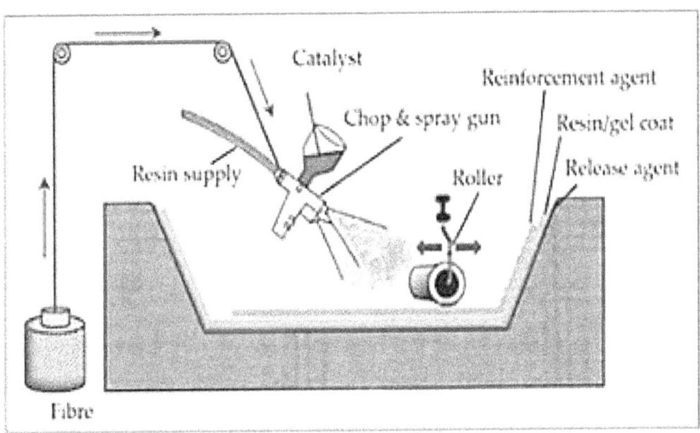

Diagram 1.23: Spray-Up Technique

If we are using gel coat then it is sprayed into the mould at a certain thickness after that gel coat is cured, and the mould is ready for process. The fibre and resin both are sprayed onto the mould with the help of chopper spray gun at a viscosity of 500–1000 cps. The basic purpose of gun is to supply fiber & resin continuously & simultaneously on the surface of the mould. At the semi-final stage the composite composition laminates is then compacted by rollers that is operated manually after that composite part is then cured at specific temperature which is depending upon the type of reinforced phase & matrix phase and allowed to cooling or solidification afterward remove the product from the mould. With the help of such a method composite of uniform coating can be easily manufactured. But we have got moderate mechanical properties of composite & also not suitable for processing reinforcing continuous fibers. The advantage of this technique is that we can produce the composite articles at low cost. Spray-up technique are used for making glass fibre composite like bathtubs, boat hulls and decks, fenders, RV components, shower stalls, spas, truck cabs, and other simple shapes products.

1.97.3. Bag-molding Process

In bag-molding process mold surface is covered with a Teflon-coated glass fabric separator which prevents sticking in the mold then upon which the prepreg plies are laid up in the desired fiber orientation or desired sequence. "Prepreg is nothing but it contains fibers in a partially cured epoxy resin. In general, a prepreg contains 42 wt% of resin. If this prepreg is allowed to cure without any resin loss, the cured laminate must contain 50 vol% of fibers. Since nearly 10 wt% of resin flows out during the molding process, the actual fiber content in the cured laminate should be 60 vol%. During the fabrication process excess amount of resin release from the prepreg in conjunction with entrapped air aswellas residual solvents, which further leads to reducing the majority of void among laminate". After that Plies are trimmed by using prepreg roller so as to achieve desired shape and size. The orientation is performed by cutting device by using mat knife or sometimes it can either be accomplished by using laser beams, high-speed water jets, or trimming dies. After that the layer-by-layer stacking of plies operation either perform manually or by numerically controlled automatic tape-laying machines. But before laying up the prepreg, the backup release film is introducing among all the plies and then somewhat compaction pressure is applied for Teflon-coated glass fabric or to the preceding ply in the layup to adhering the prepreg. Consider the following diagram regarding to the bag molding or vacuum bag auto-clave technique.

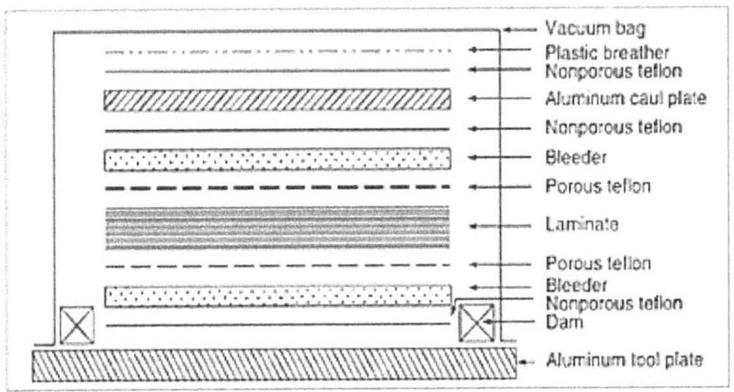

Diagram 1.24: Bag Molding Process

Afterward a porous cloth and few layers of bleeder papers are spread over the top of the prepreg stack. The basic purpose of bleeder papers is to absorb the excess resin that release from prepreg during the molding process. The complete layup is covered with another sheet of Teflon-coated glass fabric separator, a caul plate, and then a thin heat-resistant vacuum bag, which is closed around its periphery by a sealant and then a whole assembly kept an autoclave inside which it is subjected to external pressure, vacuum, and heat as a result of which consolidation and densification of separate plies takes place & transformed into solid laminate. The basic purpose of vacuum is to remove undesirable air and other volatiles while the pressure is required to accumulate & consolidation of individual layers into a laminate. As the prepreg is heated in the autoclave, the resin viscosity in the B-staged prepreg plies first decreases, attains a minimum, and then increases rapidly as the curing reaction begins and proceeds toward completion.

In order to under stand the bag moulding process, consider the case of carbon fiber–epoxy prepreg. It consists of two stage, in first stage, this cure cycle consists of increasing the temperature at a controlled rate up to 130°C and dwelling at this temperature for nearly 60 min until minimum resin viscosity does not achieve.

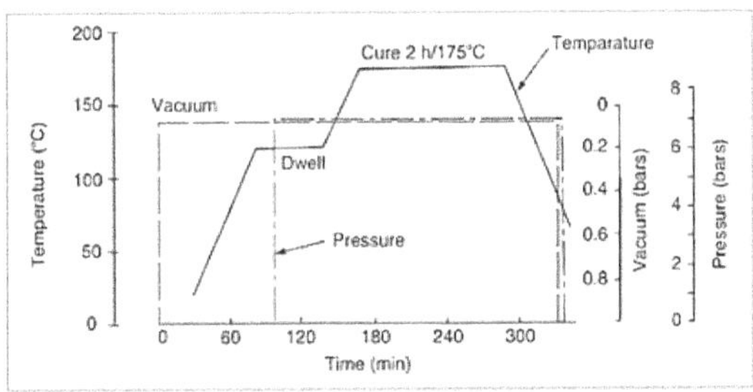

Diagram 1.25: Two Stage Cure Cycle for a Carbon Fiber – Epoxy Prepreg

During dwelling period an external pressure is applied on the prepreg stack so that excess resin may flow out into the bleeder papers. The basic advantage of resin flow is that it not only escapes entrapped air, volatiles from the prepreg but also eradicates the void from cured laminate. At the end of the temperature dwell, the autoclave temperature is increased to the actual curing temperature for the resin. The cure temperature and pressure are maintained for 2 h or more until a predetermined level of cure has occurred. At the end of the cure cycle, the temperature is slowly reduced while the laminate is still under pressure. The laminate is removed from the vacuum bag and if required, post cured can be performed at an elevated temperature in an air circulating oven.

Note

- Bag Moulding Process is of three types these are pressure bag, vacuum bag, and autoclave.

1.97.4. Filament Winding

Filament winding is known as a highly automated process which falls under the category of continuous fabrication technique. Filament winding technique is generally applied for manufacturing axisymmetric hollow parts like pipes (For chemical industry), tubes, cylinders, automotive drive shafts, pipelines for other applications and oxygen tanks. In addition to it, it can also be used to produce helicopter blades, spherical pressure vessels, conical rocket motor cases, gasoline storage tanks and apply for manufacturing prepreg sheets or continuous fiber-reinforced sheet-molding compounds, such as XMC.

Filament winding set-up consisting following elements.

1. Fiber creels
2. Resin bath tanks
3. Resin wiping device
4. Carriage
5. Mandrel

In filament-winding process, three or more creels are used that has been wounded number of fiber which pull out from these creels and then passes through a liquid resin bath tank that containing liquid resin, catalyst, and other ingredients, such as pigments and UV absorbers. Fiber unwounded process & respective tension is controlled by using the fiber guides or scissor bars which is located between each creel and the resin bath. Before entering the dry fiber into the resin bath for wetting, all the rovings are transformed into a band either passing through a textile thread board or a stainless-steel comb. After the withdrawal of band from resin tank, the resin-impregnated roving's are pulled through a wiping device whose function is to removes the unwanted excess resin from the roving's and maintaining resin thickness coating around each roving. Further resin thickness coating can be accomplished by two method.

1. A very most common wiping device using set of squeezing rollers. In this technique top roller is adjustable by which we can maintain the resin amount in addition to tension in fiber roving's.

2. In another technique for achieving wiping purpose each & every resin-impregnated roving's is passes separately through an orifice, when it passes out it will maintain the coating of resin amount aswellas tension in fiber roving's.

Consider the following diagram regarding to filament-winding technique, which shows complete overview of the process.

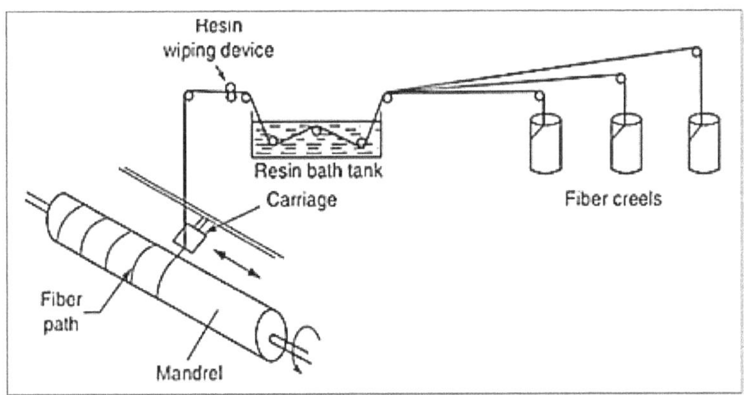

Diagram 1.26: Filament Winding Technique

After the wiping, these impregnated fiber are collected in flat band (Band former) and then accumulated on revolving mandrel. The band former is assembled with a carriage, which is free to move in traverse's direction in back and forth parallel to the mandrel whose function is to properly guide the impregnated fiber & cooperate in proper deposition around the revolving mandrel. For superior winding, traversing speed of the carriage and winding speed of the mandrel must be maintain so as to achieve desired winding angle patterns. For winding, speeds have been recommended between range of 90 to 110 linear m/min. However, for excellent & precise winding, slower speeds are always recommended.

Note

- Band formation can be achieved by using either a straight bar, a ring, or a comb.
- Filament winding is used to manufacturing an aerospace composite material.
- Now a day's computer-controlled filament-winding machines are used due to better result & low cost.
- A recommended winding speeds ranges from 90 to 110 linear M/min.

1.97.5. Conventional Filament-Winding Machine

Conventional filament-winding machine is also used to fabricating/manufacturing the composites material in which working of all the element is same except the mandrel & carriage. In this technique, motor mechanism is used to rotate the mandrel while carriage back and forth motion of carriage is perform by means of chain and sprocket setup. Remember that carriage motion always parallel to the mandrel. In driving mechanism, main sprocket is attached to the mandrel shaft through a set of gears so as to achieve carriage feed balancing with the mandrel rotation by changing the gear ratios or the sprocket size. Consider the following diagram regarding to the conventional filament-winding machine which shows a complete over view of mechanism.

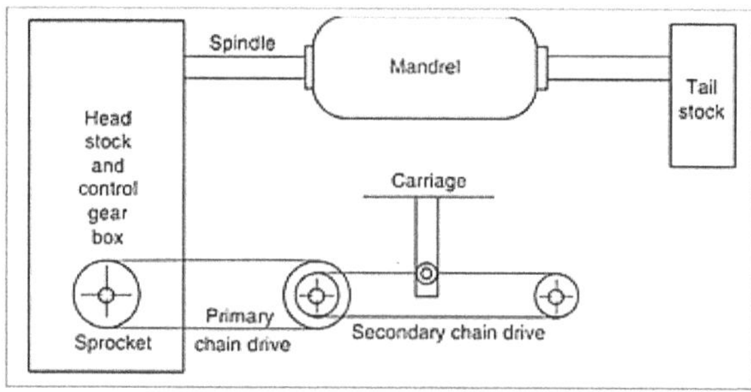

Diagram 1.27: Conventional Filament-winding Machine

Calculation of Wind Angle (Θ):- For circular mandrel which has rotational speed N revolution / minute and subjected to constant carriage feed say – V, then wind angle (Θ) may be given by the following relationship.

Where,

r = Radius of mandrel

N = Mandrel rotational speed / minute

V = Carriage feed

1.97.6. Numerical Control Filament Winding Technique

In numerical control filament winding technique, working of all the elements are similar to automated or conventional winding machine except the mechanism of mandrel motion, carriage motion, transverse feed mechanism and rotating pay out eye, consider the following diagram regarding to the numerical control filament winding which shows a complete overview of mechanism.

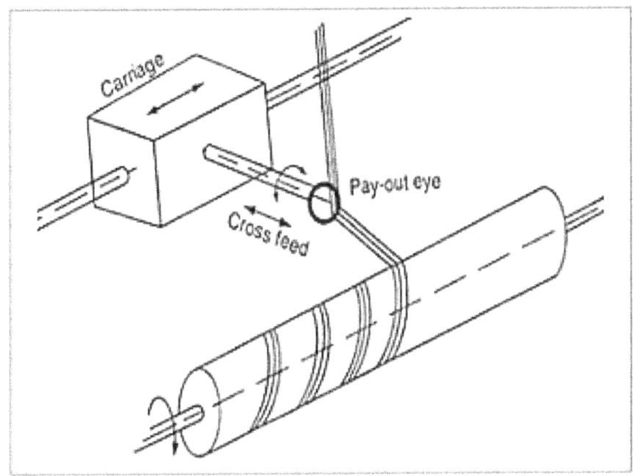

Diagram 1.28: Numerical Control Filament Winding Machine

In numerical control filament winding process, we have independent driving source for mandrel rotational motion, carriage movements, transverse feed mechanism and rotation pay out eye. In order to fulfil the desired purpose, separate hydraulic motor is installed for each motion. Consider following points.

1. Cross feed mechanism is installed on carriage which can move in & out radially.
2. Pay out eye is controlled rotational motion of horizontal axis.
3. All the movement which is related to mandrel rotation is control by numerical control mechanism.

Furthermore, combination of both i.e. cross feed mechanism & pay out eye, cooperate to prevent fiber slippages & fiber bunching on the mandrel. Eventually we can say that it imparts final shape of composites in proper manner & sequence.

Moreover, in addition to above valuable techniques, **helical windings technique & polar winding techniques** are also used for manufacturing composites articles.

- In filament winding technique, fiber tension, fiber wet out and resin amount play a dominant role for manufacturing composites articles.

- Recommended fiber tension that ranges between 1.1 to 4.4 N always desirable to maintain fiber alignment on mandrel and resin amount in the filament wound articles.

- Excessive fiber tension above the recommended value may create difference of resin in inner and outer layers, inclusion of unwanted residual stresses in finished product and responsible for large mandrel deflection.

- Fiber tension can be controlled by the following three types of fiber guides during filament winding process. Consider following diagrams.

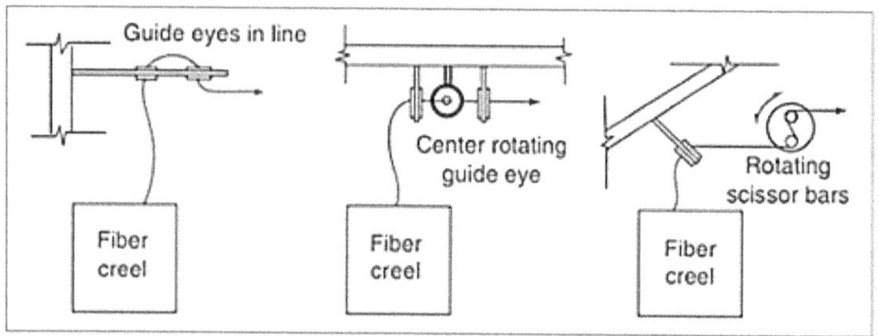

Diagram 1.29: Fiber Guides for Controlling Fiber Tension in Filament Winding Process

- Fiber wetting is a very important & essential operation of filament winding process. If fiber wetting is performed accurately then we have reduction in final filament wound articles.

- Viscosity play a vital role for wetting to fiber. resin bath composition must have low viscosity and that should be between 1 to 2 Pas, if viscosity is higher than recommended value then resin could not be properly impregnated to the fibers bonds.

- Lower viscosity of resin bath composition can be maintained by heating, resin bath as possible as at low temperature. At low temperature we can not only maintain low viscosity but also enhance pot – life.

- Fiber wetting always depending upon the winding speed, resin bath temperature, number of fiber strands in bonds, fiber tension and length of the resin bath.

- Resin contents can only be controlled by proper wiping or stripper dies, fiber tension and resin viscosity.

- Common defects in filament wound articles may appear due to voids, delamination and fiber wrinkles. voids presence due to poor fiber wetting, presence of air bubbles in resin bath, improper band width and high fiber tension on the other hand excessive time lapse between two consecutive layers responsible for delamination further wrinkles are comes on the front line due to improper winding tension & misalignment of rovings.

- Unstable fiber path are responsible for fiber slipping on mandrel which leads to fiber bunching, bridging and improper fiber orientation in wound articles.

1.98. Compression Moulding

Compression moulding is considered in the class of precision manufacturing technique by which we can produce the articles at minimum consumption of time. With the help of it, high-strength & high-quality composite parts, complex shape parts of different size can be manufactured. Now a day's compression moulding technique is widely used to composites processing for instance by thermosetting prepregs, fibre-reinforced thermoplastic and moulding compounds. Moulding compounds includes sheet moulding compound (SMC), bulk moulding compounds (BMC), and chopped thermoplastic tapes. Basically, it is not only limited up to composite processing but also equally take parts to manufacturing other metallic articles like, ribs, bosses, flanges, holes, structural automotive components, road wheels, bumpers, and leaf springs etc. The chief advantage of compression moulding is production of complex geometry shape in short periods of time without performing a secondary finishing operation like drilling, forming and welding.

Compression moulding setup consisting following components.

1. Movable platen
2. Movable upper half mold portion
3. Fixed bottom half mold portion
4. Shear edge
5. Charge
6. Ejector pin
7. Fixed platen

In compression moulding technique, composites processing material or stack of several rectangular plies known as charge, kept inside the fixed bottom half of a preheated mold cavity either by manually or robotically. The movable upper half mold portion is move downward at a constant rate by hydraulic presses so as to achieve compression force / pressure force upon the processing material and it must be continued until up to maximum recommended level of pressure force. Consider the following diagram regarding to compression moulding process which shows complete overview.

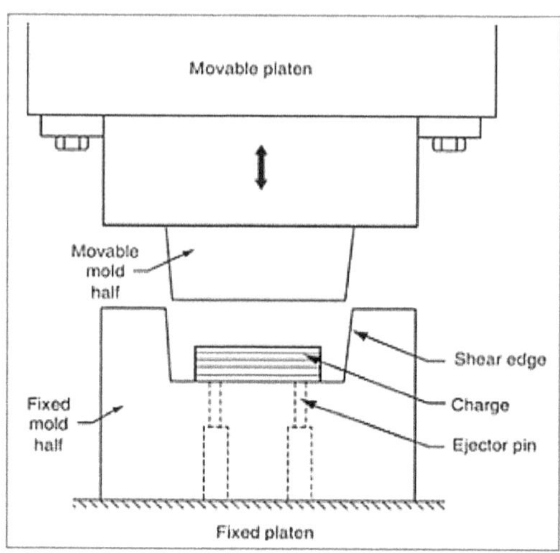

Diagram 1.30: Compression Molding Process

Further intensity of pressure force depending upon the shape/size of the articles i.e. simple or complex shape articles, length of flow, fiber contents or fiber alignment. It has been observed that high pressure force is desirable for molding parts that contain deep ribs and bosses. In general, the recommended value of moulding pressure & temperature for compression moulding process falls between 1.4 to 34.5 MPa & temperature 130ºC–160ºC respectively. After curing at a given pressure, the mold is opened and the article is withdrawn with the help of ejector pins.

Note

- Process flow of the material are always desirable for removing entrapped air from the charge & to manufacturing a void free article.
- Cycle time range depending up on the size of part and thickness of articles.

For example, Consider the case of compression moulding process of sheet-molding compounds (SMC). Compression molding is employed for transforming sheet-molding compounds (SMC) into finished products in matched molds. The compression-molding operation initiate with the placement of a pre-cut and weighed amount of SMC that can be accomplished by stacking of several rectangular plies which kept over the fixed bottom half portion of a preheated mold cavity known as charge, The ply dimensions are chosen in such a manner that it covers almost 60% to 70% of the mold surface area and thereafter mold is closed quickly after the charge placement and then movable top half of the mold is moved downward at a constant rate until the pressure value could not attain pre-set value of pressure force. After the curing under the pre-decided pressure range mold is opened to release the articles with the help of ejector pins. Consider the following time – temperature curve for SMC – compression molding operation.

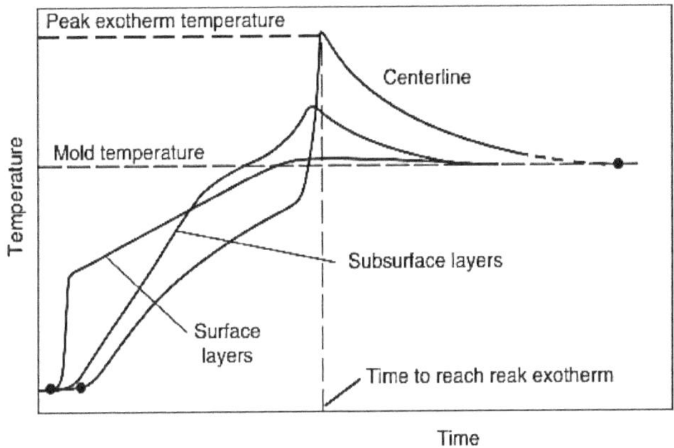

Diagram 1.31: Time – Temperature Curve Shows Distribution of Temperature with time Across the Thickness of SMC during the Compression Molding Operation

It has been concluded that,

- Time - Temperature curves (T-T Curve) analysed at the outer surface, subsurface, and centreline of thick E-glass fiber.

During molding a complex shape article heat transfer and a viscous flow phenomenon take place in the cavity. SMC moldings shows that charge attains the mold temperature rapidly and it remain uniform up to centreline temperature. However, the centreline temperature increases slowly until up to the curing reaction do not initiate at the mid-thickness of the part. The SMC material exhibits low thermal conductivity and hence the heat generated due to exothermic curing reaction in individual portion of SMC charge does not conducted to the mold surface and consequently centreline temperature increases rapidly to a peak value. When curing reaction nearer to completion, the centreline temperature decreases gradually to the mold surface temperature. Furthermore, initially surface temperature adopts a resin gel temperature and curing starts at the surface which further progresses toward, inward direction of charge. Curing achieve in fast manner at higher mold temperatures nevertheless the peak exotherm temperature may also increase. But on peak value of exotherm temperature (200°C) or higher may become a cause of burning and chemical degradation in the resin that must be avoided always.

During the compression molding process, the SMC – articles may be subjected to following defects, that is tabulated under here,

Table 1.72: Possible Defects in SMC – Articles after Compression Molding Process

Type of defects	Possible Occurrences of defect
Pinhole defect	It may be due to coarser filler particles and filler particles agglomeration
Long range waviness or ripple	It appears due to resin shrinkage and glass fiber distribution
Craters	Por dispersion of the lubricants
Sink marks	It occurs due to resin shrinkage, fiber distribution, fiber length , fiber orientation
Surface roughness	It occurs due to resin shrinkage, fiber bundle integrity, fiber distribution, strands dimensions.
Dark areas	Styrene loss from the surface
Pop-up blisters in painted parts	Subsurface voids due to trapped air & volatiles

1.99. Injection Moulding

Injection molding process is very old technique which has been used to manufacturing a articles since long period of time. It is a closed process composites manufacturing technique which has high volume production rate, fast & easy, low pressure and mostly used for manufacturing both thermoplastic and thermosetting plastic materials. It consisting following components.

1. Ladle
2. Charge
3. Nozzles
4. Mould cavity
5. Ejector pins

Consider the following schematic diagram which shows complete over-view of the injection moulding process.

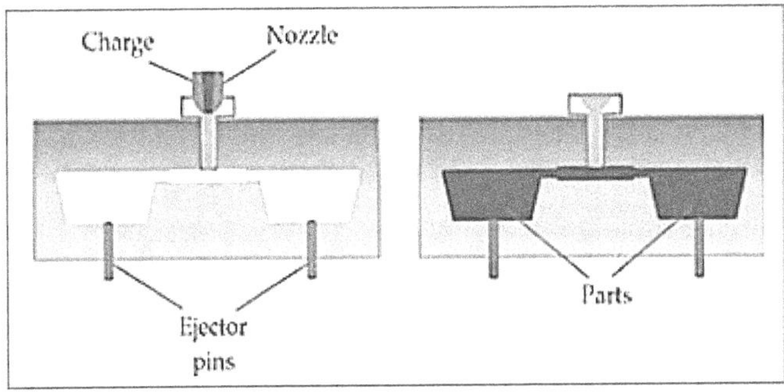

Diagram 1.32: Injection-moulding Process

Injection molding is a process that is employed for manufacturing both thermoplastic and thermosetting plastic materials articles. In this technique, properly mixed molten composition of composites is kept in ladle which equipped with a nozzle through which it is poured into mold cavity with recommended force or pressure where it cools and harden. After the cooling, upper portion of setup is removed and then plastic articles is withdrawn with the help of ejector pins. It is generally applied for manufacturing wire spools, packaging, bottle caps, automotive dashboards, pocket combs, and various other plastic products. Injection molding technique has high production rates, repeatable high tolerances, low labour cost, minimum scrap and no finishing required after molding or sometimes very little finishing if required on the contrary it is subjected to high equipment cost aswellas high running costs. But in present scenario, we have a modified form of injection molding technique, these are:

1.   Reaction Injection Molding (RIM)
2.   Reinforced Reaction Injection Molding (RRIM)

1.99.1. Reaction Injection Molding (RIM)

We know that, injection molding technique is employed for thermoplastic and thermosetting plastic materials but in case of reaction injection molding (RIM) thermosetting polymers are used instead of thermoplastic and thermosetting plastics. The manufacturing set-up is similar to injection molding technique but it has only the difference that curing reaction is accomplished within the mold cavity. Initially, the two components of the polymer are mixed properly in proper proportion and then mixture is poured into the mold under high pressure. In RIM, one processing material is polyurethane that is known as PU-RIM on the other hand second components may include either polyureas, polyisocyanurates, polyesters, poly-epoxides, or nylon 6. For polyurethane, one component of the mixture is polyisocyanate and the other component is a blend of polyol, surfactant, catalyst, and blowing agent. For example, consider following composition for reaction injection molding (RIM).

| Polyurethane = Polyisocyanate + Blend of polyol, surfactant, catalyst, and blowing agent. |
| (One Component) (Second Component) |

This method is generally employed for manufacturing, automotive bumpers, air spoilers, fenders, exterior body panels for the automotive industry, Non-E-coat polymers product (Showing excellent stiffness, impact resistance, excellent paint ability, solvent resistance and thermal resistance).

1.99.2. Reinforced Reaction Injection Molding

If reinforcing agents are added to the mixture of RIM setting then the process is known as reinforced reaction injection molding (RRIM). Common reinforcing agents include glass fibers and mica. This process is usually used to produce rigid foam automotive panels. A subset of RRIM is structural reaction injection molding (SRIM), which uses fiber meshes for the reinforcing agent. The fiber mesh is first arranged in the mold and then the polymer mixture is injected over it.

1.100. Pultrusion Process

Pultrusion process is an automated manufacturing process used for manufacturing composite materials into continuous & constant cross-section profiles. It is mostly employed for producing articles of continuous lengths from fibre-reinforced polymer that has constant cross-sections. By using this method, rods, tubes, I-beams, T-beams, frame sections, ladder rails, solid rods, hollow tubes, flat sheets, beams of various cross sections that including angles, channels, hat sections, wide-flanged sections and also used to manufacturing other long, straight structural members of constant cross-sectional area.

Pultrusion process consisting following components, these are:

1. Fibers
2. Resin bath
3. Shape performer
4. Pultrusion die
5. Puller and
6. Cutter

Pultrusion is a continuous fabrication process which is highly automated. In this process, a continuous bundle of dry fibre is pulled through a heated resin- bath for proper wetting and at the final stage to achieve finished articles a wetted composition is passes through the die rather than forced out by pressure. Consider the following schematic diagram which shows complete overview of pultrusion process.

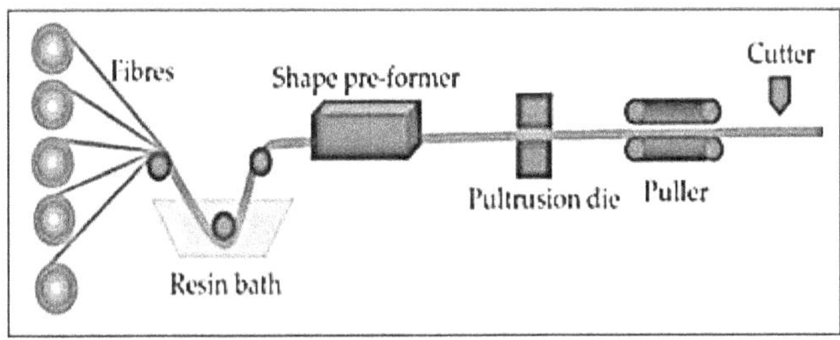

Diagram 1.33: Pultrusion Process for Composites Manufacturing

In this technique continuous fiber, strand rovings or sometimes mats are withdrawn by pull out from the creels which passes through the liquid resin bath. Liquid resin has usually composition of curing agent (Initiator), colorant, ultraviolet (UV) stabilizer, and fire retardant. For proper wetting of fiber, the viscosity of the liquid resin & submerged time of fiber in the resin bath are maintain in such a manner so that complete wetting of fibers with the resin may takes place properly. Afterward the wetted fiber–resin bundles are then passing through shape pre-formers the function of pre-formers is to ensure proper shaping, proper distribution of fiber bundles, release excess resin and impart a final configuration for material and then it is pass out through pultrusion preheated die. where final shaping, compaction, and curing take place in the die. To control the temperature of pultrusion preheated die, the entrance of the die equipped with water cooling system thereby we can reduce the chance of premature gelling and the rest of the die can be operated at recommended temperature. For curing process die is heated either by oil heaters, electric heaters or Infrared heating. A number of pulling rolls or blocks pull are installed in line whose function is to pull-out & release the cured pultruded articles from the die and then cooling it by air or water and cut into desired long-cured piece by a diamond impregnated saw at the end of the line. Moreover, one thing comes on the front line that heat generated due to exothermic curing reaction, raises the temperature in the fiber–resin stream and value of exothermic peak depends on the pulling speed of fiber–resin stream through the die. As the curing reaction approximately at completion, the exotherm temperature decreases and a cooling period begins. The rate of heat transfer from the cured material into the die walls is increased as a result of which rapid cooling of the entire cured section takes place. The temperature of the cured section is high when it exits from the die then interlaminar cracks may form within the pultruded articles.

Consider the following table, which shows mechanical properties of E – Glass polyester sheet that hs been manufacturing from pultrusion process.

Table 1.73: Mechanical Properties of Pultruded E – Glass Polyester Sheet

Total fiber contents in wt%	70 %	60%	50%	40%	30%
Continious roving (wt%)	39	29	19	19	16
Mat contents (wt%)	31.2	31.2	31.2	20.8	13.10
Roving mat ratio	1.25	0.90	0.60	0.90	1.18
Number of mat layers	2	3	3	3	2
Mat weight	1.5	1.5	1.5	1.5	1
Tesile strength in GPa					
Longitudinal	373	332.5	282	265.5	218
Transverse	86	93	94.5	84.5	68
Flexural strength in MPa					
Longitudinal	413	375	326	339	180
Transverse	204	200	220	181.5	169

Note

- The die temperature, die length, and pulling speed are controlled in such a manner so that the resin has properly cured before the pultruded member exits from the die.

- In commercial applications, polyester and vinyl ester resins, epoxies are used as the matrix material. but the major drawback is that it requires longer cure times and do not release easily from the pultrusion die. Pultrusion process has also been used with thermoplastic polymers, such as PEEK and poly-sulfone.

- A very important factor which affects the quality of finished product is depending upon the ability of wetting characteristics of fiber with the resin. Wetting always coincide to resin viscosity, residence time in the resin bath and resin bath temperature.

- Capability of wet-out can be improved by following steps.

   1. Extending, residence time of fiber bundles in resin bath

   2. By reducing the line speeds

   3. By increasing the length of resin baths

   4. Reducing the resin viscosity by resin bath temperature

- Resin penetration takes place inside the fiber bundles by capillary action and lateral squeezing between the bundles. Lateral pressure at the resin squeeze-out bushings, pre-formers, and die entrance also improves the resin penetration in the bundles. This can also be accomplished by slower line speed and lower resin viscosity.

- The fiber and resin surface energies are also responsible for absorption of resin, for instance - Kevlar 49 fibers has high surface energies thereby it accumulates large resin content from resin bath rather than either E-glass or carbon fibers under similar process conditions.

- The resin viscosity for excellent wetting of fiber bundles for pultrusion process falls between 0.4 to 5 Pas. Resin viscosities > 5 Pas responsible for poor fiber wet-out.

- Very low resin viscosities may cause of excessive resin draining from the fiber–resin stream after it leaves the resin bath. Resin viscosity can be lowered by increasing the bath temperature. Furthermore, when fiber–resin bundle enters in heated die then resin viscosity decreases, which cooperates in the wet-out of uncoated fibers.

- We know that no external pressure is required in pultrusion process, however it has been observed that at the die entrance it ranges between 1.7 to 8.6 MPa which is generated by volumetric expansion of the resin as it enters pre- heated die entrance.

- Resin viscosity, cure schedule and the line speed always based on the articles design specification which is being produced, For commercial pultrusion process line speed ranges from 50 to 75 mm/min to 3 to 4.5 m/min.

- The pulling force at the end of the pultrusion line must be maintain such that it can fulfil the following purpose.

  1. To overcome frictional force of fibers sliding, against the die wall.

  2. To maintain shear viscous force that exists among a very thin layer of resin aswellas the die wall.

  3. To adjust backflow force or drag resistance between the fibers and the backflowing resin at the die entrance.

1.101. Liquid Composites Moulding Process

In this process, a premixed liquid thermo-set resin is injected over the dry fiber preform that has been kept in the closed mould. As liquid resin is injected then it create coating over the fibers, fills the space between the fiber, remove entrapped air and thereafter curing is performed and consequently it transformed into the matrix.

In this section two method are important these are:

1.   Resin transfer moulding (RTM)

2.   Structural reaction injection moulding (SRIM)

1.101.1. *Resin Transfer Molding (RTM) Technique*

This technique consisting following components.

1.   Curative

2.   Resin container

3.   Mould cavity [Upper half mould cavity portion (Cope) & Bottom half mould cavity portion (Drag)]

4.   Mixing chamber

5.   Fiber preform

6.   Sprue (Installed at bottom portion)

7.   Resin injection point

Consider the following diagram which shows complete over-view of resin transfer moulding technique.

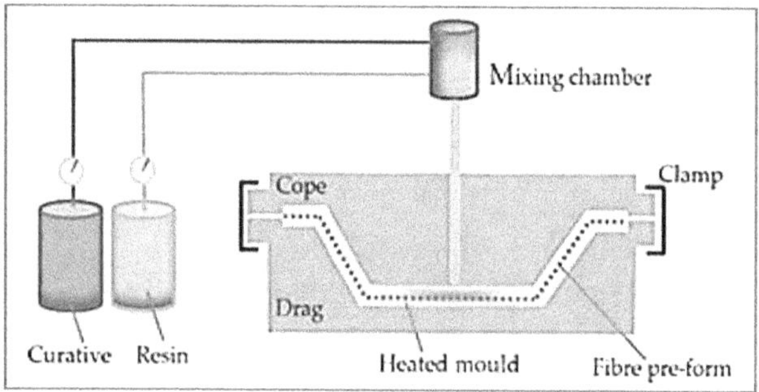

Diagram 1.34: Resin Transfer Moulding Technique

Resin Transfer Moulding (RTM) consisting a complete mould into two portion i.e. upper half portion and bottom half portion. In bottom half portion of mould, we will keep number of dry continuous strands mat, woven roving or cloth that is known as a fiber preform and then closed the mould by upper half portion. A liquid resin composition that has been properly mixed in mixing chamber (Located centrally) through which liquid resin in injected inside the heated mould cavity under recommended pressure, as resin injected it spread throughout the mould as a result of which, resin adopt a position of vacant space among the fibers, removes entrapped air and perform wetting & coating of the fibers takes place and after certain period of time after cooling ejecting the cured articles and perform trimming to achieve desired dimensions. Furthermore, now a days modified form of the RTM technique is also used, these are:

1.   **Vacuum assisted RTM (VARTM):-** Consider the following schematic diagram for VARTM.

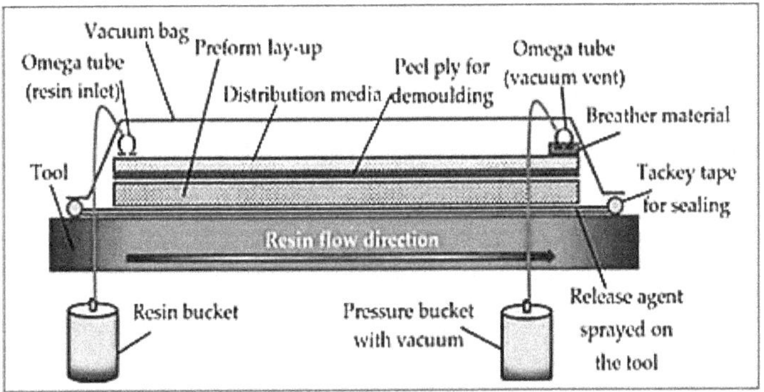

Diagram 1.35: Vacuum Assisted Resin Transfer Moulding (VARTM)

2.   **Resin Infusion Moulding (RIM):-** Resin infusion Moulding (RIM) is known as hybrid process in which a dry preform is placed in a mould on top of a layer, or interleaved with multiple layers, of high-viscosity resin film. As composition subjected to heat, vacuum, and pressure, then resin liquefies and is drawn into the preform, resulting in uniform resin distribution of high-viscosity, toughened resins in short flow distance. Consider following diagram regarding to resin film infusion (RFI).

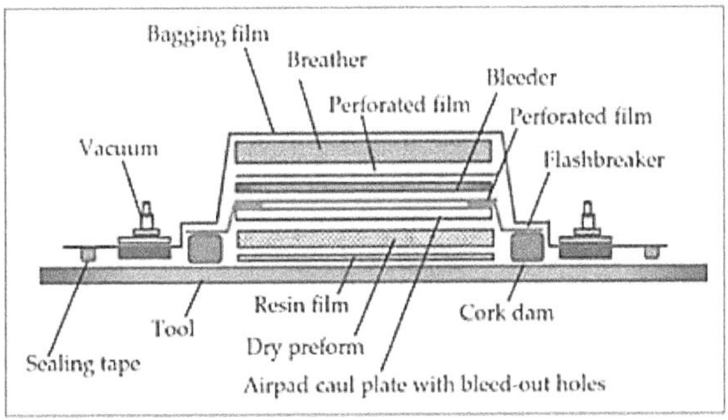

Diagram 1.36: Resin Film Infusion Process

By using resin infusion techniques, the maximum fibre volumes can handle up to 70%. Further an automated control of process co-operates in low voids and consistent preform reproduction, without the need for trimming. Resin infusion method is mostly used in boat building.

Note

- Binder are usually sprayed over the fiber so that fiber position could not be affected & maintain the preform shape.
- As compare to compression moulding process RTM has very low tooling cost.
- RTM is a low-pressure process.
- RTM technique are usually employed to manufacturing cabinet walls, chair / bench seats, hopper, water tanks, bath tubs and boat hulls.
- Generally polyurethane and epoxy are used as RTM and curing temperature for polyurethane resin ranges between 60°C to 120°C.

3. **Structural Reaction Injection Moulding (SRIM):-** Structural reaction injection moulding (SRIM) is similar to resin transfer moulding (RTM) technique except that it is based upon the resin reactivity, consider the following diagram.

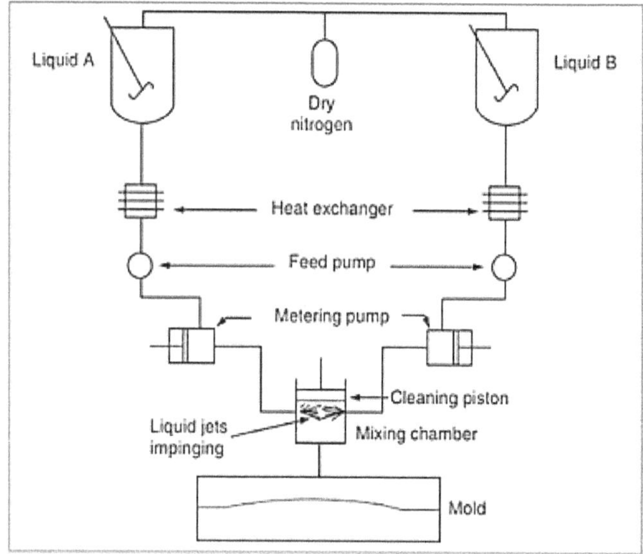

Diagram 1.37: Structural Resin Injection Moulding

To accomplished the desired resin reactivity, two low viscosity streams are supplied at high speed in mixing chamber and then this mixture is injected upon the dry fibers preform that has been previously kept in mould cavity, performed curing and thereafter ejects the article from the mould.

Note

- Curing reaction is faster in SRIM rather than RTM.
- Low viscosity is always desirable for both SRIM & RTM for good quality article manufacturing.
- Viscosity that ranges between 0.01 to 0.1 Pas are always recommended for SRIM resin.

In addition to the above fabrication or manufacturing technique, following method are also used to fulfil the same purpose. A short brief description of all those are given below.

1.102. Automated Fibre Placement

Automated fibre placement (AFP) is a newly developed advance technique used for fabricating and manufacturing of composite materials, consider following diagram regarding to automated fibre placement.

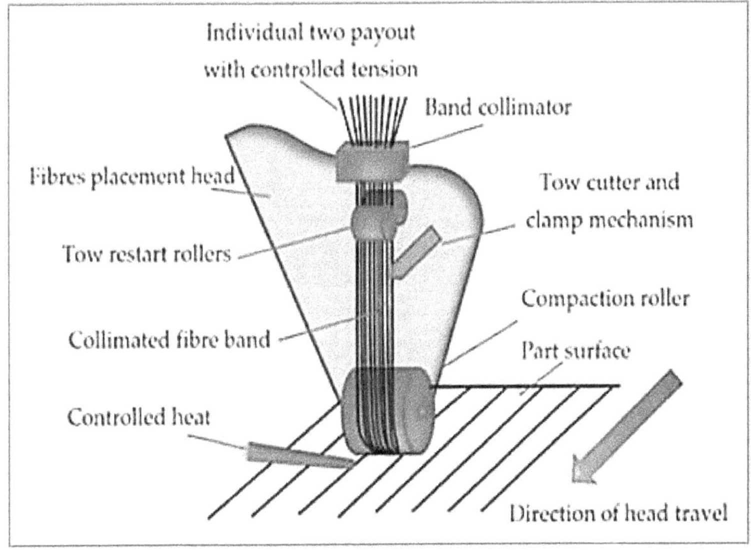

Diagram 1.38: Automated Fiber Placement Process

It is mostly used for fabricating continuous fibre-reinforced tape. A robot is use to place fibre-reinforced tape so that structure of one ply (layer) can be prepared at a time. A band of material consisting a multiple narrow strip of tape called tows. The width of such a tape, falls between range of 0.125 and 0.25 inches wide. The use of robotics facilitates the operator to perform active control for complete processes. The basic advantages of fibre placement are high processing speed, low material scrap and minimize labour costs. This process is applied for fabricating large thermoset parts that has complex shapes. In similar manner automated tape laying (ATL) is also used to manufacturing or fabricating to composites material which is also falls under the category of automated process in which prepreg tape is laid down continuously to form parts.

1.103. Additive Manufacturing

Additive manufacturing is most recently developed 3D printing technique based on the concept of rapid prototyping. This method is also used to manufacturing composite part by which we have got reduced costs of design-to-prototype product development and maintain material – labour and time-intensive area of toolmaking. In this technique, Manufacturing of composite structure is performed with a single nozzle uses, polymer composite filament and contains polymer and additives such as rubber microspheres, particles of glass or carbon fibre, wood flour, etc.

1.104. Elastic Reservoir Moulding (ERM) Technique

Elastic reservoir moulding technique are generally used to manufacturing bus roof panels, radar reflecting surface, automotive body panels and luggage carriers. Elastic reservoir moulding technique consisting following components.

1. **Mould Cavity:-** It has two portions, upper half mould portion & lower half / bottom half mould portion. A given diagram shows upper half mould portion that has protruding part of mould while lower / bottom half portion consisting depression or die like portion.
2. Resin impregnated foam.
3. Dry fibrous reinforcement at the top & bottom side foam.

Consider following diagram which shows complete over-view of elastic reservoir moulding technique.

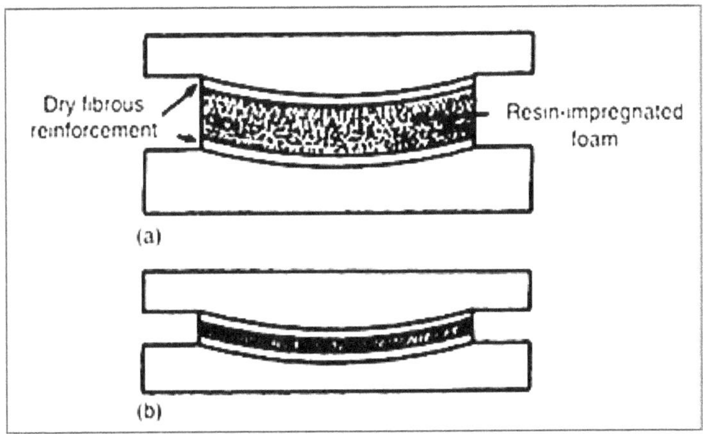

Diagram 1.39: Elastic Reservoir Moulding Process

In elastic reservoir moulding technique, sandwich shaped liquid resin impregnated foam which is supported either side by face layer of dry continuous stands mat, woven roving or cloth. A complete set is then kept inside the heated mould where it is subjected to compression force or pressure force that ranges between 500 to 1000 KPa. Here the foam is made from polyurethane which is flexible in nature and work as elastic reservoir for liquid resin, when foam has experienced a compression force resin flow takes place through face layers & wet – out it. Further when it is cured then we have got a sandwich of low-density core & fiber reinforced skins.

1.105. Tube Rolling

Tube rolling technique is generally apply for manufacturing, hollow tube, bicycle frames and space truss. In this process, pre-cut length of prepreg are rolled on to the mandrel. The uncured tube is wrapped with a heat – shrinkable film and then cured at elevated temperature in an air circulating oven. As curing starts, outer wrap shrinks due to the effect of heat and rigidly sticking over the rolled prepreg. During the curing process, entrapped air between the layers is escape through the ends and after curing mandrel is removed and hollow tube of composite material is obtained.

Note

- For better surface finishing, curing operation can be performed in a close-fitting steel tubes or split steel mould.

- It has low tooling cost with respect to filament winding technique.
- It has simple operation & ensure better control of resin contents & their distribution.
- High production rate can be achieved.
- This technique is most suitable for simple layup that containing 0^0 & 90^0 layers orientation.

Question for Discussion

1. What do you understand by composites material?. Write down some definition that has been given by the material scientists & also explain why these materials are important for diverse field of application.

2. How can be classify composites material? Explain classification on the basis of reinforced or dispersed phase.

3. How many types of matrix phases are possible for composite articles manufacturing? Explain in brief about metal matrix & ceramic matrix phase.

4. What conclusion are comes out from primary phase & secondary phase in composites materials? Is primary phase & secondary phase may be considered as a matrix phase & dispersed phase / reinforced phase respectively for any composite articles, if yes then justify.

5. Write short notes on followings.
    1. Polymer matrix composites (PMC)
    2. Carbon matrix composites (CMCs)
    3. Particles reinforcements
    4. Fiber reinforcement
    5. Property of thermosets & thermo-plastics

6. What do you understand by thermo-plastic polymer matrix phase? Explain.

7. Explain about metal matrix composites (MMC)? Provide some example regarding to metal matrix composites (MMC).

8. Write all the possible property of metal matrix composites (MMC) & also its field of application.

9. What are ceramic matrix composites (CMCs)? Explains.

10. Explain about matrix phase & dispersed phase of ceramic matrix composites? Naming all the possible matrix phase & dispersed phase.

11. Is oxides & non-oxides matrix phase may either be used to manufacturing ceramic matrix composites? Explain.

12. How many types of ceramic fibers are used to reinforcing ceramic matrix composites? Explain about silicon carbide fibers.

13. Explain various type of silicon carbide fiber say mono-filaments, multi-filaments yarns and silicon carbide whiskers that is being used for manufacturing ceramic matrix composites?

14. Write down various types of aluminum oxide fiber which is being used to manufacturing ceramic matrix composites?

15. What is the various type of property of ceramic matrix composites? Explain its field of application.

16. What are carbon matrix composites (CAMCs)? Explain advantages & possible field of application.

17. Classify the composites material on the basis of reinforcement phase? Explain various type of dispersed phase which can be used to manufacturing composites materials.

18. Write short notes on followings.
    1. Particles or particulates dispersed phase
    2. Fiber reinforcing materials
    3. Long fiber & short fiber composites

19. What do you understand by particles – reinforced metal matrix composites? Provide some possible example.

20. Explain about fiber-reinforced composites? Give some suitable example.

21. Classify fiber-reinforced composites? Is glass may be used as fiber for composites manufacturing.

22. Explain about followings.
    1. Long fiber reinforced composites
    2. Dis-continuous short fiber reinforced composites
    3. Dis-continuous short fiber reinforced metal matrix composites
    4. Short fiber rubber composites
    5. Fiber – orientation

23. Is glass fiber may be used as a dispersed phase? If yes then explain various type of glass fibers.

24. What are carbon fibers? Explain all the possible carbon fiber which is being used to manufacturing composites material.

25. Write sort notes on followings.

   1. Aramid fibers

   2. Kevlar – 49 & Kevlar – 149 fibers

   3. Nylon fibers & ceramics fibers

   4. Silicon carbide & aluminum-based fibers

26. Explain in brief about high density polyethylene fibres, Basalt fiber & Extended chain polyethylene fibres?

27. What do you understand by hybrid composites? Explain its various types.

28. Explain about green composites & Bio-composites material?

29. What are laminates composites? Explain about lamina with long fibers & lamina with discontinuous fibers.

30. What is nano-composites? Write short notes on followings.

   1. Polymer nano-composites

   2. Nano-clay composites

   3. Carbon nano-composites

   4. Natural fiber composites

31. Write down all the possible advantages & disadvantages of composites material?

32. Write short notes on followings.

   1. Common property of polymer matrix composites

   2. Physical & mechanical property of polymer matrix composites

   3. Application of polymer matrix composites

33. Explain in brief about glass fibrer – reinforced composites, Kevlar – 49 fibers reinforced composites and carbon fiber reinforced epoxy.

34. Explain in brief property of metal matrix composites like mechanical property of continuous fiber reinforced metal matrix material & also physical property.

35. Write short notes on followings.

   1. Property of discontinuous fiber reinforced metal matrix composites.

   2. Property of particulates reinforced metal matrix composites.

   3. Mechanical property of titanium carbide particulates reinforced steel & silicon carbide particulates reinforced aluminum.

   4. Physical property of silicon carbide particulates reinforced aluminum.

   5. Mechanical property of alumina particulates reinforced aluminum.

36. Explain diverse field of application of metal matrix composites?

37. Explain in brief about the property ceramic matrix composites?

38. Write short notes on followings.

    1. Mechanical property of ceramic matrix composites.

    2. Physical property of ceramic matrix composites.

    3. Application of ceramic matrix composites.

39. What do you understand by carbon-carbon composites? Explain.

40. Write short notes on followings.

    1. Mechanical property of carbon matrix composites

    2. Physical property of carbon matrix composites

    3. Application of carbon matrix composites

41. Explain in brief about application of advanced ceramic & bio- composites?

42. What do you understand by bonding in composites? Write down the various parameters upon which composites bonding depends.

43. Write short notes on followings.

    1. Mechanical bonding

    2. Dissolution & wetting bond

    3. Reaction bond & oxides bonds

    4. Diffusion bonding & plasma spray bonding

    5. Hot rolling bonding & mixed bonding

44. What do you understand by failure of composites material? Explain about failure criteria of uni-directional composites material.

45. Write short notes on followings.

    1. Quadratic interaction failure criteria

    2. Failure of composites due to residual stress

    3. Tensile property of composites

46. Explain in brief about micro-mechanics of composites material.

47. What do you understand by reinforcing material? Explain various types of synthetic & natural reinforcing material that is being used in composites materials.

48. Write short notes on followings.

    1. Glass fiber reinforcing material

    2. Carbon fiber reinforcing material

    3. Boron fiber reinforcing material

    4. Silicon carbide fiber reinforcing material

    5. Glass fiber reinforcing material

    6. Aluminum fiber reinforcing material

    7. Nylon fiber reinforcing material

49.   What is a significance of matrix material in composites material? Explain.

50.   Write short notes on followings.

    1.   Polymer matrix materials

    2.   Thermo-plastic & thermo-setting resins matrix

    3.   Metal matrix materials

    4.   Ceramic matrix materials

    5.   Carbon matrix materials

51.   Explain in brief about various types of composites manufacturing techniques?

52.   How many types of open mold & close mold techniques are used to manufacturing composites materials? Explain any one with clear sketch among both.

53.   Write short notes on followings.

    1.   Hand layup technique

    2.   Bag molding technique

    3.   Filament windings & numerical control filaments windings

    4.   Compression molding techniques

54.   What is injection molding technique? Explain with clear sketch about reaction injection molding technique & reinforced reaction injection molding technique.

55.   Explain with neat sketch about pultrusion process?.

56.   What are liquid composites molding process? Explain in brief about resin transfer molding (RTM) & Structural reaction injection molding (SRIM).

57.   Write short notes on followings.

    1.   Resin infusion molding (RIM)

    2.   Automated fiber placements

    3.   Elastic reservoir molding (ERM)

    4.   True rolling technique

References

[1]   Lubin, Hand book of composites, Van Nostarnd, New York, 1982.

[2]   Woods D.W., Ward I.M., J. Mater. Sci., 1994, 29, 2572.

[3]   Sohn M.S., Hu X.Z., Kim J.K., Polym. Compos., 2001, 9, 157.

[4]   Pigott M.R., J. Comps. Mater., 1994, 28, 588.

[5]   Hiley M.J., Plast. Rubber, Compos., 1999, 28, 210.

[6]   Rajeev R.S., Anil K. Bhowmick, De S.K Bandyopadhyay SJ. Appl. Polym. Sci., 2003, 90, 544.

[7]   Tjong S.C., Meng Y.Z., J. Appl. Polym. Sci., 1999, 72, 501.

[8]   Gonzalez P.I., Yong R.J., J. Mater. Sci., 1998, 33, 5715.

[9]   Raghavan J., Wool R. P., J. Appl. Polym. Sci., 1999, 71, 775.

[10]   Sapua S.M., Leenie A., Harimi, M Beng Y.K., Materials and Design, 2006, 27, 689.

[11]   Bipinbal P.K., Kutty S.K.N. J. Appl. Polym. Sci., 2008, 109, 1484.

[12]   Sreekumar P.A., Pradeesh Albert, Unnikrishnan G., Kuruvilla Joseph, Sabu Thomas J. Appl. Polym. Sci., 2008, 109, 1547.

[13]   Mondadori N.M.L., Nunes R.C.R., Zattera A.J., Oliveira R.V.B., Canto L.B., J. Appl. Polym. Sci., 2008, 109, 3266.

[14]   Shan Jin, Yuansuo Zheng, Guoxin Gao, Zhihao Jin, Mater. Sci. Eng., 2008, 483, 322.

[15]   Tiesong Lin, Dechang Jia, Peigang He, Meirong Wang, Defu Liang Mater. Sci. Eng., 2008, 497, 181.

[16]   Mukherjee M., Das C.K., Kharitonov A.P., Banik K., Mennig G., Chung T.N. Mater. Sci. Eng., 2006, 441, 206.

[17]   Praveen S., Chakraborty B. Cjayendran S., Raut R.D., Chattopadhyay S. J. Appl. Polym. Sci., 2009, 111, 264.

[18]   Business Communications Co, Inc, Composites: Resins, Fillers, Reinforcements, Natural Fibers and Nanocomposites (Report number RP-178).

[19]   Warner S.B., Fibre Science., Prentice Hall, Engle wood Cliffs, New Jersey, 1995.

[20]   Mallick P.K., Fibre reinforced composite materials, manufacturing and design., Marcel Dekker, Inc., Newyork, 1988, Ch. 1., p. 18.

[21]   Vincent J.F.V, Appl. Comp. Mater., 2000, 7, 269.

[22]   Pecorini T.J., Hertzberg R.W., Polym. Compos., 1994, 15, 174.

[23]   Setua D.K., De S.K., J. Mater. Sci., 1985, 20, 2653.

[24]   Bishai A.M., Ghoneim A.M., Ward A.A., Younan A.F. Int. J. Polymer. Mater., 2002, 51, 793.

[25]   Mingtao Run, Hongzan Song, Chenguang Yao, Yingjin Wang J Appl. Polym. Sci., 2007, 106, 868.

[26]   Mondadori N.M.L, Nunes R.C.R., Zattera A.J., Oliveira R.V.B., Canto L.B, J Appl. Polym. Sci., 2008, 109, 3266.

[27]   Wazzan A.A. Int. J. Polymer. Mater., 2004, 53, 59.

[28]   Rajeev R.S., De S.K., Bhowmic A.K., J. Mater. Sci., 2001, 36, 2621.

[29]   Rajesh C., Unnikrishnan G., Purushothaman E., Sabu T., J Appl. Polym. Sci., 2004, 92, 1023.

[30]   Geethamma V.G., Reethamma J., Thomas S., J Appl. Polym. Sci., 1995, 55, 583.

[31]   Sreekala M.S., Kumaran M.G., Joseph S., Jacob M., Thomas S, Appl. Compos. Mater., 2000, 7, 295.

[32]   Cao Y., Shibata S., Fukumoto I., Composites Part A, 2006, 37, 423.

[33]   Weyenberg V., Chi Truong T., Vangrimde, B., Verpoest I., Composites Part A, 2006, 37, 1368.

[34]   Wong S., Shanks R., Hodzic A., Macromol. Mater. Engg., 2002, 287, 647.

[35]   Melo J.D.D., Radford D.W.J. Compos. Mater., 2003, 37, 129.

[36]   Afaghi Khatibi A, Mai Y.W., Composites Part A., 2002, 33, 129.

[37]   Pothen L.A., Thomas S., Compos. Sci. Technol., 2003, 63, 1231.

[38]   Abdelmouleh M., Belgacem M.N., Dufresne A., Compos. Sci. Technol., 2007, 67, 1627.

[39]   Yoshitaka Uchiyama, Noriaki Wada, Tomoaki Iwai, Seiichi Ueda, Shinji Sado, J Appl. Polym. Sci., 2005, 95, 82.

[40]   Seema A., Kutty S.K.N., Int. J. Polymer. Mater., 2006, 55, 25.

[41]   Susmita S., Bhowmick A.K. J. Polym. Sci., 2005, 43, 1854.

[42]   Rajeev R.S., De S.K., Bhowmic A.K., Bandyopadhyay S. J. Appl. Polym. Sci., 2003, 90, 544.

[43]   Nugay N., Erman B., J. Appl. Polym. Sci., 2001, 79, 366.

[44]   Kozlov G.V., Burya A.I., Zaikov G.E. J. Appl. Polym. Sci., 2006, 100, 2821.

[45]   Hong Gun Kim, J. Mecha. Sci. Technol., 2008, 22, 236.

[46]   Rueda L.I., Anton C.C., Rodriguez M.C.T., Polym. Compos., 1988, 9, 198.

[47]   Das N.C., Chaki T.K., Khastgir J. Polym. Eng., 2002, 22, 115.

[48]   Arroyo M., Bell M., J. Appl. Polym. Sci., 2002, 83, 2474.

[49]   Nielsen L.E., Mechanical properties of Polymers and composites, Vol 1. Marcel Dekker, 1974.

[50]   Setua D.K., De S.K., J. Mater. Sci., 1984, 29, 3097.

[51]   Arumugam N., Selvy K.T., Rao K.V., Rajalingam P., J Appl. Polym. Sci., 1989, 37, 2645.

[52]   Czarnecki L., White J.L., J Appl. Polym. Sci., 1980, 25, 1217.

[53]   Murthy V.M., De S.K., Bhagavan S.S, Sivaramakrishnan R., Athithan S.K., J. Appl. Polym. Sci., 1983, 28, 3485.

[54]   Murthy V.M., De S.K., Bhowmick A.K., J. Mater. Sci., 1982, 17, 709.

[55]   Jana P.B., De S.K., Mallick A.K., J. Mater. Sci., 1993, 28, 2097.

[56]   Das N.C., Chaki T.K., Khastgir Chakrabarty A., J Appl Polym Sci., 2001, 80, 1601.

[57]   Pramanik P.K., Khastgir De S.K., Saha T.N., J. Mater. Sci., 1990, 25, 3848.

[58]   Roy D., Bhowmic A.K., De S.K., Polym. Eng. Sci., 1992, 32, 971.

[59]   Ibarra L., Macias A., Palma E., J Appl. Polym. Sci., 1996, 61, 2447.

[60] Abdul-Aziz, M.M., Youssef H.A., Miligy A.A., Yoshii F., Makuuchi K., Polym. Polym. Compos., 1996, 4, 259.

[61] Gabara V., in "Synthetic Fiber Materials" H. Brody, Ed., Longman, Harlow, UK, 1994, p. 239.

[62] Susan S., Taweechai A., Wiriya M., Sauvarop B.L., Polymer, 1999, 40, 6437.

[63] Senapati A.K, Nando G.B., Pradhan B., Int. J. Polym. Mater., 1988, 12, 73.

[64] Sreeja T.D., Kutty S.K.N., Int. J. Polym. Mater., 2003, 52, 239.

[65] Zhang Liqun, Zhou Yanhao, Chen Tao, Li Chen, Li Donghong Xiangjiao Gongye, 1994, 41, 267.

[66] Seema A., Kutty S.K.N., Int. J. Polym. Mater., 2005, 54, 933.

[67] Bhagwan D.A., Lawrence J.B., "Analysis and performance of fiber composites", John Wiley & Sons, New York, 1980, Cha, 2, p. 21.

[68] Monette L., Anderson M.P., Grest G.S., Polym. Comp., 1993, 14, 101.

[69] Rosen B.W., Fibre Composite Materials, American Society for metals, Metal Park, Ohio, 1965.

[70] Setua D.K., De S.K., J. Mater. Sci., 1984, 19, 983.

[71] Murthy V.M., De S.K., J. Appl. Polym. Sci., 1984, 29, 1355.

[72] Ibarra L., Chamorro A.C., J. Appl. Polym. Sci., 1991, 43, 1805.

[73] Hong Gun Kim, J. Mech. Sci. Technol., 2008, 22, 130.

[74] Akthar S., De P.P., De S.K., J. Appl. Polym. Sci., 1986, 32, 5123.

[75] Senapati A.K, Kutty S.K.N., Nando G.B. Pradhan B., Int. J. Polym. Mater., 1989, 12, 203.

[76] Rathinasamy P., Balamurugan P., Balu S., Subrahmanian V. J. Appl. Polym. Sci., 2004, 91, 1124.

[77] Chawla K.K., Composite Materials, Spriger Verlarg, New York, 1987.

[78] Richardson M.O.W., Polymer Engineering composites, Applied Science Publishers, London. 1977, p. 1.

[79] High-Performance Composites Sourcebook 2009, Gardner publications Inc.

[80] S.K. Mazumdar, Composites Manufacturing: Materials, Product, and Process Engineering, CrC press, 2002.

[81] Buchanan, G.R., Mechanics of Materials, HRW Inc., New York, 1988.

[82] Ugural, A.C. and Fenster, S.K., Advanced Strength and Applied Elasticity, 3rd ed. Prentice Hall, Englewood Cliffs, NJ, 1995.

[83] Swanson, S.R., Introduction to Design and Analysis with Advanced Composite Materials, Prentice Hall, Englewood Cliffs, NJ, 1997.

[84]   Shaw, A., Sriramula, S., Gosling, P.D., and Chryssanthopoulo, M.K. (2010) Composites Part B, 41, 446–453.

[85]   Mayer, C., Wang, X., and Neitzel, M. (1998) Composites Part A, 29, 783–793.

[86]   Avila, A.F., Paulo, C.M., Santos, D.B., and Fari, C.A. (2003) Materials Characterization, 50, 281–291.

[87]   Nicoleta, I. and Hickel, H. (2009) Dental Materials, 25, 810–819.

[88]   Bunsell, A.R. and Harris, B. (1974) Composites, 5, 157.

[89]   Mkaddem, A., Demirci, I., and Mansori, M.E. (2008) Composites Science and Technology, 68, 3123–3127.

[90]   Bednarcyk, B.A. (2003) Composites Part B, 34, 175–197.

[91]   Tabiei, A. and Aminjikarai, S.B. (2009) Composite Structures, 88, 65–82.

[92]   Huang, H. and Talreja, R. (2006) Composites Science and Technology, 66, 2743–2757.

[93]   Sriramula, S. and Chryssanthopoulos, M.K. (2009) Composites Part A, 40, 1673–1684.

[94]   Tay, T.E., Vincent, B.C., and Liu, G. (2006) Materials Science and Engineering: B, 132, 138–142.

[95]   Friedrich, K., Zhang, Z., and Schlarb, A.K. (2005) Composites Science and Technology, 65, 2329–2343.

[96]   Wakeman, M.D., Cain, C.D., Rudd, C.D., Brooks, R., and Long, A.C. (1999) Composites Science and Technology, 59, 1153–1167.

[97]   Lekakou, C. and Bader, M.G. (1999) Composites Part A, 29, 29–37.

[98]   Paul, S.A., Boudenne, A., Ibos, L., Candau, Y., Joseph, K., and Thomas, S. (2008) Composites Part A, 39, 1582–1588.

[99]   John, M.A., Francis, B., Varughese, K.T., and Thomas, S. (2008) Composites Part A, 39, 352–363.

[100]   Bledzki, A.K. and Gassan, J. (1999) Progress in Polymer Science, 24, 221.

[101]   Bergeret, A. and Bozec, M.P. (2004) Polymer Composites, 25, 12.

[102]   H.E. Deve and C. McCullough, "Continuous-Fiber Reinforced Al Composites: A New Generation," JOM, July 1995, pp. 33–37.

[103]   T. Donomoto et al., "Ceramic Fiber Reinforced Piston for High Performance Diesel Engines," SAE Technical Paper No. 830252, 1983.

[104]   G.F. Hawkins, J.P. Nokes, R. Zaldivar, J. Newman, and C. Zweben, "Measuring the Glass-Transition Temperature of Composites in the Field," In Proceedings, SAMPE 2002, Long Beach, CA, May 12–16, 2002.

[105]   T. Hayashi, H. Ushio, and M Ebisawa, "The Properties of Hybrid Fiber Reinforced Metal and Its Application for Engine Block," SAE Technical Paper No. 890557, 1989.

[106]   D.B. Marshall and A.G. Evans, "Failure Mechanisms in Ceramic-Fiber/Ceramic-Matrix Composites," J. Am. Ceramic Soc., 68(5), 225–231, May 1985.

[107]   Military Handbook—5F, Metallic Materials and Elements for Aerospace Vehicle Structures (approved for public distribution).

[108]   L.H. Miner, R.A. Wolffe, and C. Zweben, "Fatigue, Creep and Impact Resistance of Kevlar® 49 Reinforced Composites," In Composite Reliability, ASTM STP 580, American Society for Testing and Materials, Philadelphia, PA, 1975, pp. 549–559.

[109]   R. Morrell, Handbook of Properties of Technical & Engineering Ceramics, Part 1: An Introduction for the Engineer and Designer, Her Majesty's Stationery's Office, London, 1985.

[110]   J.C. Norman and C. Zweben "Kevlar® 49/Thornel® 300 Hybrid Fabric Composites for Aerospace Applications," SAMPE Quarterly, 7(4), 1–10, July 1976.

[111]   W.H. Pfeifer, J.A. Tallon, W.T. Shih, B.L. Tarasen, and G.B. Engle, "High Conductivity Carbon-Carbon Composites for SEM-E Heat Sinks," Paper presented at the Sixth International SAMPE Electronic Materials and Processes Conference, Baltimore, MD, June 22–25, 1992.

[112]   G. Savage, Carbon-Carbon Composites, Chapman & Hall, London, 1993.

[113]   K.A. Schmidt and C. Zweben, "Advanced Composite Packaging Materials," Electronic Materials Handbook, Vol. 1: Packaging, ASM International, Materials Park, OH, 1989, pp. 1117–1131.

[114]   D.L. Smith, K.E. Davidson, and L.S. Thiebert, "Carbon-Carbon Composites (CCC): A Historical Perspective," In Proceedings, 41st International SAMPE Symposium, March 24–28, 1996, pp. 32–41.

[115]   R. Warren (Ed.), Ceramic-Matrix Composites, Chapman and Hall, New York, 1992.

[116]   C. Zweben, "Thermomechanical Properties of Fibrous Composite Materials: Theory," B. Bever (Ed.), Encyclopedia of Materials Science and Engineering, Pergamon, Oxford, 1986.

[117]   C. Zweben, "Mechanical and Thermal Properties of Silicon Carbide Particle Reinforced Aluminum,".

[118]   L.M. Kennedy, H.H. Moeller and W.S. Johnson (Eds.), Thermal and Mechanical Behavior of Metal Matrix and Ceramic Matrix Composites, ASTM STP 1080, American Society for Testing and Materials, Philadelphia, PA, 1989.

[119]   C. Zweben, "The Future of Advanced Composite Electronic Packaging," In D.D.L. Chung (Ed.), Materials for Electronic Packaging, Butterworth-Heinemann, Oxford, 1995a.

[120]   C. Zweben, "Simple, Design-Oriented Composite Failure Criteria Incorporating Size Effects," In Proceedings, Tenth International Conference on Composite Materials, ICCM-10, Whistler, British Columbia, Canada, August 1995b.

[121]   C. Zweben, "Overview of Composite Materials for Optomechanical, Data Storage and Thermal Management Systems," In Proceedings of the SPIE 44th Annual Meeting and Technical Display, The International Symposium on Optical Science, Engineering, and Instrumentation, Denver, CCO, July 18–23, 1999.

[122]   C. Zweben, "Metal Matrix Composites, Ceramic Matrix Composites, Carbon Matrix Composites and Thermally Conductive Polymer Matrix Composites," In C.A. Harper (Editor-in-Chief), Handbook of Plastics, Elastomers and Composites, 4th ed., McGraw-Hill, New York, 2002, Chapter 5.

[123]   C. Zweben, "Thermally Conductive Composites," JEC Composites Mag., January 2009.

[124]   C. Zweben, "Advanced Thermal Management Materials for Electronics and Photonics," Adv. Microelectron., 37(4), July/August 2010.

[125]   C. Zweben, "Advanced Thermal Management Materials for LED Packaging," LED Professional Review, January/February 2010.

[126]   C. Zweben, "Advanced Thermal Management Materials," Chip Scale Review Tech Monthly, May 2011.

[127]   M. Kolybaba, L.G. Tabil, S. Panigrahi, W.J. Crerar, T. Powell, Wang, Biodegradable Polymers: Past, Present, and Future.

[128]   A.V. Ratna Prasad K. Murali Mohan Rao and G. Nagasrinivasulu "Mechanical properties of banana empty fruit bunch fiber reinforced polyester composites" Indian journal of fiber and textile reasearch, Vol. 34, 2009.

[129]   Lina Herrera, Selvum Pillay and Uday Vaidya "Banana fiber composites for automotive and transport applications" Department of Matrial Science & Engineering, University of Alabama at Birmingham, Birmingham, AL 35294.

[130]   Panthapulakkal S, Sain M. Injection-molded short hemp fiber/glass fiber reinforced polypropylene hybrid composites – mechanical, water absorption and thermal properties. J Appl Polym Sci., 2007; 103: 2432–41.

[131]   Arbelaiz et al," Influence of matrix/fiber modification, fiber content, water uptake and recycling", Composites Science and Technology, 2005; 65: 1582–92.

[132]   Harris, B., Engineering Composite Materials, The Institute of Metals, London, 1986.

[133]   Hull, D. and T.W. Clyne, An Introduction to Composites Materials, Cambridge University Press, 1996.

[134]   Jones, R.M., Mechanics of Composite Materials, McGraw-Hill, New York, 1975.

[135]   Powell, P.C, Engineering with Polymers, Chapman and Hall, London, 1983.

[136]   Roylance, D., Mechanics of Materials, Wiley & Sons, New York, 1996.

[137]   Chen, J.; Hamon, M.A.; Hu, H.; Chen, Y.; Rao, A.M.; Eklund, P.C. & Haddon, R.C. (1998). Solution properties of single-walled carbon nanotubes. Science, Vol. 282, pp. 95-98.

[138]   Chen, Y.L.; Liu, B.; Huang, Y.; & Hwang, K.C. (2011). Fracture toughness of carbon nanotube-reinforced metal- and ceramic-matrix composites. Journal of Nanomaterials, Vol. 2011, Article ID 746029.

[139]   Cho, J.; Boccaccini, A.R. & Shaffer, M.S.P. (2009). Ceramic matrix composites containing carbon nanotubes. Journal of Materials Science, Vol. 44, pp. 1934–1951.

[140]   Ding, W.Q.; Calabri, L.; Chen, X.Q.; Kohhaas, K.M. & Ruoff, R.S. (2006). Mechanics of crystalline boron nanowires. Composites Science and Technology, Vol. 66, pp. 1112–1124.

[141]   Ebbesen, T.W.; Lezec, H.J.; Hiura, H.; Bennett, J.W.; Ghaemi, H.F. & Thio, T. (1996). Electrical conductivity of individual carbon nanotubes. Nature, Vol. 382, pp. 54–56.

[142]   Evans, A.G. (1990) Perspective on the development of high-toughness ceramics. Journal of the American Ceramic Society, Vol. 73, pp. 187–206.

[143]   Hull, D & Clyne T.W. (1996). An Introduction to Composite Materials (Second edition). Cambridge University Press, 0521388554, The Edinburgh Building, Cambridge CB2 2RU, UK.

[144]   Japanese Industrial Standards (JIS). (1995). R 1607. Kita, J.; Suemasu, H.; Davies, I.J.; Koda, S. & Itatani, K. (2010). Fabrication of silicon carbide composites with carbon nanofiber addition and their fracture toughness. Journal of Materials Science, Vol. 45, pp. 6052-6058.

[145]   Ma, R.Z.; Wu, J.; Wei, B.Q.; Liang, J. & Wu, D.H. (1998). Processing and properties of carbon nanotubes–nano-SiC ceramic. Journal of Materials Science, Vol. 33, pp. 5243-5246.

[146]   Miyahara, N.; Yamaishi, K.; Mutoh, Y.; Uematsu, K. & Inoue, M. (1994). Effects of grain size on strength and fracture toughness in alumina, JSME International Journal, Vol. 37, pp. 231-237.

[147]   Rice, R.W. (1996). Grain size and porosity dependence of ceramic fracture energy and toughness at 22°C, Journal of Materials Science, Vol. 31, pp. 1969-1983.

[148]   Sheldon, B.W. & Curtin, W.A. (2004). Nanoceramic composites: tough to test. Nature Materials, Vol. 3, pp. 505–506.

[149]   Sun, L.; Gao, L. & Li, X. (2002). Colloidal processing of carbon nanotube/alumina composites. Chemistry of Materials, Vol. 14, pp. 5169–5172.

[150]   M.; Yokomizo, K.; Hashida, T. & Adachi, K. (2008). Structural characterization and frictional properties of carbon nanotube/alumina composites prepared by precursor method. Materials Science and Engineering B, Vol. 148, pp. 265–269.

[151]   Yao, W.; Liu, J.; Holland, T.B.; Huang, L.; Xiong, Y.; Schoenung, J.M. & Mukherjee, A.K. (2011). Grain size dependence of fracture toughness for fine grained alumina, Scripta Materialia, Vol. 65, pp. 143–146.

[152]   Yu, M.F.; Lourie, O.; Dyer, M.J.; Moloni, K.; Kelly, T.F. & Ruoff, R.S. (2000). Strength and breaking mechanism of multiwalled carbon nanotubes under tensile load. Science, Vol. 287, pp. 637-640.

[153]   R.M. Jones, Mechanics of Composite Materials, 2nd Ed., Taylor & Francis, Philadelphia, PA (1999).

[154]   C.T. Herakovich, Mechanics of Fibrous Composites, John Wiley & Sons, New York, NY (1998).

[155]   K.K. Chawla, Composite Materials, 2nd Ed., Springer-Verlag, New York, NY (1998).

[156]   B.D. Agarwal, L.J. Broutman, and K. Chandrasekharan, Analysis and Performance of Fiber Composites, 3rd Ed., John Wiley & Sons, New York, NY (2006).

[157]   I.M. Daniel and O. Ishai, Engineering Mechanics of Composite Materials, 2nd Ed., Oxford University Press, London (2005).

[158]   A.K. Kaw, Mechanics of Composite Materials, 2nd Ed., CRC Press, Boca Raton, FL (2006).

[159]   S.W. Tsai and H.T. Hahn, Introduction to Composite Materials, Technomic Publishing Co., Lancaster, PA (1980).

[160]   D. Hull and T.W. Clyne, An Introduction to Composite Materials, 2nd Ed., Cambridge University Press, London (1997).

[161]   M.R. Piggott, Load Bearing Fibre Composites, 2nd Ed., Kluwer Academic Press, Dordrecht (2002).

[162]   K.H.G. Ashbee, Fundamental Principles of Fiber Reinforced Composites, Technomic Publishing Co., Lancaster, PA (1989).

[163]   D.F. Adams, L.A. Carlsson, and R.B. Pipes, Experimental Characterization of Advanced Composite Materials, 3rd Ed., CRC Press, Boca Raton, FL (2003).

[164] J.R. Vinson and R.L. Sierakowski, The Behavior of Structures Composed of Composite Materials, 2nd Ed., Kluwer Academic Publishers, Dordrecht (2002).

[165] P.K. Mallick and S. Newman (eds.), Composite Materials Technology: Processes and Properties, Hanser Publishers, Munich (1990).

[166] T.G. Gutowski (ed.), Advanced Composites Manufacturing, John Wiley & Sons, New York (1997).

[167] G. Lubin (ed.), Handbook of Composites, Van Nostrand-Reinhold Co., New York (1982).

[168] P.K. Mallick (ed.), Composites Engineering Handbook, Marcel Dekker, New York (1997).

CHAPTER 2

INTRODUCTION TO POLYMER MATERIAL

2.1. Introduction

Polymer exists in human life from the birth of human being which can be observed in modern age. At very initial stage by the medical science researchers invented that it comes on the front line in plants & animals, in the form of DNA, RNA, proteins and polysaccharides aswellas it also plays a prominent role in human body because our body is a constituent of large number of polymers such as proteins & enzymes etc. Further in broad sense we can say that polymer having two types:

1. Natural polymers
2. Synthetic polymers

2.1.1. *Natural Polymers*

It has been observed that many materials found in nature are polymers, say plant & animals, these polymers are known as natural polymers. In this class following are considered such as silk, shellac, rubber, bitumen, cellulose, wood, leathers, DNA, RNA, proteins, polysaccharides, horns of animals, tortoise shells, rosin (Pine tree), asphalts, tar (from organic materials) and starch. Cellulose are generally made from glucose, rubber from isoprene and protein from amino-acids.

2.1.2. *Synthetic Polymers*

Synthetic polymer is nothing but it is man-made polymer that is being used to fulfil the pre-decided goal or objectives, backlite is the first synthetic polymer that had been produced in year around 1910. In this class following are important say – plastic, rubber, fiber material, rubber resin, adhesives, surface coating material, polyethylene (PE), PP, polyvinyl chloride (PVC), nylons, SBR, BR, Bakelite, Teflon, PMMA, PAN, adhesives tapes, rayon, thermo-plastic elastomers foams, paint & sealants, poly-carbonates (PC), polystyrene (PS), silicon rubbers, Dacron, polyesters, epoxy, vinyl polyurethane, silicone, Lucite, boat resin and so forth.

The basic purpose of utilization of both natural polymers & synthetic polymers are to acquire, comfort & facilitation in human life for instance, medications, nutrition's, communication, transportation, container manufacturing, clothing, and construction work (say building & highways). Furthermore, it is not a matter of surprise that we live in an age of polymer because in fact we are all surrounded by the polymer articles or product without which human life could not be run smoothly or without which we are unable to imagine our life. The reason is that polymer product application, starts from housed-hold product to engineering field or we can say that it has wide span of application for example – house hold utensils, cloths, furniture, automobiles, space air craft, thermal & electrical applications, decorations, shelters, tools, weapons, writing materials, paints, construction work that includes flooring, windows, claddings, rain water, pipes, membranes, seats, glazing, insulation & sign gauge, clothing made from synthetic fibers, polyethylene cups, fiber glass, nylon bearings, plastic bags, polymer based paints, epoxy glue, polyethylene foam cushion, silicon heart valve, Teflon coated cookware, plastic table ware (say spoon, fork & knife), billiards balls, bi-cycle helmets, dices, wheels (plastic hubs & rubber tire), plastic milk cartoons and so many others. Polymer has also been used in industrial application since long period of time due to cost effectiveness, easy accessibility, malleability & non-toxic nature. Consider the following table which shows some common & very famous polymers which is widely used to manufacturing polymer articles.

Table 2.1: Some common type of polymer

Polymer name	Structure
Commodity thermoplastics	
Polyethylene	$+CH_2-CH_2+$
Polystyrene	$+CH_2-CH+$ (pendant phenyl ring)
Polypropylene	$+CH_2-CH+$, CH_3
Polyvinyl chloride	$+CH_2-CH+$, Cl
Polymers in electronic applications	
Polyacetylene	$+CH=CH+_n$
Poly(p-phenylene vinylene)	$+\langle\bigcirc\rangle-CH=CH+_n$
Polythiophene	thiophene ring repeat unit
Polyphenylene sulfide	$+\langle\bigcirc\rangle-S-\langle\bigcirc\rangle+_n$
Polyanilines	$+CH_2-CH+_n$, NH_2
Biomedical applications	
Polycarbonate (diphenyl carbonate)	$+\langle\bigcirc\rangle-C(CH_3)_2-\langle\bigcirc\rangle-O-\overset{O}{C}-O+_n$
Polymethyl methacrylate	$+CH_2-C(CH_3)+_n$, $COOCH_3$
Silicone polymers	$+Si(CH_3)_2-O+_n$
Specialty polymers	
Polyvinylidene chloride	$+CH_2-CCl_2+$
Polyindene	$[-CH-CH-]_n$ (fused phenyl ring with CH_2)
Polyvinyl pyrrolidone	$+CH_2-CH_2+$ (pyrrolidone ring: N, CH_2, $C=O$, CH_2-CH_2)
Coumarone polymer	$[-CH-CH-]$ (benzofuran ring, O)

The importance of polymers can be easily understood by the following table which shows all the possible types of polymers with their respective applications.

Table 2.2: Some Important Polymers & their Respective Application

Polymer	Applications
	Thermoplastics
Amorphous	
Polystyrene	Mass-produced transparent articles, thermoformed packaging, foamed packaging, and thermal insulation products, etc.
Polymethyl methacrylate	Skylights, airplane windows, lenses, bulletproof windows, automotive stop lights, etc.
Polycarbonate	Helmets, eyeglass lenses, CD's, hockey masks, bulletproof windows, blinker lights, head lights, etc.
Unplasticized polyvinyl chloride	Tubes, window frames, siding, rain gutters, bottles, thermoformed packaging, etc.
Plasticized polyvinyl chloride	Shoes, hoses, roto-molded hollow articles such as balls and other toys, calendered films for raincoats and tablecloths, etc
Semi-crystalline	
High density polyethylene	Milk and soap bottles, mass production of household goods of higher quality, tubes, paper coating, etc.
Low density polyethylene	Mass production of household goods, squeeze bottles, grocery bags, etc.
Polypropylene	Goods such as suitcases, tubes, engineering application (fiberglass-reinforced), housings for electric appliances, etc.
Polytetrafluoroethylene	Coating of cooking pans, lubricant-free bearings, etc.
Polyamide	Bearings, gears, bolts, skate wheels, pipes, fishing line, textiles, ropes, etc.
	Thermosets
Epoxy	Adhesive, glass fiber reinforced automotive leaf springs, carbon fiber reinforced bicycle frames, aircraft wings and fuselage etc.
Melamine	Decorative heat-resistant surfaces for kitchens and furniture, dishes, etc.
Phenolics	Heat-resistant handles for pans, irons and toasters, electric outlets, etc.
Unsaturated polyester	Toaster sides, iron handles, satellite dishes, glass fiber reinforced breaker switch housings and automotive body panels, etc.
	Elastomers
Polybutadiene	Automotive tires (blended with natural rubber and styrene butadiene rubber), golf ball skin, etc.
Ethylene propylene rubber	Automotive radiator hoses and window seals, roof covering, etc.
Natural rubber (polyisoprene)	Automotive tires, engine mounts, etc.
Polyurethane elastomer	Roller skate wheels, sport arena floors, ski boots, automotive seats (foamed), shoe soles (foamed), etc.
Silicone rubber	Seals, parts for medical applications, membranes, heat resistant kitchen containers, etc.
Styrene butadiene rubber	Automotive tire treads, etc.

Consider another table which shows name of some polymers with their commercial acronyms.

Table 2.3: Some Important Polymer Representing by their Acronyms

Polymer	Abbrev. (Type)	Polymer	Abbrev. (Type)
Acrylonitrile-butadiene-styrene	ABS (TP)	Polyether ether ketone	PEEK (TP)
Butadiene rubber	BR (E)	Polyether sulfone	PES (TP)
Cellulose acetate	CA (TP)	Polyethylene terephthalate	PET (TP)
Cellulose acetate butyrate	CAB (TP)	Polyimide	PI (TP)
Cellulose nitrate	CN (TP)	Polyisobutylene	PIB (TP)
Epoxy	EP (TS)	Polymethyl methacrylate	PMMA (TP)
Ethylene-propylene-diene rubber	EPDM (E)	Polyphenylene sulfide	PPS (TP)
Expanded polystyrene	EPS (TP)	Polyphenylene sulfone	PPSU (TP)
High density polyethylene	PE-HD (TP)	Polypropylene	PP (TP)
Impact resistant polystyrene	PS-HI (TP)	Polypropylene copolymer	PP-CO (TP)
Linear low density polyethylene	PE-LLD (TP)	Polystyrene	PS (TP)
Linear medium density polyethylene	PE-LMD (TP)		
Liquid crystalline polymer	LCP (TP)	Polysulfone	PSU (TP)
Low density polyethylene	PE-LD (TP)	Polytetrafluoroethylene	PTFE (TP)
Melamine-formaldehyde	MF (TS)	Polyurethane	PUR (TS)
Metallocene catalyzed polyethylene	mPE (TP)	Polyvinylidene acetate	PVAC (TP)
Natural rubber	NR (E)	Polyvinyl alcohol	PVAL (TP)
Olefinic thermoplastic elastomer	TPO (TP)	Polyvinyl carbazole	PVK (TP)
Phenol-formaldehyde (phenolic)	PF (TS)	Polyvinyl chloride	PVC (TP)
PF mineral filled moldings	PF-mf (TS)	Polyvinylidene fluoride	PVDF (TP)
PF organic filled moldings	PF-of (TS)	Rigid PVC	PVC-U (TP)
Plasticized PVC	PVC-P (TP)	Silicone	SI (E)
Polyacetal (polyoxymethylene)	POM (TP)	Silicone rubber	SI (E)
Polyacrylate	PAR (TP)	Styrene-acrylonitrile copolymer	SAN (TP)
Polyacrylonitrile	PAN (TP)	Thermoplastic elastomer	TPE (TP)
Polyamide 6	PA6 (TP)	Thermoplastic poly-urethane elastomer	TPU (TP)
Polyamide 66	PA66 (TP)	Unsaturated polyester	UP (TS)
Polybutylene terephthalate	PBT (TP)	Urea-formaldehyde	UF (TS)
Polycarbonate	PC (TP)	Vinyl ester resin	VE (TP)

Where,

TP – Stand for Thermo-Plastics

E - Stand for Elastomers

TS – Stand for Thermo-Sets

Moreover, polymer material also plays a dominants role in the civil engineering construction field. They are generally applied in flooring, windows, claddings, rain water, pipes, membranes, seals, glazing and insulation etc. Consider the following tables which shows application of polymers material in civil engineering construction works.

Table 2.4: Application of Various Polymers in Civil Engineering Construction Works

Polymer Type	Applications
Epoxy resins	Solid resin and Terrazzo flooring, Anchor fixings, Adhesives
Ethyl vinyl acetate (EVA)	Solar panel encapsulants
Expanded polystyrene (EPS)	Concrete moulds, Insulation, Packaging
Polycarbonate	Lighting housings, Fittings in hot water systems, Glazing
Polyester (thermosetting)	FRP Bridge sections, Cladding Panels, Sinks, Surfaces, Coatings
Polyethylene	Foam underlay, Damp-proof membranes, Coatings
Polyisobutylene (PIB)	Glazing sealants, Waterproof membranes
Polymethylmethacrylate / Acyrlic (PMMA)	Surfaces, Sinks
Polypropylene (PP)	Sound insulation, Water pipes, Waste pipes
Polyurethane (PU)	Sealants, Concrete jointing
Polyvinylchloride (PVC)	Sealants, Concrete jointing
Rubber	Bridge bearings, Flooring

Furtherore, application & use of polymer is not limited like a sky for example – Polyprpene used in industires (say textiles, pacakagings, stationary, plastics, air-craft, constriction, rope & toys etc.), polysterene (used in pacakagings industries & also use to manufacturing bottles, toys, containers, trays, disposable glass, pltes, TV cabinates and insulators), polyvinyl chloridres (PVC) [Used to manufacturing sewage pipes, applied as insulators in electrc cables, but now a days it is used to manufacturing cloting, furniture, doors, vinyl flooring and window], Urea-formaldehyde resins (Used to manufacturing adgesives, moulds, laminated sheets and unbreakable container), glyptol (used for making paints, coatings, and lacquers). Bakelite is a first polymer which is invented in year 1910, is being used to manufacturing swithches, kitchen products, toys, jewellery, fire arms, insulators and computer disces etc.

Consider following tables which shows a brief description of various polymers, monomers and their used to manufacturing a polymer products.

Table 2.5: Showing Various Polymers their Monmers & Resprctive Uses

Polymers	Monomers	Use of polymers
Rubber	Isoprene (1, 2-methyl 1 – 1, 3-butadiene)	Making tyres, elastic materials
BUNA – S	(a) 1, 3-butadiene (b) Styrene	Synthetic rubber
BUNA – N	(a) 1, 3-butadiene (b) Vinyl Cyanide	Synthetic rubber
Teflon	Tetra Flouro Ethane	Non-stick cookware – plastics
Terylene	(a) Ethylene glycol (b) Terephthalic acid	Fabric
Glyptal	(a) Ethylene glycol (b) Phthalic acid	Fabric
Bakelite	(a) Phenol (b) Formaldehyde	Plastic switches. Mugs. bucket
PVC	Vinyl Cyanide	Tubes, Pipes
Melamine Formaldehyde Resin	(a) Melamine (b) Formaldehyde	Ceramic plastic material
Nylon-6	Caprolactum	Fabric

2.2. Definition of Polymer

Basically **"polymer"** is found in greek dicitionary. If we splits up the whole word then we have got a complete meaning, say,

$$\boxed{\begin{array}{ccc} \text{Polymer} = & \text{Poly} & + & \text{Mer} \\ & \text{(Many)} & & \text{(Part)} \end{array}}$$

Here,

Poly stand for **"Many"**

Mer stand for **"Part"**

From the above formula polymer may be defined as,

"A matter consisting long macro-mlecules or large molecules having a high molecular weight in which a lot of small repeating molecules are linked together so as to form a long chain".

Small molecules or sub units are known as monomers. These sub units / small molecules are linked together by means of covalent bonds on the contrary long molecules joint together by means of either vanderwalls bonds, hydrogen bonds or covalent cross – links. Furthermore, polymer are accumulated units of number of monmers whose quantity may be either hundreads, thousands or in millions howeever property of polymer always different from monmers. Starch, polyvinyl chloride, polyethylene, nylon - 66 are monmers while plastics, rubbers, fibers, thermo-plastic elastomers, adhesives foam, paints, sealants, surface coating materilas, PE, PVC, Nylons and bakelites etc. are all polymers. Further the process by which monomers are transformed into polymers is known as polymerization process. The first synthetic polymer was invented in year 1910 that is known as Bakelites. Bakelite is used to manufacturing billiards balls. First synthetic rubber is methyl rubber & first synthetic fiber was rayon which was being replaced of silk. Furthermore, the definition of polymer may be easily understood by the following example – polythene is a polymer that formed by plenty of molecules of ethene (C_2H_4). Here polythene is a polymer & ethene (C_2H_4) is a monomers. Further a molecular weight of polymer generally falls between 5000 to 20,0000 dalton. The number of repeating units of polymer is known as degree of polymerization. Consider the following diagram that is dedicated to polyethylene (PE), poly-vinyl chloride (PVC) and polypropylene (PP) which shows complete over view of structure.

Table 2.6: Showing Polymerization of Polymers, Polymer & Monomers Structure

Polymerization process
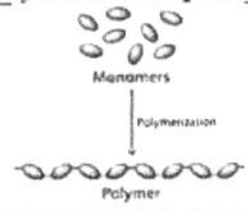 Monomers ↓ Polymerization Polymer

	Monomers
Polyethylene (PE)	repeat unit H H H H H H ·C—C—C—C—C—C· H H H H H H Polyethylene (PE)
Poly-vinyl chloride (PVC)	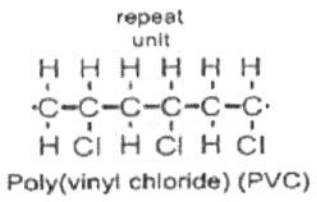 repeat unit H H H H H H ·C—C—C—C—C—C· H Cl H Cl H Cl Poly(vinyl chloride) (PVC)
Polypropylene (PP)	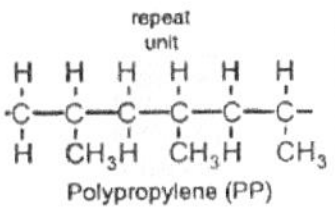 repeat unit H H H H H H ·C—C—C—C—C—C— H CH₃H CH₃H CH₃ Polypropylene (PP)

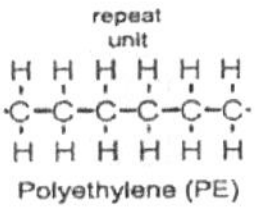

Note

- Polymers can either be organic compounds or inorganic compounds.

- Monomers are joint together or linked together by means of covalent bonds.

- Mnomers can have a quantity of hundreads, thousands or in millions population.

- Molecular weight are measured in dalton.

- A long units of small molecules which bonded together by means of covalent bond is known as monomers.

2.3. Polymer a Class of Engineering Materials

We have lot of material that uses in diverse field of engineering application such as copper, steel, cast iron, mild steel, aluminum, silver, titanium, gold etc. in conjunction with other alloys. In addition to it, we also have other metals & non-metals like ceramic, composites which is being widely used in engineering as well as other applications. But evolvements of human wants & their requirements/necessity, engineering materials also expanded their span, among those engineering materials, polymer material play a dominant role in various field of application because polymers are promising material for some specific engineering application due to their fascinating characteristics like - high strength, modulus to weight ratios, toughness, mechanical strength, resilience, resistance to corrosion, lack of conductivity (heat as well as electrical), colour, transparency. In addition to it easy processing and cost effectiveness also favoured. For example, in case of plastic polymer material, low density and high weight savings, enhanced their utility for automotive application, marine and aerospace applications. Further polymer may be used in structural or non-structural application. In case of structural application mechanical properties are always imperative like tensile strength, stiffness, impact strength, and chemical resistance whereas nonstructural application required surface finishing, ease of painting, effect of humidity & ultraviolet radiation and cracks creation on polymer material. A brief summary of polymer application such as, sky lights, eye glass lenses, air plane window, bullet proof window, helmets, head lights, tubes, engineering applications (fiber-glass reinforced), electric applications, lubricant free bearings, bearings, gears, bolts, skate wheels, pipes, textile ropes, adhesives, glass fiber reinforced materials, automotive leaf springs, carbon fiber reinforced materials, bi-cycle frames, air craft wings & fuselages, heat resistance surfaces, electric outlets, glass fibers reinforced breaker switch, tires, engine mounts, seals, part of medical appliances, membranes, automotive tire trades, tools, weapons, flooring, window, claddings, rain waters, pipes, seals glazing, insulation, signage, solid resin, terrazzo flooring, anchor fixing, adhesives, solar panels encapsulants, concreate molds, insulation, packaging, lighting house, fitting of hot water system, FRP bridge section, cladding panels, form underlays, damp proof membranes, coating, water proof membranes, glazing sealants, water & waste pipes, sound insulation, concreate joining, bridge bearings, nylon bearings, silicone heat valve, communication, transportation, buildings, high ways, air craft applications, construction application, insulators, vinyl flooring, electrical switches, computer discs, tubes, pipes, plastic hubs, automotive components, electronic industries, aviation industries, bio-medical application, automotive application, marine application & so forth. Here I am going to describe some possible field of polymers material in engineering applications, these are:

1. Polymer material used in mechanical engineering field
2. Polymer material used in electrical & electronic engineering field
3. Polymer material used in civil engineering field
4. Other field of application such as medical field

2.3.1. Polymer Material Used in Mechanical Engineering Field

It has been observed that polymer material showing wide range of application in the field of mechanical engineering due to their, high strength, light weight, low modulus to weight ratio, stiffness, toughness, resilience, low cost, tensile strength, compressive strength, flexural strength, hardness, creep resistance, fatigue resistance and impact resistance. Consider following points regarding to mechanical engineering applications.

- Like another engineering material mechanical properties depending upon the structure of constituent's elements & their arrangements.

- Property of plastic polymers like – low density, weight saving and high strength always desirable for automotive application, marine engineering and aero-space engineering applications.

- In case of structural-mechanical engineering application, some properties are always imperative like tensile strength, stiffness, impact resistance and strength.

- Thermoplastic polymers are generally used to applied engineering design applications due to their high thermal stability, high dimensional stability, high rigidity, light weight, resistance to creep and deformation under load.

- Mechanical properties of thermoplastics say low stiffness, low tensile strength and low hardness than that of metals & ceramics, and property of greater ductility can be used in different field of mechanical engineering applications.

- Extreme ductility or pliability characteristics, very large molecular structure, low density, moderate temperature stability, low strength, low melting temperature of plastic polymer facilitates easy transformation of complicated or intricate shaped mechanical components.

- Polymer application not only limited up to mechanical engineering application but also cultivation of polymer properties moves our journey towards air-craft engineering & aero-space engineering applications.

- In case of amorphous polymers, if alignment of all the polymer chain is performs in one direction then we have got enhancement in glass transition temperature as a result of which both stiffness & strength can be increased in the direction of molecular chain and hence we can use this stiffness & strength in various field of mechanical engineering applications.

- Property of polymer can be improved by increasing crystallinity, as crystallinity increase then we have got improvements in some specific characteristics like – density, stiffness, strength, toughness, heat resistance capacity. These properties play a vital role in different mechanical application.

- High crystalline polymer showing brittleness on the other hand amorphous polymer indicate plastic deformation hence we can say that high crystallinity polymer can be used in engineering application where hardness is desirable.

- Strength & stiffness of polymer can be augmented by transforming it into reinforced composites by adding filler particles, whiskers, short fibers (Discontinuous fiber) and long fiber (Continuous fiber) to polymer matrix say epoxies, unsaturated polyester and vinyl chlorides. It has been observed that polymer composites consisting 50% to 70% of glass, carbon or polyaramid in thermo-plastic or thermo-setting then new polymer material indicate light weight than that of steels. If impact strength & toughness is to be improved by polymer blending then resulting polymer can be used in automotive applications.

- It has been observed that mechanical properties of polymer depending upon the temperature, for example - At low temperature polymer of glass represent brittle behavior & showing rubber like characteristics at high temperature.

Consider the following table which shows complete over view of thermo-plastic polymer, thermo-setting polymer and elastomers which is being widely used in mechanical engineering application.

Table 2.7: Contribution of Thermo-plastic Polymer Material in Mechanical Engineering Application

Thermo-plastic (Polymer) mechanical engineering applications

Name of Thermo-plastic (Polymer)	Properties	Applications
Acrylonitrile-butadiene-styrene (ABS)	Superior strength and toughness, dimensional stability, rigid, impact resistance, abrasion resistance, low temperature resistance	Refrigerator lining, automotive components (wheel covers, head light bezels), tool holders,
Acrylics (poly-methyl-methacrylate) [PMMA]	Moderate strength & wear resistance	Drafting equipment's
Fluorocarbons	Low coefficient of friction, resistance to high temperature	Low friction surface, Maximum operating temperature 260°C, bearings, seals, valves, anti-adhesive coating, gaskets.
Polyamides (Nylons)	Good mechanical strength, abrasion resistance, toughness, low coefficient of friction, abrasion resistance, self – lubricating characteristics	Bearings, gears, cams, bushes.
Aramids (aromatic polyamides)	High tensile strength , Good stiffness	Used as fibers for reinforced plastic polymer
Polycarbonates	Superior impact resistance & ductility	Load bearing electrical components, business machine components etc.
Nylon-6,6	Tough, good abrasion resistance and self-lubricating property.	Ball bearing cages, for manufacturing various machine components, conveyor belts, gears, bearings, bushes, Nylon 6.10 (Used as bristles, brushes)
Glass reinforced Nylon plastics		Radiator parts of automobile and for relay coil formers
Polyethylene available in three forms - LDPE, HDPE, UHMWPE	Tough, low strength, low coefficient of friction, UHMWPE – Indicate high impact resistance, toughness resistance, abrasive wear resistance	LDPE & HDPE (Used for manufacturing bumpers), UHMWPE (Used in artificial knee and hip joints)
Polyimides	Good mechanical properties at elevated temperature, Good creep resistance, low friction and wear characteristics.	Pump components (bearings, seals, valve seats, piston rings), used manufacturing Aerospace and automobile engineering parts
Polystyrene	Excellent moisture resistant polymer, brittle in nature, resistance to tearing	For manufacturing refrigerator parts, Automotive components,
Acetals	Good strength, stiffness, resistance to creep, heat and abrasion	Used for manufacturing high performance parts (long period service period) , for manufacturing bearings, cams, gears, bushings, rollers, wear surfaces, valves etc.
Polyvinyl Chloride (PVC)	Water resistant, tough and hard	Used to manufacturing turbine pumps and seals, Flexible PVC (Gaskets and Seals)
Unsaturated Polyesters reinforced with either glass or other fibers	Good mechanical strength	Automotive body
Polyesters (acid and alcohol)	Showing good mechanical, chemical Resistance, Abrasion resistance and low friction	Used for manufacturing gears, cams, rollers, load bearing members, pumps, electro-mechanical components.

Table 2.8: Contribution of Thermo-setting Polymer Material in Mechanical Engineering Application

Consider another table which shows property and application of thermo-setting polymer materials,

Name of thermosetting polymer	Property	Application
Epoxies	Excellent mechanical properties, corrosion resistance, resistance to high heat chemicals, water, various solvents, acids, alkalis, good strength and dimensionally stable	Used to manufacturing components of aircrafts and automobiles, tools, dies and adhesives.
Fiber Reinforced Epoxies	Excellent mechanical properties	Pressure vessels, rocket motor casings, tanks and other structural components, used in aero-nautical engineering applications
Phenolics	Excellent thermal stability, brittle & rigid in nature and dimensionally stable, Resistant to heat, water and chemicals	Used to prepare motor housings, Auto-distributors, used as bonding materials to hold abrasive grains together in grinding wheels
Polyimides	Good mechanical at elevated temperature, Good creep resistance, low friction and wear characteristics.	Pump components (bearings, seals, valve seats, piston rings), used manufacturing Aerospace and automobile engineering parts
Phenol formaldehyde resins	Hard, rigid, strong materials, excellent heat & moisture resistance, good chemical resistance & abrasion resistance, excellent bonding strength and adhesive properties.	Used as adhesives for grinding wheels and brake linings of automobiles

Table 2.9: Contribution of Elastomer (Polymer) Material in Mechanical Engineering Application

Applications & property of elastomers,

Name of elastomers	Property	Applications
Synthetic natural rubber, butyl, styrene-butadiene, and ethylene propylene	Superior resistance to heat and chemical resistance	Shock absorbers and seals, Butyl rubber (Conveyor belts), styrene-butadiene rubber (Gaskets, moulded mechanical goods)
Polyurethane	High strength, stiffness, hardness, Resistance to abrasion, cutting and tearing	Used for manufacturing Seals, gaskets and auto body parts
Chloroprene	Weather resistant	Shock mounts, seals
Nitrile rubber	High resistance to heat, good abrasion resistance	Automobile engineering parts and components of high-altitude air-craft hose, Conveyor belts, Gaskets.
Rubber	High friction properties, Good resistance to abrasion and fatigue	Used as a vibration and noise absorber, Engine Mount
Silicone resins	High temperatures stability, water resistance, good oxidation resistance, non-toxic	Used as high temperature lubricants, heat transfer media, greases (used as lubricants where temperatures fluctuation occurs), In solid form used as seals, Gaskets, heat seals, water proof materials, heat shield for space shuttle
Thiokol	Resistance to chemical	To manufacturing engine gaskets, Containers for transporting solvents and used as solid propellant fuels for rockets
TEFLON OR Poly tetra fluoro-ethylene	Extreme high toughness, high softening point, high chemical-resistance, high density, low coefficient of friction, good mechanical properties (Machined, punched and drilled)	Used for manufacturing gaskets, pump parts, used as non-lubricating bearings & non-sticking stop-cocks, tanks & tank linings.
EPDM- Copolymer	Superior weathering resistance	Used as auto door, decklid seals and foamed seals

2.3.2. *Polymer Material Used in Electrical Engineering Field*

Polymer materials are also applied in the field of electrical engineering application due to light weight & high electrical insulating property. For example – plastic polymer has density and flexibility. Now a days it is being widely used in electrical engineering application as an insulator due to deficiency of electrical conductivity. In present time polymer are used in electrical appliances, electric out-lets, lighting house, electrical switches, computer discs, insulating the wire & electrical equipment's etc. Furthermore, in the category of non-conducting materials polymer play a dominating role which has been applied for storage of electrical charge, when it is used for storing charge then polymer material is known as die-electric materials. It has been observed that if any material showing good die-electric property then it would be superior insulators. Polymer is a high die-electric strength material because it has capacity to store electrical charge when experiencing an electric field. A property of collecting & storing highest value of current is known as die-electric strength. Die-electric strength of polymer may be high as high 1000 Mv/m and upper die-electric strength limit always depending upon the ionization energy of electrons in covalent bonds throughout the polymer structure. We know that polymer is insulator but it may become an electrical conductor which is completely depending upon the crowd of free electrons as well as mobility of free electrons when it is subjected to electrical field. For example – Polymer in glassy state showing electrical conductivity that ranges between 10^{-13} to 10^{-19} Ohm^{-1}. Electrical conductivity of polymers depending upon the temperature, if temperature increases then conductivity of polymer would be raised. Electrical conductivity of any polymers can be expressed by the following relationship.

$$E_C = A \cdot e^{(-\Delta U/RT)}$$

Where,

E_C = Electrical conductivity

A = Coefficient depending upon the temperature

ΔU = Activation energy

T = Temperature

R = Resistance for current flow

Note

- Basically, polymer material showing low electrical conductivity.
- Polymer material showing low thermal conductivity.
- Polymer (Thermo-plastic) material has high coefficient of thermal expansions, it is about 5 times from metals and 10 times from ceramics.

- Polymer not only indicates low electrical conductivity but also represent non-magnetic behavior.

- Polymer unique die-electric strength property which always imperative to electric & electronic engineering application.

- Now a days PTFF, PE, PVC, EP and MF polymers are most widely used in electrical application due to its superior electrical performance.

Modification in polymer may develop electrical conduction characteristics in polymer material. For example - Polyaniline (PANI) is high electrically conductive polymer which is being used in rechargeable batteries, sensors, electronic devices, light emitting diodes and so forth. Another example of polymer under this category is Teflon which has been widely used as insulating material for cables, wires and transformers. Furthermore, Bakelite which is a first synthetic rubber has excellent electrical insulating property used for making electrical insulating parts such as switches, plugs, switch boards, heater handles and telephones. Now let be consider the following table which shows complete over view of thermo - plastic polymer, thermo-setting polymer and elastomers which is being widely used in electrical engineering application.

Table 2.10: Contribution of Thermo-plastic Polymer Material in Electrical Engineering Application

Thermo-plastic (Polymer) used in Electrical Engineering Applications

Name of Thermo-plastic (Polymer)	Properties	Applications
Acrylonitrile-butadiene-styrene (ABS)	Good electrical property i.e. high electrical resistance, dimensionally stable and rigid,	Electronic goods, head light bezels, telephones application, decorative panels.
Acrylics (poly-methyl-methacrylate) [PMMA]	Good electrical resistance	lighted signs, displays, lighting fixtures
Fluorocarbons	Good electrical property i.e. high electrical resistance	High temperature electronic parts, electrical insulation for high temperature wire and cables, gaskets.
Polyamides	High die-electric strength, Good electrical properties at elevated temperature	Jacketing for wires and cables, electrical connectors for high temperature use
Polycarbonates	Exhibits Electrical properties	Load bearing electrical components, electrical insulators,
Nylon-6,6	High temperature stability	Electro-insulating elements
Polyethylene	Good electrical property i.e. electrically insulating property	Battery parts, preparation of insulator parts ,
Polystyrene	Excellent electrical properties, highly electric insulating,	Battery cases, indoor lighting panels, used as insulators, making radio components, TV and refrigerator components

Table 2.11: Contribution of Thermo-setting Polymer Material in Electrical Engineering Application

Consider another table which shows property and application of thermo-setting polymer materials for electrical engineering applications.

Name of thermosetting polymer	Property	Application
Epoxies	Good electrical properties.	Electrical mouldings, protective coatings, epoxy resins mould used for the production of aircrafts and automobiles components, used to manufacturing electrical components of good strength and high insulation
Phenolics	Resistant to electricity, electrical insulation characteristics	Used to prepare motor housings, electrical fixtures, Auto-distributors, telephones, electronic components (insulators, wiring devices),
Bakelite (Phenol formaldehyde resins)	Electrical insulation characteristics	Phenol formaldehyde resins used for making Domestic plugs, switches and electrical insulation and protective coatings.

Property & Application of other polymers material that is used in Electrical Engineering Applications.

Table 2.12: Contribution of other Polymer Material in Electrical Engineering Application

Name of elastomers	Property	Applications
Silicone resins	Indicate excellent electrical property over wide range of temperature and humidity	Electrical components requiring strength at high elevated temperatures, oven gaskets, heat seals and electric insulations.
TEFLON OR Poly tetra fluoro-ethylene	Extremely good electrical	Used as insulating material for motors, transformers and cables, wires
EPDM- Copolymer	Weathering resistance & electrical insulator	Used in wire insulation
Polyester	Good mechanical electrical and chemical Resistance	Electro-mechanical components
Polyvinyl Chloride (PVC) - Flexible PVC	Electrical insulator	Used for wire and cable coatings for insulation
Alkyds	Possess good electrical insulating properties	Used to manufacturing electrical and electronic components
Aminos resin (Urea)	Hard, rigid, resistant to abrasion, creep and electrical arcing	Used in electrical and electronic components,
Butyl rubber	Good electrical insulation properties	Used for Insulating high voltage wires and cables
Conducting polymers	Flexible & good electrical insulation properties	Rechargeable Light weight batteries, wiring in aircrafts and aerospace components, telecommunication systems, electronic devices such as transistors and diodes, solar cells. In photovoltaic devices, In molecular wires and molecular switches.
Polymer coating	Adhesion, barrier properties, scratch and abrasion resistance, chemical resistance, wettability.	Solar cells, lithium-sulphur batteries, membrane, Light Emitting Diodes,
Bakelite	Excellent electrical insulating character, rigid, hard, scratch-resistant, infusible, water-resistant, insoluble solids,	Used for manufacturing electric insulator parts like switches, plugs, switch-boards, heater-handles, telephone parts, cabinets for radio and television.
Polyaniline (PANI)	Highly conductive polymer, low density, mechanical flexibility, easy processability and high environmental stability.	Rechargeable batteries, sensors, electronic devices, light emitting diodes.

Conducting polymers (CPs)	Electroluminescent and semiconducting properties	Organic light-emitting diodes, organic field-effect transistors and organic solar cells.
Polymers (PTFE, PE, PVC, EP and MF)	Electrical insulator	Use as insulants in electrical and electronic engineering, ease of processing and high electrical performance.

2.3.3. *Polymer Material Used in Civil Engineering Field*

According to previous history, we know that at very initial stage timber & masonry are chiefly used for construction work by human being and these material are known as traditional or conventional building material. But about two hundred year ago, introduction of steel, concrete & cement has been invented which evolved the span of construction field. An accumulated form of steel work & RCC (Reinforced cement concrete) can be considered as new material or advanced material for construction industries. However, it has been observed that after service of certain period / few year such a newer material subjected to degradation. But a question arises here how such a degradation can inhibit? Answer of above question definitely be "Application of polymer material" in civil construction work. The utilization of polymer material not only decline degradation but also enhance strength, flexibility, durability, cost effectiveness and reducing corrosion resistance. Polymer material can be used for structural and non-structural application both. Ropes, girders, rebars-fiber reinforced plastic (FRP), reinforced bar for concrete, pre-stressing tendon for concrete members and FRP sheets falls under the category of structural application. Basically, application of polymer material in construction work including flooring, windows, cladding, rainwater, pipes, membranes, seals, glazing, insulation etc. Further inclusion of polymer improves material property, structure strength, tensile strength and resilience. Furthermore, a very interesting aspect is that polymer composition with concrete are available in three forms which has been widely used in civil engineering construction work, these are:

1.   Polymer impregnated concrete (PIC)
2.   Polymer concrete (PC)
3.   Polymer modified concrete (PMC)

All those composition augmenting, compressive strength, durability and reducing corrosion resistance, now we will discuss in brief all the above polymer concrete that is being widely used in construction work.

1.   **Polymer Concrete:-** Polymer concrete is prepared by using polymer instead of binder (lime - cement).

   According to American concrete institution polymer concrete may be defined,

   "Polymer concrete is a composite material in which aggregate is bound together in a matrix with a polymer binder in which Portland cement can be used as an aggregate or filler"

Polymer concrete composition = Polymer + Aggregate (Portland Cement)

2. **Polymer Modified Concrete (PMC):-** Polymer modified concrete (PMC) is nothing but it is a composition of polymer & Portland cement & such a composition is known as Polymer modified concrete (PMC) or Polymer cement concrete (PCC).

> **Polymer modified concrete (PMC) composition = Polymer + Portland cement**

Polymethylmethacrylate (PMMA) has been excellent polymeric material with UV resistance, PMMA is widely used in window and door profiles, canopies, panels, façade design, etc. It also facilitates light transmission and provides good heat insulation, hence a suitable choice of building green houses. PMMA is also used to build aquariums and marine centres. Now a day's basalt fiber-reinforced polymer (BFRP) is most widely used to strengthen structural elements such as reinforced concrete (RC) beams because of extensively low cost & stellar mechanical performance. Moreover, it has been observed that strain hardening cementitious composite (SHCC) plate reinforced which with Carbon Fiber Reinforced Polymer (CFRP) sustain higher durability for structure and higher protection to the FRP component. As a result of which we can get high temperatures stability, vandalism, delays and inhibiting a detachment of concrete substrate. A study reveals that if fiber reinforced polymers, consisting polyester or epoxy resins - and non-metallic reinforcing fibres which embedded or firmly fixed in materials like glass, aramid or carbon fibers then they develop high strength, low self-weight, high corrosion resistance and fatigue resistance.

Note

- Alumino-silicate geo-polymer are used in construction application. In such a case mortar has following composition.

> **Mortar = Silica Sand (40%) + Alumino-Silicate geo-polymer (Cement)**

This composition imparts low grade refractoriness in addition to it, it also developed desired level of compressive strength & thermal stability throughout the construction.

- Polymer composites are generally used to constructing foot-bridge.
- Composition of polymer improves workability due to increase in plasticity with concrete.
- If we use impregnated polymer then we have got reduction in water cement ratio.
- Utilization of polymer in construction protected corrosion due to weathering (rain & UV light), improve strength, durability and enhanced useful life of construction.

- Polystyrene (Thermo-setting plastic polymer) is generally used to manufacturing wall tiles.

- Epoxies are used to manufacturing sinks and for preparing skid resistant surface for highways.

- Polyvinyl chloride (PVC) is used in construction industries (For manufacturing pipes & turbine pumps).

- Amino polymer is used to manufacturing a small housing application & toilets top.

2.3.4. *Limitation of Polymer in Engineering Applications*

We have been studied number of application of polymer in mechanical engineering, electrical engineering applications and civil engineering applications. But few attributes can limit its circumstances of applications, these are low strength, low stiffness, service temperature, visco-elastic property, or degradation of polymer due to the effects of sunlight & other radiations. In addition to, engineering applications, polymer material also has an application in the field of medical field.

2.3.5. *Application of Polymer in Medical Fields*

Remarkable property of polymer material not only utilized in various field of engineering application but also these material are equally participates in medical field or medical devices. whenever any material is applied in medical field or devices then it termed as bio-materials. Bio-material is defined as,

> A material used within human body either as artificial organs/limbs, bone cements, dental cements, ligaments, pacemakers, or contact lenses etc. which showing compatibility with human body elements say bloods, cells, proteins and tissues etc. in such a manner that functioning of human body does not affect or perform work satisfactorily.

We have number of materials which is being widely used in medical field among those polymeric material is one which is extracted from either animals or plants for example - cellulosics, chitosan, dextran, agarose, and collagen etc. For medical application both types of polymer material may be used say Natural polymer and Synthetic polymer. Natural polymer includes - cellulosics, chitosan, dextran, agarose, and collagen on the other hand synthetic polymer or man-made polymer also used in medical field due to their bio-compatibility or inertness with human body. For example - Poly-siloxane, polyurethane, poly-methyl-methacrylate, polyacrylamide, polyester, and polyethylene oxides etc.

Consider following table which shows bio-medical application of some common polymer materials.

Table 2.13: Contribution of Polymer Material in Bio-medical Applications

Polymer material	Bio-medical applications
Polycarbonates	Lenses light globes, optical lenses , medical apparatus
Polystyrene	For lenses manufacturing and medical devices
Polyethylene (UHMWPE)	Parts requiring high impact toughness and abrasive wear resistance used for manufacturing artificial knee and hip joints.
Nylon 6.6	Mouldings of Nylon 6.6 is used in the application of medicine and pharmacy because of sterilis-ability.
Conducting polymers	Used in drug delivery system for human body
Smart polymers	Used for controlling delivery of drugs and genes
Composites with polymers (PVC and PMMA)	Found applications in medicine,
Silicone rubber and PU/PVC copolymer membranes	Used for detecting sodium ions in body fluids.
Polyaniline (PANI)	Used for detecting calcium ions in body fluids by transforming ionic reaction to electrical signals
Poly-vinyl chloride (PVC), Poly(3,4-ethylenedioxythiophene) (PEDOT) and Poly ethylene glycol (PEG)	Used in medical devices due faster rate of electrical signal generation & biocompatibility.
Stress sensors	Used in medical devices in cardiac recordings for measuring blood pressures & heart rate
Biosensors	Used in medical diagnoses and environmental pollution control
Polymer coatings	Used in bio-medical devices
Conducting polymer poly(3,4-ethylenedioxythiophene) (PEDOT).	Orthopaedic implantable devices
Titanium alloys device coated with biodegradable and antibacterial polymers (Chitosan and PLGA)	Orthopaedic implantable devices application

Further, consider another table which is dedicated to application of bio-sensors in medical field.

Table 2.14: Contribution of Polymer Bio-sensor material in bio-medical Applications

Polymer	Application	Properties
cellulose membranes with bacterial origin	glucose sensor	increasing and prolonging stability
poly(vinyl chloride)	analysis of creatine in urine	polymer membrane with natural electrically inert lipids applied as plasticizers

Polymer	Application	Properties
Polyaniline	glucose, urea, triglyceride sensor	polymer deposition and enzymatic immobilization controlled electrochemically
poly(o-aminophenol)	glucose biosensors	amperometric sensor on platinized glassy carbon electrode
Polypyrrole	glucose detection	electrode immobilization of enzymes by electropolymerization of pyrrole
Polyamine	L-amino acid sensor	Immobilization of enzymes by electropolymerization
redox polymers capable of crosslinking	biosensors for enzymes	use of polymers with an ability to cross-link

2.4. Comparative Study Among Metals, Ceramics & Polymers

We have plenty of solid materials that has been used in many engineering as well as non-engineering applications since long period of time due to their distinct features. These solid martials may be either metals, ceramics, composites, polymers, non-metals, alloys of metals and many others. All those can be distinguished not only on the basis of their specific field of application but also their property. Basically, applications of material always govern to property. Property of metals, ceramic and polymer create possible differentiation, whose attribute is based upon the mechanical characteristics, chemical characteristics, electrical characteristics, optical characteristics, physical characteristics, nuclear characteristics and electronic characteristics etc. On the basis of above characteristics, we can briefly describe about metals, ceramics and polymer, which is given below.

2.4.1. Metals

Metallic materials or its alloys are widely used in various engineering or non-engineering applications. These materials known for stiffness, strength, ductility, resistant to fracture, brittleness, electrical conductivity, magnetic behaviour, heat conductivity, transparent to visible light, lustrous appearance after polishing, opacity, malleability, toughness, creep resistance, fatigue resistance, ability to transform in various shapes, weldability etc. In metallic category, iron, aluminium, copper, titanium, gold, nickel, zinc & its alloys, aluminium alloys, stainless steel, mild steel, cast iron, steels alloys and other metals are very famous. Consider the following points regarding to metals which differentiate also ceramics & polymers.

- If we consider atomic arrangement then we found that in metals & its alloys, atoms are arranged in specific order with respect to ceramic & polymer materials.

- Metallic material has crowed of free electrons so that it indicates high electrical conductivity & heat conductivity rather than ceramic materials & polymer materials.

- Metallic material always indicates crystalline structure or nature and showing strength & formability after heat treatment or casting, consequently it has large application in various fields with respect to ceramic materials & polymer materials. However ceramic materials are used in specific applications where metals don't fulfil the desired purpose like metal cutting operation of very hard metals.

- Ceramic & polymer materials offers resistance to corrosion resistance, chemical compounds and other severe environmental conditions on the contrary metal does not inhibit to it.

Consider the following bar charts in which density are given for metals, ceramics polymer & composites.

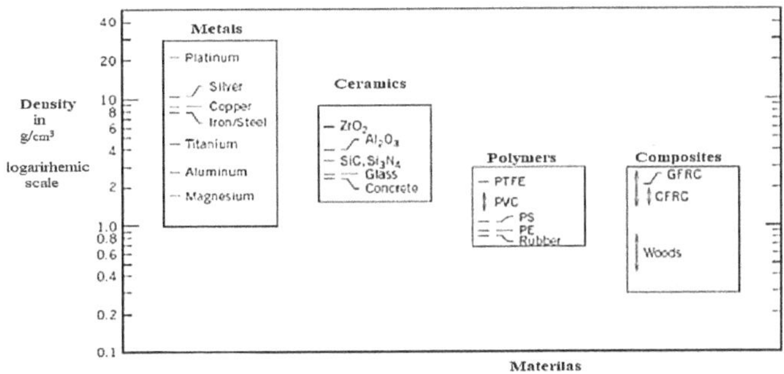

Curve 1: Showing, Density at Room Temperature of Metals, Ceramics, Polymers and Composites Materials

From the above curve it has been observed that value of density at room temperature for metals are always higher, further the decreasing order of density of metals, ceramics, polymers and composites are as given below.

Metals > Ceramics > Polymers > Composites materials

Moreover, consider the following curves which shows stiffness, tensile strength, fracture toughness / fracture resistance and electrical conductivity of metals, ceramics, polymers, and composites material respectively.

(i) Stiffness Curve

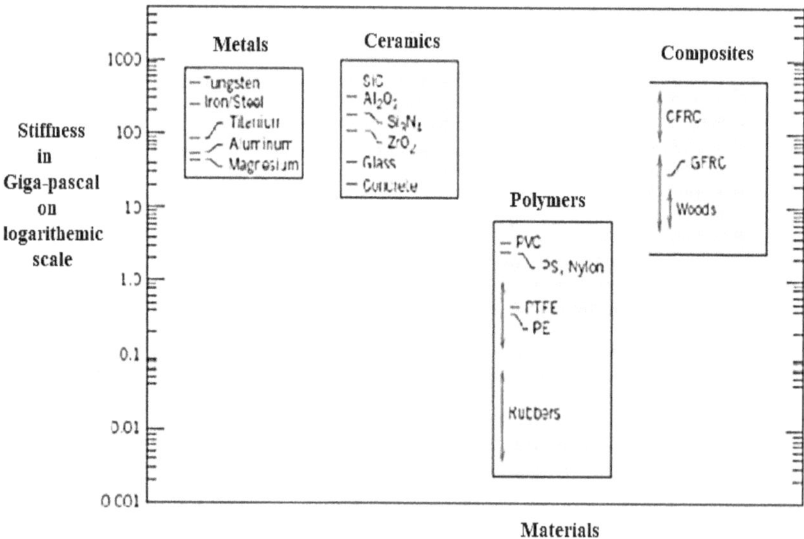

Curve 2: Showing, Stiffness at Room Temperature of Metals, Ceramics, Polymers and Composites Materials

From the above curve it has been observed that stiffness of ceramic material is somewhat higher than that of metals. The decreasing order of the stiffness are given below.

Ceramics > Metals > Composites > Polymers

(ii) Strength Curve

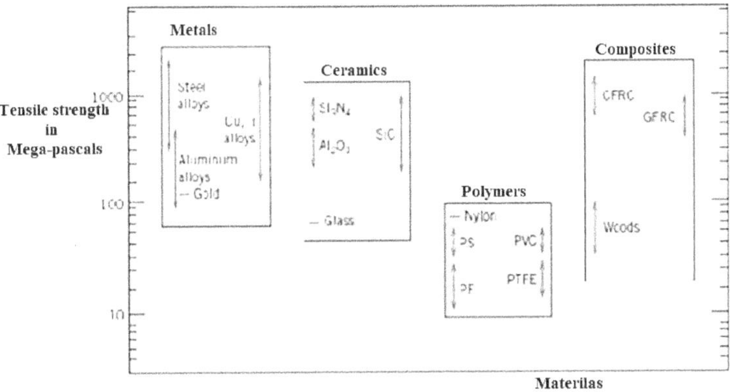

Curve 3: Showing, Tensile Strength at Room Temperature of Metals, Ceramics, Polymers and Composites Materials

From the above curve it has been observed that strength of metal is higher than that of other material. The decreasing order of the stiffness are given below.

$$\text{Metals} > \text{Composites} > \text{Ceramics} > \text{Polymers}$$

(iii) Fracture Toughness or Fracture Resistance Curve

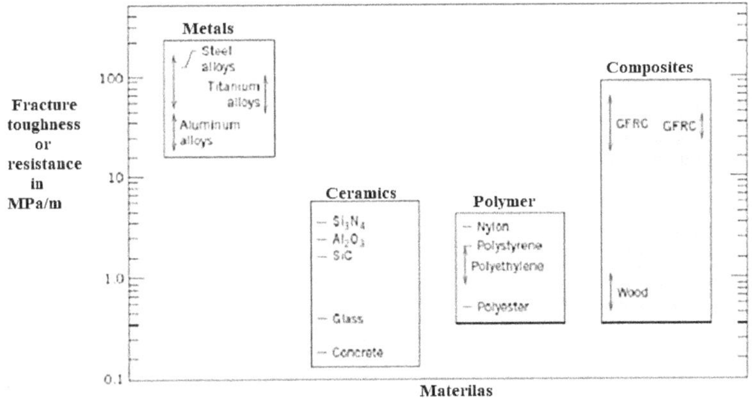

Curve 4: Showing, Fracture Toughness at Room Temperature of Metals, Ceramics, Polymers and Composites Materials

From the above curve it has been observed that fracture toughness or fracture resistance of metal is higher than that of other material. The decreasing order of the Fracture toughness or fracture resistance is given below.

Metals > Composites > Ceramics > Polymers

(iv) Electrical Conductivity Curve

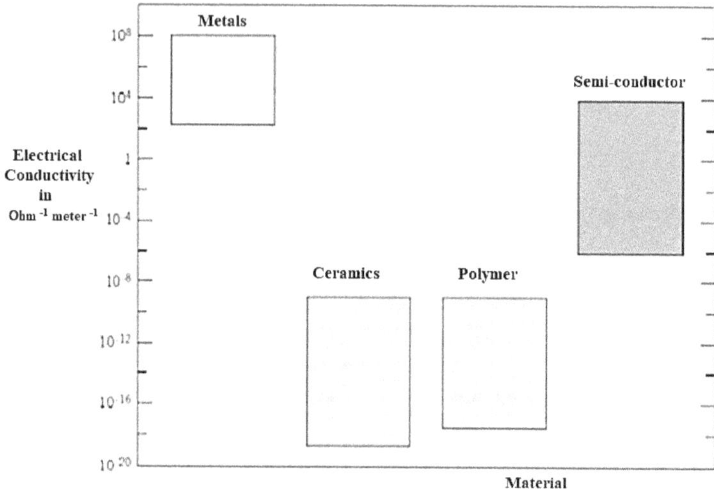

Curve 5: Showing, Electrical Conductivity Range of Metals, Ceramics, Polymers and Semi-conductor Material

From the above curve it has been concluded that electrical conductivity of metal is higher than that of other material. The decreasing order of the electrical conductivity are given below.

Metals > Semi-Conductor material > Ceramics > Polymers

2.4.2. Ceramics

The word ceramics extracted from the Greek word keramikos which means "potter's clay" or pottery. In deep sense keramikos can be understand by the word "To burn off". A ceramic is an inorganic, metallic or non-metallic solid material that is made by compacting a powder under high pressure or powder suspension into a body which is then sintered at high temperature or in other word we can say that it is neither a metal, a semiconductor nor a polymer. Ceramics may be a compound of oxides, nitrides, and carbides. Ceramic material has fascinating property that attribute as, lower electrical and thermal conductivity, resistant to high temperatures, corrosion resistant, good wear resistance, higher stiffness, high hardness, abrasion resistance, brittleness, chemical inertness. good resistance to corrosive environments, and lower fracture toughness than metals. For manufacturing a ceramic product or articles, a mixture of ceramic powders, water, and binder materials is taken in proper proportion and mixed properly & then poured into mold so that we can achieve desired shape as per the design specification. This semi-finished product is called "green bodies". Afterward dry it to remove moisture contents and perform sintering as a result of which binder oxidize, leaving the ceramic powder particles to bond to each other during the high temperature baking. Ceramic materials may be classified into two categories i.e. oxide and non-oxide ceramics. furthermore, it can also be classified as traditional ceramics & technical / engineering / specific ceramic. Traditional ceramics composed of clay minerals (i.e., porcelain), as well as cement, and glass. On the other hand, specific ceramics are synthetically producing or we can say it is manmade. ceramic material showing, superior mechanical properties, electrical & thermal insulating properties, piezo-electric properties, optical properties (Transparent, translucent, or opaque), magnetic properties (for example - Fe_3O_4), nuclear property, superior chemical properties etc. We have plenty of ceramic compound for instance - include aluminium oxide (or alumina, Al_2O_3), silicon dioxide (or silica, SiO_2), silicon carbide (SiC), silicon nitride (Si_3N_4), cement, glass and graphite etc. Glass is grouped in the class of amorphous material (Non-crystalline material).

- Now a days we have number of applications in specific field & non-specific field due to their remarkable property where metals & polymer does not perform the task satisfactorily.
- Metal is conductor of electricity & heat on the contrary ceramic & polymer are work as insulator for both.
- Ceramic materials are relatively showing high stiffness, strength, wear resistance, and abrasion resistance as compare to metals & polymer.
- Ceramic compounds indicate high resistant to temperatures and hostile environments than that of metals and polymers.
- Ceramics exhibits optical property like transparency, translucency and opacity on the contrary metals & polymer does not having optical property.
- Metals & ceramic material showing magnetic behaviour (For example- oxide ceramics Fe_3O_4) on the other hand polymer does not show magnetic property.

- Ceramic showing various kind of atomic structure in the compound rather than metals (BCC, HCP, FCC etc.) & polymer materials.

- Scissors, china tea cup, bricks, floor tiles, glass ware, cutting tools, insulating lining of furnaces and so many other articles/product are manufactured from ceramic compounds furthermore it has wide span of applications.

2.4.3. Polymer

Polymers are organic compounds that is derivative of carbon, hydrogen and including other non-metallic elements. Polymers have very long chain like molecular structures. Polymers includes plastics and rubber materials. It can be broadly classified into three cadre thermo-plastic polymers, thermo-setting polymers and elastomers (rubbers). For example - Polyethylene (PE), nylon, nylon-66, polyvinyl chloride (PVC), polycarbonate (PC), polystyrene (PS), polyesters, phenolics and silicone rubber etc. are very common in polymer material. Consider following points regarding to polymer.

- Polymer material have low densities, high flexible rather than metals & ceramics and widely used in insulators application for both thermal and electrical.

- Mechanical characteristics of polymer are dissimilar to the metallic and ceramic materials and these materials neither stiff nor strong like other engineering material.

- Polymers are very light in weight, having good stiffness & strength as compare compared to ceramics, metal and glasses.

- Most of the polymers are showing ductile nature and pliability which means they can easily be transform into complex or intricate shapes rather than ceramics material and metals.

- Polymers are relatively chemically inert and unreactive to hostile environments as compare to metals and ceramics materials.

- Polymer showing low electrical, thermal conductivities and nonmagnetic characteristics on the contrary metals indicate high electrical and thermal conductivities. Further ceramic material showing electrical, thermal insulating properties however it exhibits magnetic behaviour.

- Plastic table ware (spoons, fork and knife), billiard balls, bi-cycle helmet, two dice, plastic hub and rubber tires, plastic milk cartons are usually manufactured from polymer materials.

2.5.   Classification of Polymer Materials

Polymers are high molecular weight material which has different chemical structure, physical properties, mechanical behaviour and thermal characteristics etc. We have number of ways on the basis of that polymer material can be classified like – Tacti-city, Crystallinity, Polarity, Mechanism of polymerization, types of mesomeres involved in the polymers, Forces between polymer chains, molecular structures, their technological applications, Organic polymers, Inorganic polymers etc. In addition to it, Commodity plastic, Intermediate plastic, Engineering plastic, Advanced / high performance plastic may also be included in the classification. Consider following diagram which shows a complete over-view of classification of polymer material.

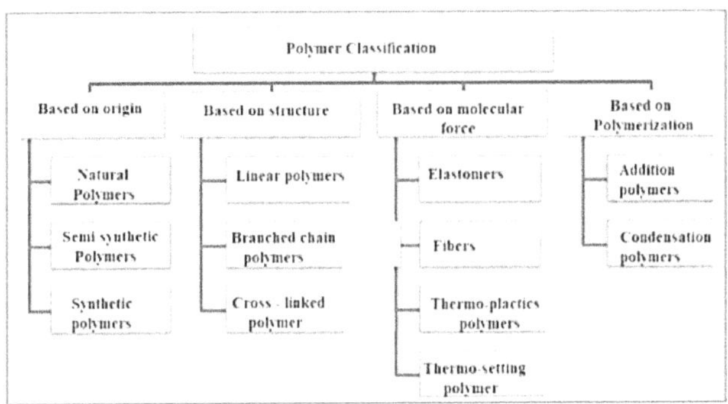

Diagram 2.1: Classification of Polymer Materials

Furthermore, natural polymer and synthetic polymer may be classified as, Consider following diagram.

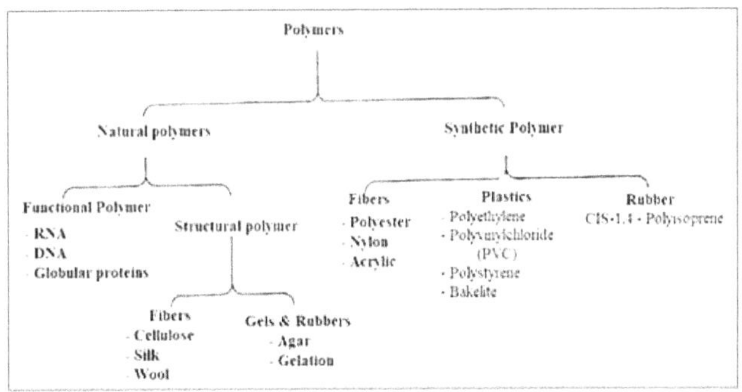

Diagram 2.2: Classification of Natural Polymer and Synthetic Polymer

A possible brief description of all those types are given under here.

2.5.1. *Classification of Polymer on the Basis of Origin*

In this classification we will coincide the existence or extraction of polymer means birth point of polymer. For example, natural polymer is achieved from plant & animals say rubber, silk, starch, protein, shellac, bitumen, cellulose, wood, leather, DNA, RNA, polysaccharides, isoprene, horns of animals, tortoise shells, asphalt and tar etc. are very famous in this category. Similarly, semi-synthetic that is also known as a modified natural polymer. Furthermore, synthetic polymer that is manufactured by synthetic methods by human being & known as man-made polymer. eventually we have following points.

- Natural polymer, extracted from nature or we can say it is originated in nature.

- Semi-synthetic polymer is nothing but it is modification of natural polymer but after all it is also originated from nature. & only modification has been performed by human being.

- Synthetic polymer is a man-made polymer that is manufactured or originated by human being in laboratory.

A brief summary is given below.

1. **Natural Polymer:-** As name implies that the source or occurrences of such a polymer is nature & hence it is known as natural polymer or bio-polymer. For example - natural rubber, natural silk, cellulose, starch, proteins and resins etc.

2. **Semi-synthetic Polymer:-** Semi-synthetic polymer is chemically modified natural polymer. The basic purpose of such a modification is to develop some properties or augmenting the property of natural rubber. hydrogenated, natural rubber, cellulosic, cellulose nitrate, methyl cellulose, Starch, silicones and cellulose acetate (rayon) are some common example which falls under the category of semi-synthetic polymer.

3. **Synthetic Polymer:-** It is explicit from their name that such a polymer is prepared or manufactured synthetically in laboratory and known as synthetic polymer. Such a polymer is known as man-made polymer. Poly-vinyl alcohol, polyethylene, polystyrene, poly-sulfone, nylon, polyethene, synthetic rubber, PVC, Teflon, plastic (polythene), synthetic fibres (nylon 6,6) and synthetic rubbers (Buna - S) etc. are very famous in this cadre.

2.5.2. *Classification based on the Structure of Polymers*

We know that **"Polymer is a matter** that **consisting long macro-mlecules or large molecules having a high molecular weight in which a lot of small repeating molecules are linked together so as to form a long chain"**, Small molecules or sub units are known as monomers. These sub units / small molecules are linked together by means of covalent bonds on the contrary long molecules joint together by means of either vanderwalls bonds, hydrogen bonds or covalent cross – links.

On the basis of molecular structure polymers material may be of following two types.

- Linear-chain polymer
- Branched-chain polymer
- Network or gel polymer (Cross Linked Structure)

1. **Linear-chain Polymer:-** In linear polymer, all the sub units / small repeating molecules known as monomers which is bonded together by means of strong covalent bonds are joined in linear chain like structure, and form a long straight chain. Such type of polymer is called linear chain polymer. For example – Polyethylene, PVC, nylons, polyester, acrylics, high density polyethylene (HDPE), poly-styrene. Such type a chain like structure are found in thermoplastics. Consider the following diagram regarding to it.

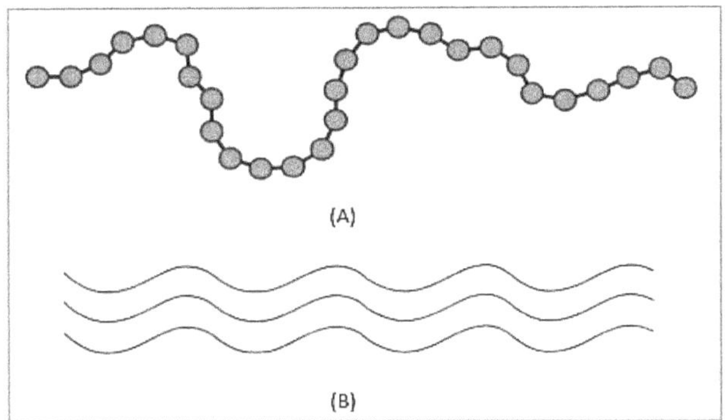

Diagram 2.3: Linear Chain Polymer

Since in linear polymer structure (chain like structure), molecules are closely packed to each other as a result of which it showing high density & tensile strength. Furthermore, mostly it has been observed that linear structured polymer showing thermoplastic nature, for clarification, we know that repeated units / sub units are strongly bonded by covalent bond on the other hand molecules are connected through weak secondary bonding. Whenever linear polymer experienced heat or thermal effect then randomly motion of molecules takes place which always try to overcome weak secondary forces, as secondary force has been completely overcome the molecules moves around the polymer melts. Eventually such a mechanism only occurs in the thermo-plastic polymer. In addition to it, polymer that has a long linear chain & inter-molecular forces that exists between that chains and whenever material subjected to pressure then molecules will tend to moves past one another and this mechanism is called plastic flow. Such kind of polymer may be dissolved in other solvent that has similar chemical structure which inhibit not only motion of the molecules but also arrangement of molecules past one another. Consequently, it appears in the solid form at room temperature. This characteristic is also justified to thermo-plastic nature.

Note

- In linear chain structure (Thermo-plastic polymer), monomers have covalent bonds & secondary bond exists among the chains.

- Thermo-plastic polymer can have linear or branched macro-molecules.

- Structure of polymer always cooperate to evaluate properties of polymers, for instance – Rigidity of polymer molecules, Intensity of forces between chains, Degree of crystallinity in domain of the chains, Degree of cross-linking between the chain etc.

- Structure of polymer is only a single mode by which we can evaluate mechanical property like – Tensile strength, toughness and elasticity. In addition to it, with the help of structure we can also determine intensity of inter-molecular forces.

- It is essential for polymers formation that, monomers must have reactive functional groups or double or triple bonds. The functionality of monomer is depending upon quantity of these functional groups. Furthermore, double bonds showing functionality of 2 & triple bond has a functionality of 4. In order to form a polymer, the monomer must be at least bifunctional, whenever it is bifunctional, the polymer chains are always linear.

2. **Branch Chain Polymer:-** Branch chain polymer consisting two components i.e.
    - Long linear chain of polymer (Linear backbone).
    - Branch of monomers on either side of back-bone.

Eventually we can say that **"Branch chain polymer is defined as a polymer that contains chain of monomers on one side or either side of linear back-bone chain (Chain of molecules)".**

OR

In other word we can say that a linear long chain that consisting side branches called branched chain polymer, such type of structure is found in low density polyethylene, polypropylene, amylopectin and glycogen and thermo-plastic polymer etc.

$$-\overset{\displaystyle H}{\underset{\displaystyle H}{C}}-\overset{\displaystyle H}{\underset{\displaystyle H}{C}}-\overset{\displaystyle H}{\underset{\displaystyle H}{C}}-\overset{\displaystyle H}{\underset{\displaystyle H}{C}}-$$

Diagram 2.4: Polyethylene Polymer

The basic structure of branch chain polymer can be understood by following diagram,

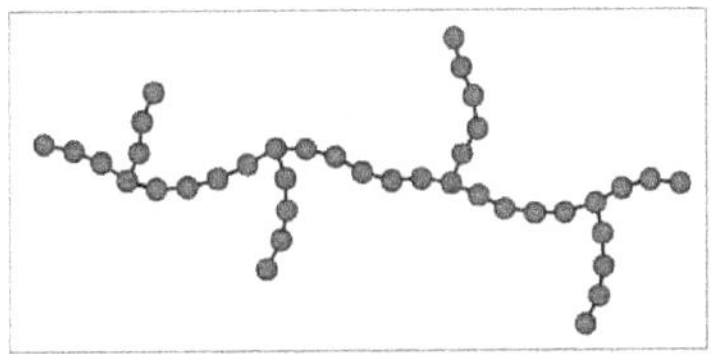

Diagram 2.5: Basic Building Block of Branch Chain Polymer

Basically, branch chain polymer is associated with low conversion of monomers & hence polymer having low molecular weight. For example – low density polyethylene (LDPE) in which ethylene molecules are monomers which linked to long chain polymeric molecules & such a structure is known as branched chain polymers. Since molecules are packed in irregular pattern so therefore such type of polymer has low density, low tensile strength & low melting point.

Note

- To manufacturing branch polymer, monomer must have a capability of growing in more than two direction and hence it is imperative that starting monomers should have functionality of greater than two.
- For manufacturing branched polymers of high molecular weight there is a requirement of polymerization of multifunctional monomer so that large conversions may take place as a result of which we have got polymer which has three-dimensional network called a gel.

3. **Cross – Linked Polymer:-** Cross linked polymer structure is also known as three dimensional structure (Network polymer). In cross linked polymer, two linear chains are joint together by covalent bond and overall structure can be appear in three dimensions.

Cross linking polymer usually manufactured by heating resins in Bakelite mould or mixture of resins with suitable cross-linking agent say amine, epoxies, vulcanized rubber, phenolics, unsaturated polyester resins and amine resins (Ureia & melamine) by curing at low temperature. Basically, cross-linked polymers are generally formed from bi-functional & tri-functional monomers which contains strong covalent bonds between various linear polymers chains. Bakelite, rubber, casting (boat) resin, cross-linked polyester (Glyptal), melamine, formaldehyde resins, vulcanised rubber, Phenol-formaldehyde resin, urea-formaldehyde resin, epoxy resins (araldite), melamine and elastomers product are generally showing cross-linked structure. Further cross – linked structure may be of two types.

- Loosely cross – linked structure that is found in elastomers. Consider following diagram.

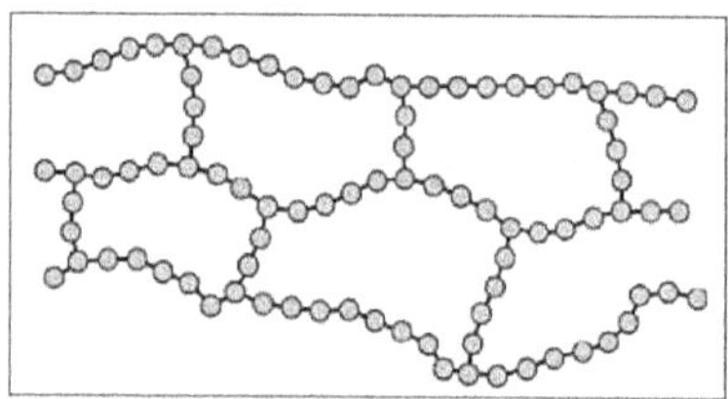

Diagram 2.6: Loosely Cross – Linked Structure in Elastomer

In such a structure primary bonding exists among branches & other molecules are connected at a certain point, as explicit from above diagram.

- Tightly cross-linked structure that is generally found in thermosetting polymer. Consider following diagram which shows a complete over-view of tightly cross-linked network structure.

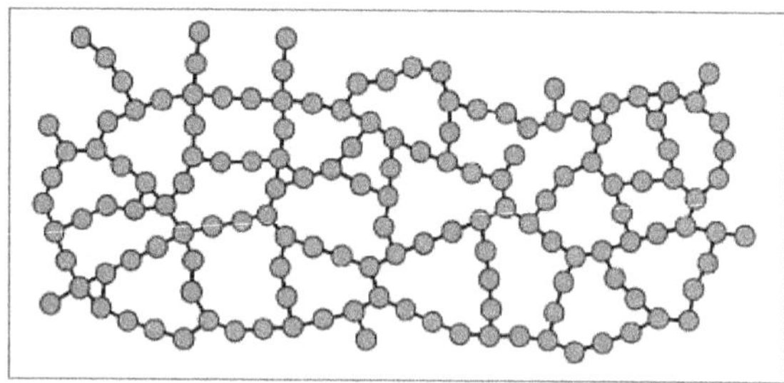

Diagram 2.7: Tightly Cross – Linked Structure in Thermosetting Polymer

Furthermore, thermo-setting polymers are generally manufactured from low molecular weight substance (say – Semifluid substance). Whenever such a lower molecular polymer is heated in mould then high cross-linked structure is formed in polymer that has three-dimensional network structural bonding which interconnecting that polymer chains. Further, cross linked structured (Thermosetting polymer) indicates hardness, rigidity, infusibility, insolubility and they do not get become melts & soft after processing and could not be reshaped or reuse again. Thermosets (Say - Formaldehyde resins, urea-formaldehyde resin, epoxy resins (araldite), melamine and Bakelite) falls under this category. Moreover, Consider the following diagram, which gives an over-view of transformation of highly cross-linked polymer from uncrossed linked thermo-setting polymer.

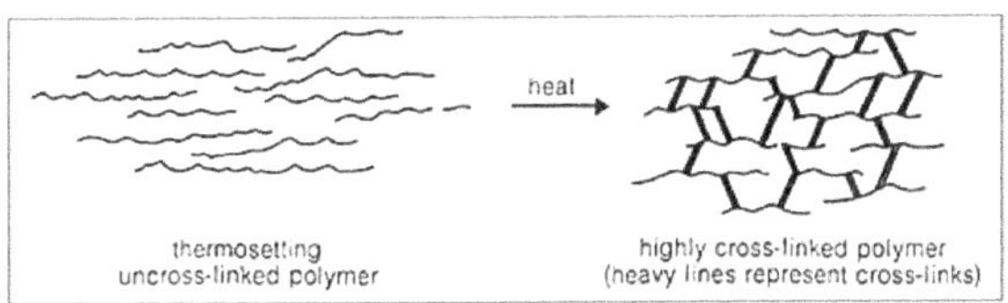

Diagram 2.8: Transformation of Uncrossed Linked into Highly Cross-linked Polymer

From the above diagram it has been concluded that whenever uncross-linked two-dimensional structured polymer is heated then it transformed into highly cross-linked three-dimensional network structure.

Note

- Cross link structure showing only chemical bonds between polymer chains and other ends. During cross linking enhancement of molecular weight takes place & try to maintain translation motion of chains with respect to one another. Cross link also cooperates to evaluate the physical property of polymer product. Cross link structure showing absorption of solvent instead of solubility in solvent, if degree of cross-linking reduces on the other hand if degree of cross-linking is high then solvent molecules could not be penetrated the chains.

- Whenever lightly – cross linked structure is formed then complete entire mass behave as one gigantic macro-molecules.

- Degree of cross linking is nothing but it is a number of junction point / unit volume.

- Cross linked structure of polymer is hard, brittle and rigid due to their network structure and these polymer does not dissolve in solvents due to chains are covalently bonded but they can absorb the solvents only.

2.5.3. *Classification of Polymer on the Basis of Thermal Behaviour*

On the basis of thermal behaviour polymer may be two types i.e. thermoplastics polymer and thermosetting polymer. In this we will have to observe the response of both the polymer when they experience a thermal effect.

2.5.3.1. Thermo-plastic Polymer

On the basis of thermal behaviour we will identify the reusability or recycling capability of plastic polymer without degrading their characteristics.

OR

In other word we can say that,

"A characteristic of thermoplastic polymer by virtue of which it can reuse, recycle and reshape again in desired shape after heating without affecting its attributes".

We know that thermo-plastic polymer can be heated from solid state to molten state thereafter molten form may be transformed into desired design specification as per requirement then allowed to solidify. A major advantage of such a thermo-plastic polymer is that heating & cooling can be performed several times without affecting the polymer characteristics. The basic reason behind such a thermal behaviour is noticed due to existence of linear macro-molecules that inhibit cross linking during the heating operation as a result of which reheating, reshaping could be performed many times and hence a polymer that become soft on heating and stiff on cooling known as thermo-plastic polymer. Furthermore, it has been observed that if thermo-plastic is heated it become soft and further heating gives a molten form because of broken down of weak non-covalent bonds. This molten form is used to obtain desired shape after moulding process. In subsequent process if plastic is cooled it become stiff & brittle, the temperature at which brittleness attain known as glass transition temperature. In order to minimize number of secondary bonds, plasticizers are mixed with thermo-plastics whose function is not only enhancing easiness during shaping the product or articles and imparting desired level of brittleness or hardness in the plastic but also reducing the glass transition temperature of thermoplastic. Thermoplastic polymer may be of two types i.e.

- Commodity thermo-plastic polymer
- Engineering thermo-plastic polymer

Commodity thermo-plastic polymer are used to manufacturing those products which is being widely used in daily home service like bucket, jugs, glass and many more other items.

Polyethylene (LDPE, LLDPE, HDPE), Polypropylene (PP), Polyvinylchloride (PVC), Polystyrene (GPPS, HIPS), Styrene acrylonitrile (SAN) and Polymethyl-methacrylate (PMMA) are most common commodity thermo-plastic polymer. On the contrary engineering or commercial thermo-plastic are generally used to manufacturing fibers, lenses, films, sheets, packaging materials, paints and varnishes. Polyamide (Nylon-6 & Nylon-6,6), acrylonitrile butadiene styrene (ABS), poly-acetal, poly-carbonate (PC), thermo-plastic polyesters (PET, PBT), poly-phenylene oxides (PPO), and thermo-plastic poly urethane (TPU) are very famous.

Note

- Thermo-plastic may dissolve in solvents & can be retained in original form after evaporation.
- Most common thermo-plastic polymers are polyolefins, nylons, linear polyester, PVC, sealing wax, polyethylene, acrylics, polypropylene, polystyrene, polymethyl-methacrylate (Used for manufacturing plastic lenses or Perspex), poly-tetrafluoro-ethylene (PTFE or Teflon) and plexiglass.
- Thermo-plastic consisting either linear or somewhat branched macromolecules in which molecules has attracted to each other by inter-molecular forces.
- Thermo-plastic polymers are hard, rigid and brittle at room temperature but it can reuse, recycle again after heating.
- Thermo-plastic polymer exhibits low stiffness, low tensile strength, low hardness than metals and ceramics, moderate ductility, low density than metals and ceramics, higher coefficient of thermal expansion, lower melting point and work as insulators.
- Plasticizers are included in thermo-plastic polymer so that glass transition temperature can reduce that is removed automatically in further heating process.
- It is mostly used to manufacturing wool fibers, polyethylene jug (thermo-plastic jug), polycarbonate lenses (thermo-sets) and plastic lenses or Perspex etc.
- Thermo-plastic polymer does not have any cross-linked structure. It has long molecular chains these are bonded together by secondary bonds.
- Manufacturing of product or articles is accomplished by simultaneous application of heat & pressure both during fabrication process.
- Reheating, reusing, recycling and reshaping is possible due to the absence of cross-linking structure formation during heating process.
- Thermo-plastic consisting only linear polymer chain & indicate solubility in organic solvents.

2.5.3.2. Thermosetting Polymer

We know that thermo-plastic polymer can reuse, recycle and reshape again by reheating it several times on the contrary thermosetting polymer does not reheat again once it transformed into desired shape if so degradation of characteristics would be achieved. Mostly it has been found that it has poor response for heating and further processing does not possible once they set and transformed into rigid and hard product / articles due to the formation of tightly cross-linked structure of molecules. Basically, whenever thermoset polymer subjected to heating process, chemical reaction takes place as a result of which hard, rigid and infusible mass would be produced that has cross linking of the polymer chain among molecules. Phenolic, resins, urea, epoxy resins, polycarbonates, Bakelite, urea-formaldehyde resin, phenol-formaldehyde resin, melamine are some example of thermo-set polymer.

Thermosetting polymer may be defined as,

> A polymeric substance which when subjected to heat it become soften & transformed into infusible mass due to the effect of chemical reaction and after molding that is performed by application of pressure it takes a desired shape. Once it cools, they become rigid, hard, showing insolubility and unable to further reheating aswellas reprocessing because of highly cross-linked structure, such a compound is called thermosetting polymer.

Further, thermosetting polymer is to be molded at room temperature or above the room temperature, such a polymer usually manufactured by low molecular weight substance whenever these substances are heated & molded then we have highly cross-linked three-dimensional network structure which is a responsible factor for hardness, infusibility, reheating and insolubility of thermosets polymer. Moreover, manufacturing of thermosets product not only required heat but also demanded adequate amount of recommended pressure for completing the process. The purpose of heat is to initiate chemical reaction and transformed it into moldable form and thereafter molding is performed by application of pressure to obtain a pre-decided permanent shape. Moreover, thermosetting polymer play a vital role in engineering design field due to remarkable & fascinating characteristics like - High thermal stability, high dimensional stability, high rigidity, light weight, high electrical & thermal insulating properties, creep resistance, resistance to deformation under load, non-softening characteristics, non-swelling nature, hardness and brittleness etc.

Note

- Thermosetting polymer is known for their high cross-linked three-dimensional structure & covalent bonding among molecules.

- Heat response for thermosetting polymer, means it does not re-melt, reform and recycle, if it is heated intentionally then it becomes decompose / burns instead of softening or melting.
- Mostly thermosets composed of long chain which has strong cross-linked to one another as a result of which it has strong & hard nature with respect to thermoplastic polymer.
- Cross linking reaction can be achieved by two method i.e. heating the resins in suitable mold (Say – Bakelite Mold) or mixing a suitable cross linking agent with resins & then cured at low temperature.
- Cross linking agent may either be amine, epoxies, vulcanized rubber, phenolics, unsaturated polyester resins or amino resins.
- We can have desired shape of thermosets during processing i.e. heating and molding. Once after cooling they could not be softened again by supplying heat, if heated then burning of thermosets would take place.

2.5.4. Classification of Polymer on the Basis of Molecular Forces

Mechanical properties of polymers like tensile strength, toughness, elasticity etc. depends upon the intermolecular forces like Vander Waals forces and hydrogen bonding. On the basis of these forces they are classified as,

2.5.4.1. Elastomers

These are rubber – like solids with elastic properties. In these elastomeric polymers, the polymer chains are held together by the weakest intermolecular forces. These weak binding forces permit the polymer to be stretched. A few 'crosslinks' are introduced in between the chains, which help the polymer to retract to its original position after the force is released as in case of vulcanised rubber. The examples are Buna-S, Buna-N, neoprene, rubber and spandex etc.

$$\left(CH_2-\underset{\underset{Cl}{|}}{C}=CH-CH_2\right)_n$$

Neoprene

Elastomer consisting long chains of molecules which has cross-linked structure. The degree of cross-linking always less than thermosets. Further all the molecules are tightly linked to each other in unstressed condition but whenever elastomers experienced a low stress value then large elastic deformation may be achieved.Consider the following diagram, which shows basic building blocks of elastomers, in case of unsressed condition & stressed condition (tensile loading application).

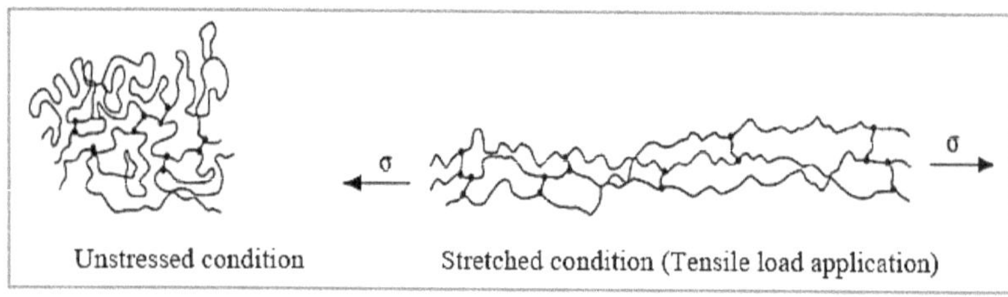

Diagram 2.9: Basic Building Block of Elastomer in Stressed & Unstressed Condition

It has been observed that from above diagram, whenever elastomer is streched by application of tensile load application that acts upon molecules as a result of which it tends to lenghthen due to elasticity characteristics. If load is further increases then covalent bonds which exists among cross-linked molecules showing increasing value of elastic modulus aswellas stiffness. Elastomer exibhits high molecular weight, long flexible chain, weak inter-molecular forces, tensile strength and elongation.

Note

- Elastomer indicate elastic deformation characteristics i.e. when external force is applied then elastomer fiber stretches but when this force is relieved then elastomer fiber spring back to original form. In actual practise elastomer molecular structure including chains which tangled to each other. If they are stretched, a restoring force comes in action which try to pulls them back into initial & tangled position.

2.5.4.2. Fibres

Fibres are the thread forming solids which possess high tensile strength and high modulus. These characteristics can be attributed to the strong intermolecular forces like hydrogen bonding. These strong forces also lead to close packing of chains and thus impart crystalline nature. It is used in textile industries the examples are polyamides (Nylon 6, 6), polyesters etc.

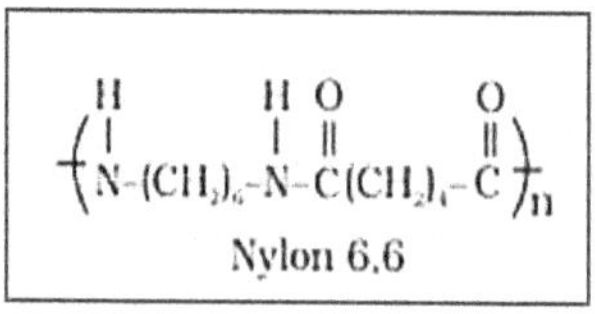

$$\left\{ N-(CH_2)_6-N-C(CH_2)_4-C \right\}_n$$

Nylon 6,6

Diagram 2.10: Fibers (Nylon-66)

Fiber attributed as long chain polymer having secondary forces and showing low elasticity than plastic & elastomers, high tensile strength, moisture absorption capacity and light weight.

2.5.4.3. Thermoplastic Polymers

These are the polymers having intermolecular forces between elastomers and fibres and become soft on heating and hard on cooling at room temperature. They may have either linear or branched chain polymers structure further these polymers can be recycled many times. Example-, Polythene, polystyrene, PVC.

2.5.4.4. Thermosetting Polymers

This polymer is hard and infusible on heating could not be soften on re-heating and unable to reuse or recycle. These polymers having cross linked or heavily branched molecules for Example - Bakelite, urea-formaldehyde resins, etc.

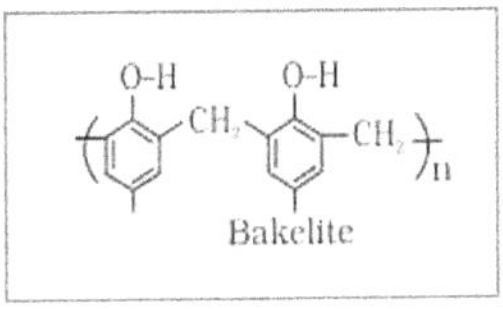

Diagram 2.11: Bakelite (Thermosetting Polymer)

2.5.5. Classification of Polymer on Basis of Polymerization or Formation

Polymer material can also be made by synthetic method which has been invented by human being, such a mode of polymer formation is known as polymerization process. Polymerization consisting two technique i.e. addition polymerization & condensation polymerization. In addition, polymerization process, all the atoms of the monomer's molecules take parts in polymerization on the other hand in condensation polymerization, among all the atoms few atoms of the monomer split during the reaction which transformed into water, alcohol, ammonia or carbon dioxides. A brief description of addition technique & condensation technique is given below.

2.5.5.1. Addition Polymerization Technique

Addition polymerization technique sometimes also referred as a chemical reaction polymerization in which repeated monomers units (Multifunctional units) are connected or added together in such a manner that they may form a long linear chain or branched chain with back-bone due to the effect of chemical reaction to manufacturing a polymer. Since polymer is formed by the addition of repeated monomers without removal of any by products are known as addition polymers. Consider the following building blocks which shows inherent structure of monomers throughout the polymer.

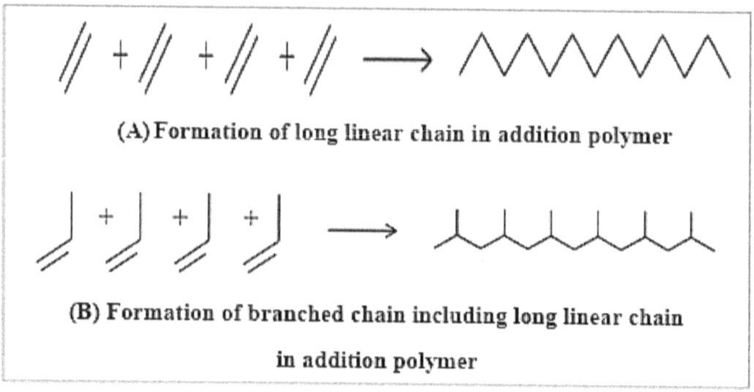

Diagram 2.12: Block Diagram of Chain Structure in Addition Polymer

Furthermore, monomer contains carbon-carbon double bonds. Now a days most of the polymer are generally made from addition polymerization technique and it has been observed that alkenes or alkynes are play an inevitable role for manufacturing addition polymer. polyethylene, polypropylene, polystyrene, poly- 1,3-cyclopentadiene, alkene polymers, poly-alka-dienes, poly-fluoroalkenes, poly-methanol, poly-propene, polypropylene, polyvinyl chloride (PVC) and polytetrafluoroethylene (Teflon) are some common example of addition polymers.

In addition to understand the addition polymerization, consider the example of polyethylene. In case of polyethylene, a plenty of ethylene molecules called monomers are linked together in long chain by breaking the π-bonds & to establishing two new single bonds between the monomer's units. Consider the following diagram which shows complete over-view of polyethylene addition polymers.

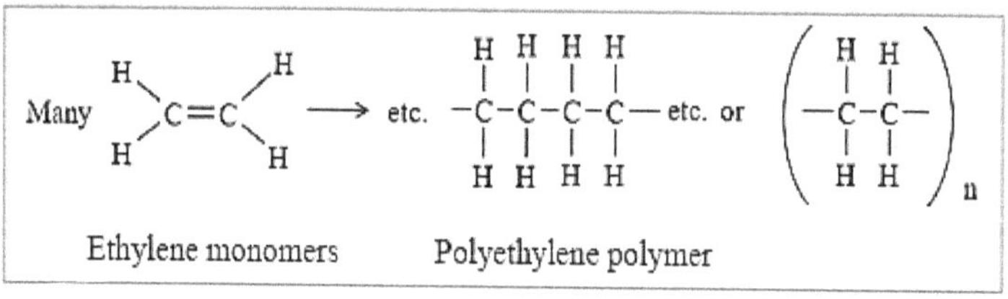

Diagram 2.13: Polyethylene Addition Polymers

In order to fulfil the desired purpose three things are required heat, pressure and chemical catalyst throughout the reaction. We can have variation in chain length aswellas chain structure, means chain of monomers may either be linear or branched. Polyethylene that has linear structure facilitate to easy close packing of monomers and referred as high-density polyethylene which indicate rigid nature on the contrary low-density polyethylene consisting branched aswellas cross-linking of molecules in the chain and hence these are more flexible than that of high-density polyethylene.

Consider another example of addition polymer that is polypropylene. In this case monomers are propylene and final addition polymer consisting branched methyl on alternate carbon atoms of the chain. Consider the following diagram which shows complete over-view of polypropylene addition polymer.

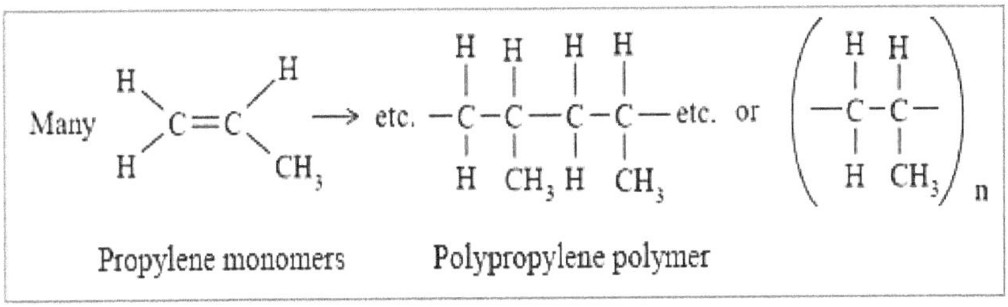

Diagram 2.14: Polypropylene Addition Polymers

2.6. Stages in Addition Polymerization

It is explicit that in addition polymerization, molecules of same monomers or different monomers are added together to form a polymer. This polymerization contains carbon-carbon double bond that is mostly a derivative of unsaturated compounds like- alkenes which take part in chain reaction. Chain reaction consisting following three stages.

- Initiation stage
- Propagation stage
- Termination stage

1. **Initiation Stage:-** At initial stage our basic motive is to generate chemical reaction. For that initiator is added with monomers & then heating is performed to composition so that molecules could decompose & chemical reaction may take place to achieve active species. These active species usually recognized as a free radical, a cation or an anion. Further whenever these achieve active species interacting with carbon-carbon double bonds of monomers then polymerization initiation occurs. This reaction takes place in such a manner that new free radical, a cation or an anion must be generated.

2. **Propagation Stage:-** Propagation stage is nothing but it indicates continuation of process. In this stage newly developed free radical, a cation or an anion play a vital role so that subsequent polymerization can automatically run in smooth manner likewise an initiation stage and this process repeated again & again until up to final step does not achieve or we can say that it continue up to or before the termination occurrence.

3. **Termination Stage :-** In termination stage, we have got termination of develop chain from another linked develop chain due to the decomposition of active site. But sometimes it has also been noticed that in some cases anions remain adhere in polymer and does not terminate hence such a polymer is known as living polymer. Consider the process of polyethylene.

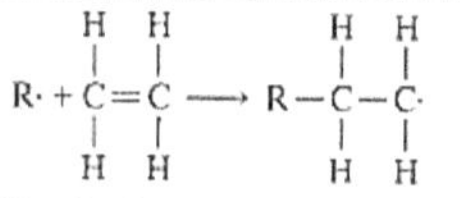

$$
R\cdot + \underset{\underset{H}{|}}{\overset{\overset{H}{|}}{C}} = \underset{\underset{H}{|}}{\overset{\overset{H}{|}}{C}} \longrightarrow R - \underset{\underset{H}{|}}{\overset{\overset{H}{|}}{C}} - \underset{\underset{H}{|}}{\overset{\overset{H}{|}}{C}}\cdot
$$

Termination stage polyethylene

Diagram 2.15: Termination Stage of Polyethylene

Where,

R – Showing active initiator

Propagation showing linear growth of the molecules of monomers which attached to one another in such a manner that they produce successive chains of monomers. Consider following diagram.

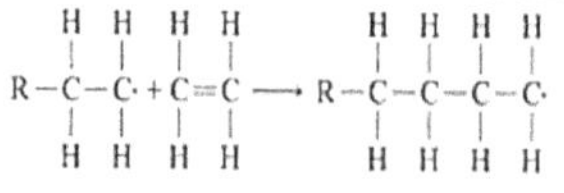

$$
R - \overset{H}{\underset{H}{C}} - \overset{H}{\underset{H}{C}}\cdot + \overset{H}{\underset{H}{C}} = \overset{H}{\underset{H}{C}} \longrightarrow R - \overset{H}{\underset{H}{C}} - \overset{H}{\underset{H}{C}} - \overset{H}{\underset{H}{C}} - \overset{H}{\underset{H}{C}}\cdot
$$

Propagation stage polyethylene

Diagram 2.16: Propagation Stage Polyethylene

Sometimes, if we want to maintain or controlled molecular weight in polymer then polymerization can be terminated in some stages, for that it is imperative, active ends of propagating chain may react or linked together in such a manner so that it forms a non-reactive molecule. Consider following diagram.

$$
R\cdots C - C - C - C\cdot + \cdot C - C - C - C\cdots R \longrightarrow R\cdots C - C - C - C - C - C - C - C\cdots R
$$

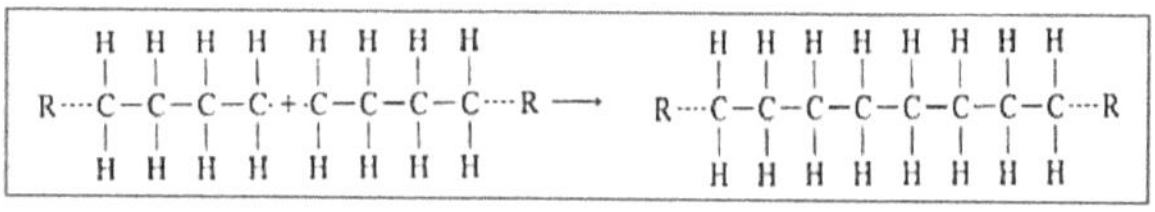

From the above diagram, termination of growth of chain or an active chain end may react with in initiator or other chemical species that has a single active bond, Consider following diagram.

$$\cdots\overset{\displaystyle H}{\underset{\displaystyle H}{C}}-\overset{\displaystyle H}{\underset{\displaystyle H}{C}}-\overset{\displaystyle H}{\underset{\displaystyle H}{C}}-\overset{\displaystyle H}{\underset{\displaystyle H}{C}}+\cdot R \longrightarrow -\overset{\displaystyle H}{\underset{\displaystyle H}{C}}-\overset{\displaystyle H}{\underset{\displaystyle H}{C}}-\overset{\displaystyle H}{\underset{\displaystyle H}{C}}-\overset{\displaystyle H}{\underset{\displaystyle H}{C}}-R$$

Eventually, from the above diagram, we can completely inhibit the chain growth.

2.7. Application of Addition Polymer

We have number of polymers among those most of them polymer are made from addition technique or process. For example – polystyrene is an addition polymer which is being widely used to manufacturing containers like – viscos-elastic cases, compact disk jewel boxes, table ware (Fork, knives, spoons, and cafeteria trays), coffee cups, grocery store meat trays and building insulations. In addition to it, consider the following table which gives an information of some other addition polymer. In this table we have addition polymer, structure of polymer, monomers and their respective uses.

Table 2.15: Some Most Important Addition Polymer

EXAMPLE	MONOMER(S)	POLYMER	USE
Polyethylene	$CH_2{=}CH_2$	$-CH_2{-}CH_2-$	Most common and important polymer. Bags, insulation for wires, squeeze bottles
Polypropylene	$CH_2{=}CH$ $\quad\ \ \|$ $\quad\ \ CH_3$	$-CH_2{-}CH-$ $\qquad\ \|$ $\qquad\ CH_3$	Fibers, indoor-outdoor carpets, bottles
Polystyrene	$CH_2{=}CH$ (phenyl)	$-CH_2{-}CH-$ (phenyl)	Styrofoam, inexpensive household goods, inexpensive molded objects
Polyvinyl chloride (PVC)	$CH_2{=}CH$ $\qquad\|$ $\qquad Cl$	$-CH_2{-}CH-$ $\qquad\ \|$ $\qquad\ Cl$	Synthetic leather, clear bottles, floor covering, phonograph records, water pipe
Polytetrafluoroethylene (Teflon)	$CF_2{=}CF_2$	$-CF_2{-}CF_2-$	Nonstick surfaces, chemically resistant films
Polymethyl methacrylate (Lucite, Plexiglas)	$\qquad CO_2CH_3$ $\qquad\quad\|$ $CF_2{=}C$ $\qquad\ \|$ $\qquad\ CH_3$	$\qquad CO_2CH_3$ $\qquad\quad\|$ $-CH_2{-}C-$ $\qquad\ \|$ $\qquad\ CH_3$	Unbreakable "glass", latex paints
Polyacrylonitrile (Orlon, Acrilan, Creslan)	$CH_2{=}CH$ $\qquad\|$ $\qquad CN$	$-CH_2{-}CH-$ $\qquad\ \|$ $\qquad\ CN$	Fiber used in sweaters, blankets, carpets
Polyvinyl acetate (PVA)	$CH_2{=}CH$ $\qquad\|$ $\qquad OCCH_3$ $\qquad\ \ \|\|$ $\qquad\ \ O$	$-CH_2{-}CH-$ $\qquad\ \|$ $\qquad\ OCCH_3$ $\qquad\quad\|\|$ $\qquad\quad O$	Adhesives, latex paints, chewing gum, textile coatings
Natural rubber	$\qquad CH_3$ $\qquad\ \|$ $CH_2{=}CCH{=}CH_2$	$\qquad CH_3$ $\qquad\ \|$ $-CH_2{-}C{=}CH-CH_2-$	The polymer is cross-linked with sulfur (vulcanization)
Polychloroprene (Neoprene rubber)	$\qquad Cl$ $\qquad\|$ $CH_2{=}CCH{=}CH_2$	$\qquad Cl$ $\qquad\|$ $-CH_2{-}C{=}CH-CH_2-$	Cross-linked with ZnO, resistant to oil, gasoline
Styrene butadiene rubber (SBR)	$CH_2{=}CH$ (phenyl) $CH_2{=}CHCH{=}CH_2$	$-CH_2CH-CH_2CH{=}CHCH_2-$ (phenyl)	Cross-linked with peroxides. Most common rubber. Used for tires. 25% styrene. 75% butadiene

2.8. Characteristics of Addition Polymer Technique

1. It has been observed that at the beginning of initiation stage, formation of polymer chain rapidly takes place.

2. Since amount of active species is very low and hence polymerization process consisting mixture of newly formed polymer & unreacted monomers.

3. The existence of carbon-carbon double bond in the monomers always transformed into single carbon-carbon bonds in the polymer during reaction which leads to energy liberation and hence cooling is required.

Note

- Generally, benzoyl peroxide, acetyl peroxide, tert-butyl peroxide are used as an initiator.

- Mechanism of an addition polymerization mostly accomplished by free radicals or ionic polymerization. Furthermore, if we consider ionic polymerization then it can be achieved either by cationic polymerization or anionic polymerization.

- Addition polymer are generally formed from olefinic, diolefin, vinyl or by other related monomers.

2.9. Free Radical Polymerization

Whenever compound like benzoyl peroxide, acetyl peroxide, tert-butyl peroxide, etc. are used as a catalyst or initiator for beginning the polymerization of alkenes or dienes then free radicals are generated. These free radicals are responsible for polymerization & also cooperate to successive polymerization process. Furthermore, free radicals are nothing but it is an intermediate compound whose function is to catalysing the process, they consisting electrons that neither have any charge carrying opportunity nor freedom to movement. In order to understand the phenomenon of free radical's polymerization consider the case of polyethene.

- Ethene work as monomers which mixed with catalyst or initiator say benzoyl peroxide and then heating is performed.

- During the initiation stage, free radicals of phenyls is formed. Further addition of phenyls free radicals to peroxide to the ethene double bond also responsible for generating new larger free radicals throughout the process.

- During the second stage or propagation stage, free radicals react with next molecules of ethene that further formed new large size of radicals. This process repeated again & again and chain propagation has achieved. Consider the following diagram.

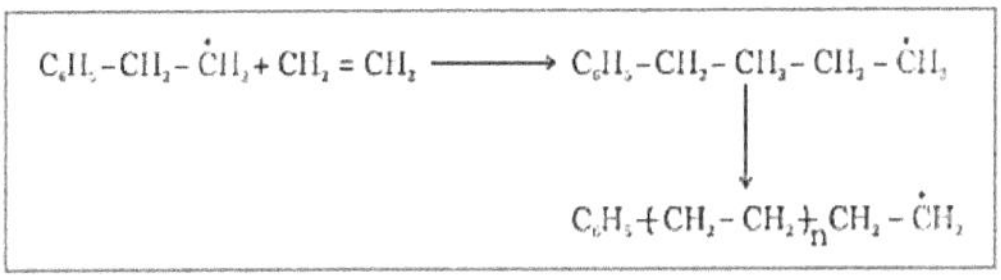

$$C_6H_5-CH_2-\overset{\bullet}{C}H_2 + CH_2 = CH_2 \longrightarrow C_6H_5-CH_2-CH_2-CH_2-\overset{\bullet}{C}H_2$$

$$\downarrow$$

$$C_6H_5 \cdot (CH_2 - CH_2)_n \, CH_2 - \overset{\bullet}{C}H_2$$

Diagram 2.17: Chain Propagation of Ethene Monomers

- During final stage (known as termination stage), free radical reaction with another monomer is terminated or stopped and finally we have polymer product. Consider the following diagram which shows final polyethene product.

$$C_6H_5 \cdot (CH_2 - CH_2)_n \, CH_2 - \overset{\bullet}{C}H_2$$
$$C_6H_5 \cdot (CH_2 - \overset{\bullet}{C}H_2)_n \, CH_2 - \overset{\bullet}{C}H_2$$
$$\longrightarrow C_6H_5 \cdot (CH_2 - CH_2)_n \, CH_2 - CH_2 - CH_2 \cdot (CH_2 - CH_2)_m C_6H_5$$

Polythene

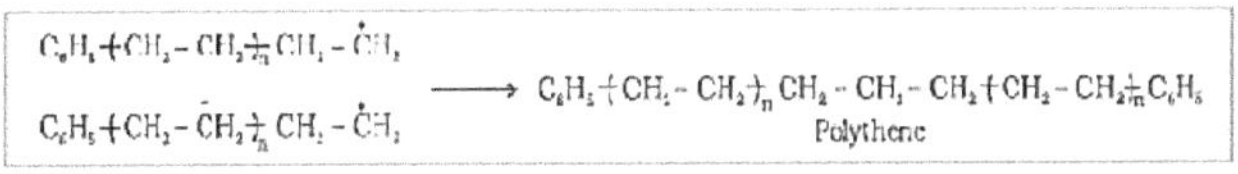

Diagram 2.18: Chain Termination Stage & Final Polyethene Product

2.10. Ionic Polymerisation

In addition, polymerization, anions & cations are also used as initiator. During the polymerization process anions & cations work as initiators. In this class initiator either may be anions or cations, A brief description of cationic polymerization and anionic polymerization is given blow.

2.11. Cationic Polymerization

In cation polymerization, we will used those compound which not only generate positive ions but also effectively cooperate to perform subsequent polymerization process. Mostly cations generating group are, **(Aluminium chloride + water) or (boron trifluoride + water).** These cations group are generally very suitable to polymerization with methyl & phenyl. Mainly used to manufacturing propylene & styrene.

- Initiation stage showing decomposition of initiator, the reaction is given below.

$$BF_3 + H_2O \text{ -------- } H^+ + BF_3(OH^-)$$

This reaction liberates H^+ cations & it further joint with carbon-carbon double bond of alkene and consequently we have got stable carbocation. Consider following diagram regarding to initiation stage.

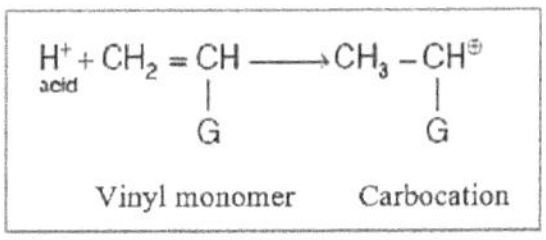

$$\underset{\text{acid}}{H^+} + CH_2 = \underset{\underset{G}{|}}{CH} \longrightarrow CH_3 - \underset{\underset{G}{|}}{CH^{\oplus}}$$

Vinyl monomer   Carbocation

Diagram 2.19: Initiation Stage

- In propagation stage (Second stage), carbo-cations further adhere to carbon-carbon double bond of next monomer molecules & this process repeat again & again, consequently we have got a propagation in chains. Consider the following diagram which shows chain propagation stage.

$$CH_3 - \overset{\oplus}{C}H + CH_2 = CH \longrightarrow CH_3 - CH - CH_2 - \overset{\oplus}{C}H$$
$$\underset{G}{|} \qquad \underset{G}{|} \qquad\qquad \underset{G}{|} \qquad \underset{G}{|}$$

$$CH_3 - CH - CH_2 - \overset{\oplus}{C}H + CH_2 = CH \xrightarrow{\text{Repeated}} CH_3 - CH - CH_2 - CH - CH_2 - \overset{\oplus}{C}H$$
$$\underset{G}{|} \quad \underset{G}{|} \quad \underset{G}{|} \qquad\qquad \underset{G}{|} \quad \underset{G}{|} \quad \underset{G}{|}$$

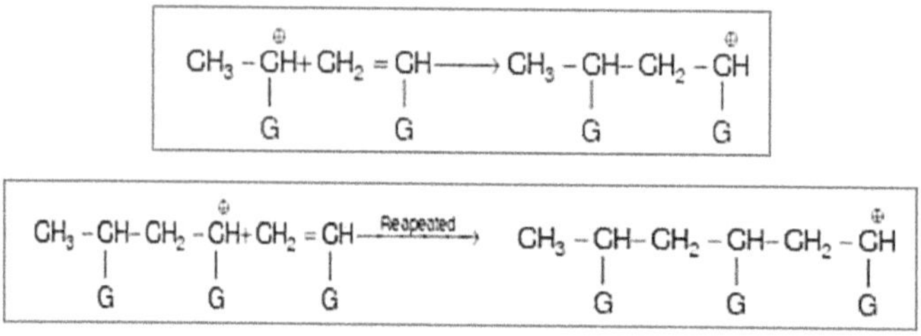

Diagram 2.20: Chain Propagation Stage

- In termination stage carbocations (Positive ions) joint to negative ions and finally we have got termination of propagated chain. Consider the following diagram which shows termination of chains.

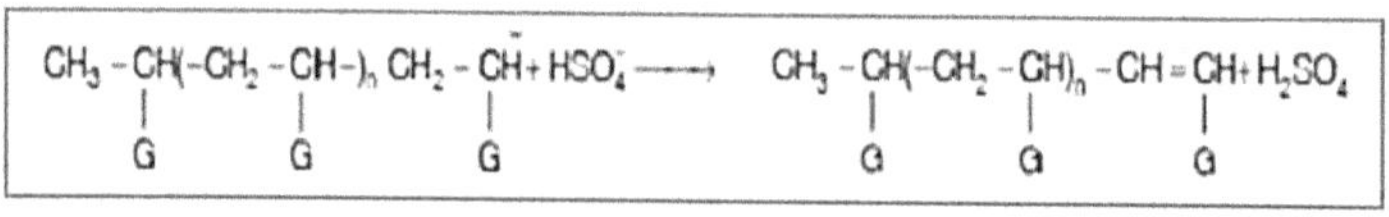

$$CH_3 - CH(-CH_2 - CH-)_n CH_2 - \overset{-}{C}H + HSO_4^- \longrightarrow CH_3 - CH(-CH_2 - CH)_n - CH = CH + H_2SO_4$$
$$\underset{G}{|} \quad \underset{G}{|} \quad \underset{G}{|} \qquad\qquad \underset{G}{|} \quad \underset{G}{|} \quad \underset{G}{|}$$

Diagram 2.21: Termination Stage

2.12. Anionic Polymerization

In this polymerization process anions or negative ions are used as initiators. To form negative ions (anions) following initiator or catalysts are used these are, **(Sodium + Liquid ammonia), Alkali, metals, alkyls, Grignard's reagents and tri-phenyl-methyl sodium.** These group are very suitable for nitrile, chloride. For example- acrylonitrile, vinyl chloride, methyl methacrylate etc.

- In initiation stage, potassium amide added to carbon-carbon double bonds of alkene as a result of which stable carbonations would be formed.

- In propagation stage, this carbonation is again added to carbon-carbon bonds of next monomer molecules & hence new carbanions would be formed. This process repeated again & again to form a chain of monomers.

- It has very special case in which polymerization technique does not have any chain termination reaction & hence it is called living polymerization technique.

2.13. Condensation Polymers

According to medical science researcher polymer exists in human life from their birth, specially condensation polymer. For example - proteins, polypeptides, starch, cellulose and DNA molecule etc. Proteins are gigantic polymeric substances formed from monomer units of amino acids, Starch and cellulose like polymeric material consisting sugar monomer of glucose, A DNA molecule is made up of the sugar deoxyribose which linked with phosphates to form the backbone of the molecule. All the above polymeric substance belongs to category of condensation polymers and known as natural condensation polymer. But now a day's human wants and evolvements of necessity moves their journey towards manufacturing of advanced polymer by synthetic method among all those synthetics process, condensation technique is one. Generally, condensation polymers are manufactured by the reaction between either between two bi-functional monomers, poly-functional monomeric units/molecules, two different bi-functional, between tri-functional monomeric units or between bi-functional & ploy-functional monomeric units. Furthermore, in such kind of polymerization, small molecular weight by product are formed like – water, ammonia, hydrogen chloride, alcohol (methanol), gases and salt which is eliminated during the process.

If we consider the case of two bi-functional monomers condensation polymerization then repetitive condensation reaction takes place between two bi-functional monomers molecules and successive condensation create bi-functional species again & again. Since each step product has distinct functional species which independent to each other therefore it is also known as a step growth polymerization. After condensation process we have got high molecular weight condensation polymer. Basically, the reaction & property of finished product depending upon **"How many functional groups of monomers involved in the reaction"**. Generally, nylon-6, Dacron, polyurethane, polyimides, polyesters, phenol-formaldehyde, nylons-6, polycarbonates nylon-66 etc are some common product which is manufactured by condensation polymerization. Consider the following diagram which shows reaction of condensation polymerization,

$$\text{H}{-}\square{-}\text{X} + \text{H}{-}\square{-}\text{X} \longrightarrow \text{H}{-}\square{-}\square{-}\text{X} + \text{HX}$$

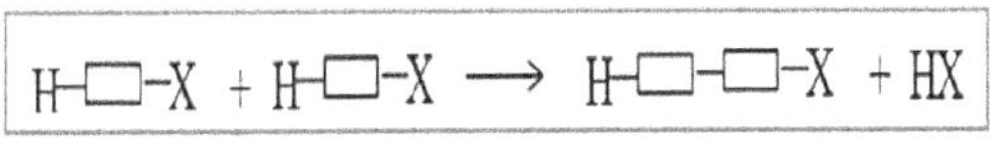

Diagram 2.22: Reaction in Condensation Polymerization

Moreover, let be consider some example of polymer product that is manufactured from condensation polymerization.

1. Dacron or poly-ethylene-terephthalate (PET) is a polyester, can be prepared by condensation polymerization in which a dicarboxylic acid reacts with a bi-functional alcohol. Reaction process for poly-ethylene-terephthalate (PET) is given below.

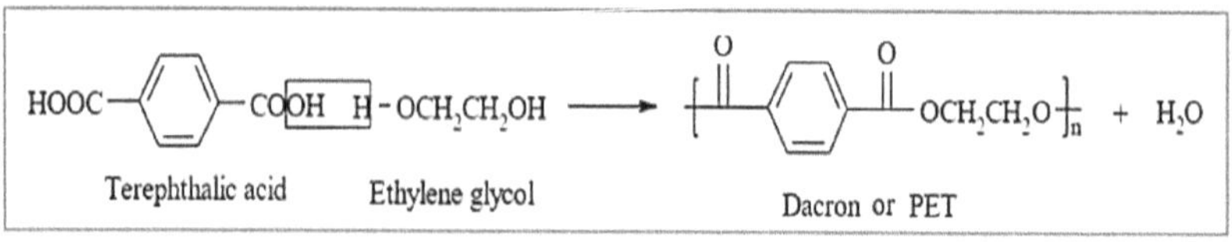

Diagram 2.23: Synthesis of Dacron / PET by Condensation Polymerization

During the polymerization of dimethyl terephthalate and ethylene glycol react then polyester by-product, methyl alcohol would be produced that is condensed off as a result of which two monomers combine to each other & producing a larger molecule.

2. Nylon 6-6 is a polyamide which can be prepared industrially by condensation polymerization in which dicarboxylic acid react with a bifunctional amine or by hexamethylene diamine with adipic acid. Consider following reaction diagram that is dedicated to manufacturing Nylon 6-6.

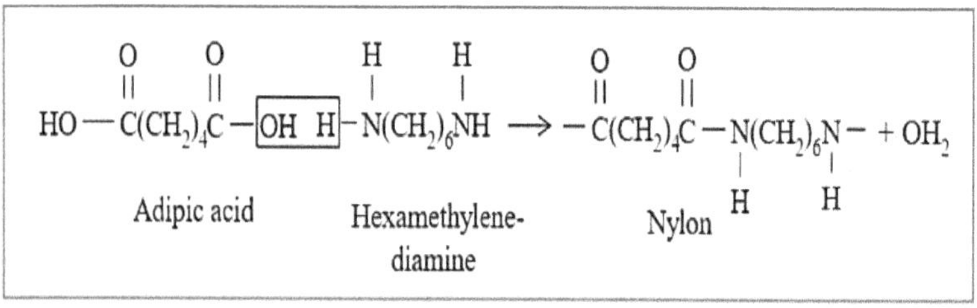

OR

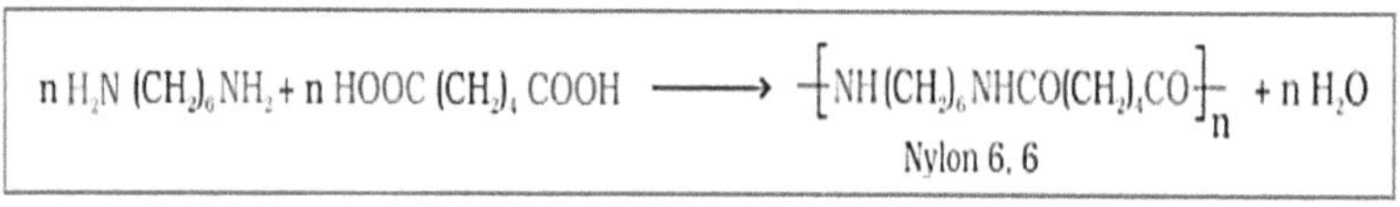

Diagram 2.24: Synthesis of Nylon 6-6 by Condensation Polymerization

In this condensation polymerization, water is formed during the reaction stage which would be a by-product & eliminated it.

3. Consider another example which shows the formation of a polyester from the reaction between ethylene glycol and adipic acid.

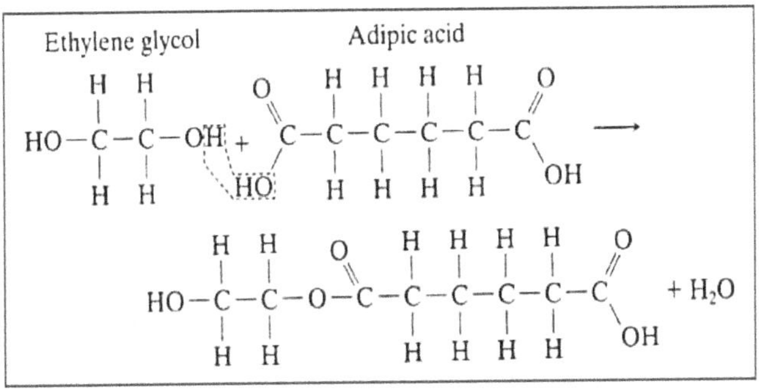

Diagram 2.25: Synthesis of Polyester by Condensation Polymerization

In this case stepwise process is successively repeated again & again and producing a long linear molecular chain & it has been observed that reaction times for condensation are larger than that of addition polymerization.

4. Polyurethane is a foamed condensation polymer that is manufactured from two liquid i.e. its part 1st has polymeric diol or triol (Glycerol), a blowing agent, a silicone surfactant, and a catalyst on the other hand 2nd part contains a poly-isocyanate (Diphenyl-methane-diisocyanate). During the polymerization reaction, presence of small amount of water decomposed to some of the isocyanate and carbon dioxide is formed. When carbon dioxide interacts with the blowing agent then it creates gas bubbles in the viscous mixture as the foam sets into a rigid mass. Further, blowing agent is a low melting point liquid which become vaporize during the reaction when it experienced a heat. The controlling of cell size & structure of the foam mostly accomplished by the silicone surfactant. Consider following general equation of reaction.

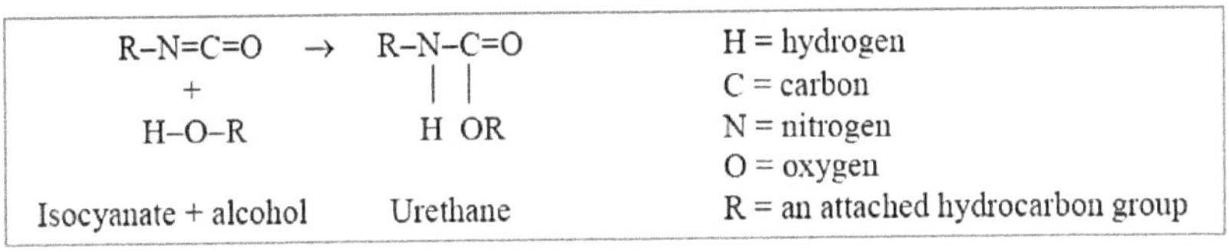

R–N=C=O → R–N–C=O	H = hydrogen
+ | |	C = carbon
H–O–R H OR	N = nitrogen
	O = oxygen
Isocyanate + alcohol Urethane	R = an attached hydrocarbon group

Diagram 2.26: A General form of Reaction for Polyurethane Manufacturing

Further the basic reaction which takes place between poly-isocyanate (Diphenyl-methane-diisocyanate) & Glycerol is given below.

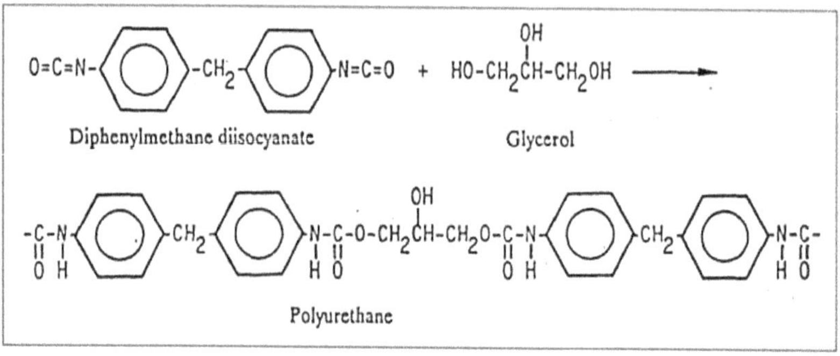

Diagram 2.27: Synthesis of Polyurethane by Condensation Polymerization

The formation of carbon dioxide is achieved by the following reaction, which is given below.

Diagram 2.28: Reaction Showing Formation of Carbon Dioxide

5. Polyamide is prepared by amide linkages between monomers, which involve the functional groups carboxyl and amine (An organic acid and an amine monomer).

6. Nylon-6 can also be manufactured by the condensation polymerisation of hexamethylenediamine with adipic acid under high pressure and at high temperature. The reaction equation for nylon-6 is given below.

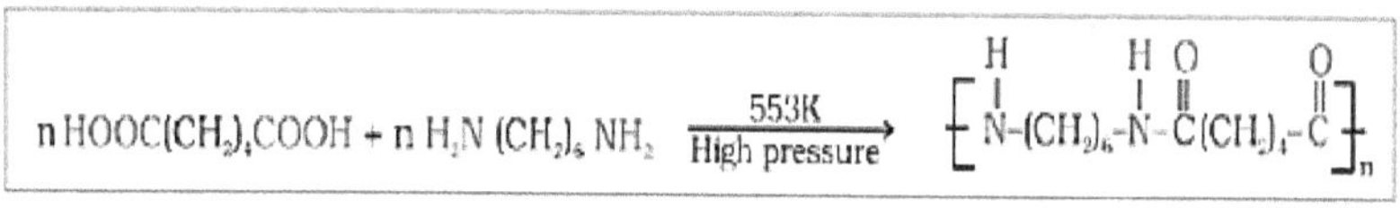

Diagram 2.29: Synthesis of Nylon-6 by Condensation Polymerization

In addition to it, addition polymerization & condensation polymerization another polymerization process may also be employed to polymer manufacturing. These processes are described briefly under here.

2.14. Coordination Polymerization

Coordination polymerization technique also falls under the category of addition polymerization technique. In this process Ziegler-natta catalysts or initiator is used to accomplishing the polymerization technique. This catalyst is mostly used to polymerization of alkene. When Ziegler-natta catalyst interacting with transition metals compound like chlorides of titanium with organic metal compound then active species imitates the polymerization among monomers of carbon-carbon double bond. For example – polymerization of propene by which we can produce poly-propene.

Note

- Brach chain structure is not found in coordination polymerization & hence formation of free radicals does not achieve.

- Long linear chain consisting active sites of the growing chain in which carbon-carbon atoms directly bonded to the metals.

2.15. Co-polymerization

It is a polymerisation reaction in which a mixture of more than one monomeric species is allowed to polymerize and form a copolymer. The copolymer can be made not only by chain growth polymerisation but also by using step growth polymerisation a. It contains multiple units of each monomer that is used in the same polymeric chain. For example, a mixture of styrene and 1, 3 – butadiene can form a copolymer called styrene butadiene rubber (SBR). A reaction in which a mixture of two (or) more monomers is allowed to undergo polymerisation is known as copolymerization. The polymer is known as copolymer.

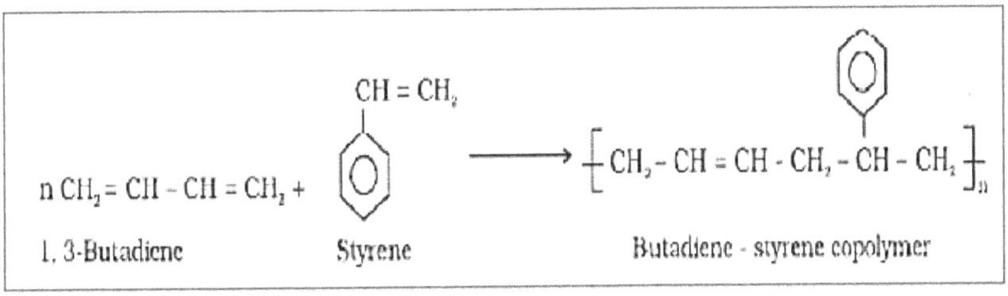

$$n \, CH_2 = CH - CH = CH_2 + \underset{\text{Styrene}}{\overset{CH = CH_2}{\bigcirc}} \longrightarrow \left[CH_2 - CH = CH - CH_2 - CH - CH_2 \right]_n$$

n CH₂= CH – CH = CH₂ +   (Styrene)   ⟶   Butadiene - styrene copolymer

1. 3-Butadiene   Styrene   Butadiene - styrene copolymer

Diagram 2.30: Copolymer-styrene Butadiene Rubber (SBR)

2.16. Application of Condensation Polymer

We have plenty of condensation polymer that has been used in diverse field of application. For example - Polyurethane is used to manufacturing a casting of objects, insulating material or sound proofing material etc. and nylon is used to manufacturing parachutes etc. Since we have so many types of condensation polymer. Consider following table which has summary of few important condensation polymer. This table also shows type of condensation polymer, their monomers and respective uses.

Table 2.16: Application of Some Common Condensation Polymer

EXAMPLE	MONOMERS	POLYMER	USE
Polyamides (Nylon)	$HOC(CH_2)_aCOH$ $H_2N(CH_2)_aNH_2$	$-C(CH_2)_aC-NH(CH_2)_aNH-$	Fibers, molded objects
Polyesters (Dacron, Mylar, Fortrel)	$HOC-\!\langle\ \rangle\!-COH$ $HO(CH_2)_aOH$	$-C-\!\langle\ \rangle\!-C-O(CH_2)_aO-$	Linear polyesters Fibers, recording tape
Polyesters (Glyptal resin)	(phthalic anhydride) $HOCH_2CHCH_2OH$ OH	$-COCH_2CHCH_2O-$	Cross-linked polyester Paints
Polyesters (Casting resin)	$HOCCH=CHCOH$ $HO(CH_2)_aOH$	$-CCH=CHC-O(CH_2)_aO-$	Cross-linked with styrene and peroxide. Fiberglass boat resin
Phenol-formaldehyde resin (Bakelite)	(phenol) $CH_2=O$	(cross-linked phenol-methylene network)	Mixed with fillers. Molded electrical goods, adhesives, laminates, varnishes
Cellulose acetate*	CH_2OH ... OH ... OH CH_3COOH	CH_2OAc ... OAc ... OAc	Photographic film
Silicones	$Cl-Si-Cl$ H_2O CH_3 / CH_3	$-O-Si-O-$ CH_3	Water-repellent coatings, temperature-resistant fluids and rubbers (CH_3SiCl_3 cross-links in water)
Polyurethanes	CH_3 $N=C=O$ $N=C=O$ $HO(CH_2)_aOH$	CH_3 $NHC-O(CH_2)_aO-$ $NHC-O(CH_2)_aO-$	Rigid and flexible foams, fibers

2.17. Characteristics of Condensation Polymerization Technique

The following are some important features of condensation polymerization technique.

- The polymer chain forms slowly, it requires several hours or several days.
- All of the monomers are quickly converted to oligomers, thus, the concentration of growing chains is high.
- Since most of the chemical reactions employed have relatively high energies of activation, the polymerization mixture is usually heated to high temperatures.
- Step-reaction polymerizations normally afford polymers with moderate molecular weights, i.e., <100,000.
- Branching or crosslinking does not occur unless a monomer with three or more functional groups is used.

Note

- Whenever polymerization process takes place among the monomers of one functional group then one reaction group can terminate a growing chains and final finished polymers product showing lower molecular weight.
- If reaction between monomers of two reactive group takes place then linear structure polymer would be formed on the contrary if such a reaction achieved more than two end group then cross-linked three-dimensional polymer would be formed.
- Condensation polymerization only occurs in those monomer molecules that has reactive functional group, such as, - OH, - COOH, - NH_2, - NCO etc.
- If condensation polymerization including more than one monomer species then it is known as step growth polymerization.
- Usually monomer molecules, oligomer molecules and high molecular weight intermediate that has complementary functional end units are take parts in poly - condensation reaction.
- For manufacturing high molecular weight polymer, then high yield condensation reaction always imperative.

2.18. Molecular-weight & Molecular Weight Distributions

Molecular weight & molecular weight distribution in polymer perform a dominating role not only in polymer synthesis but also cooperate to evaluate the physical property, chemical property and mechanical property of polymer material. We have number of synthetic process by which polymer product or polymer material can be manufactured. These polymer materials have remarkable property like toughness, tensile strength, adherence and environmental resistance, impact strength or melt viscosity which completely governs to molecular weight & molecular weight distribution. In actual practise polymer is nothing but it is mixture of molecules that has different molecular weight so therefore we can evaluate only average accumulated value of molecular weight rather than fixed conformed value. Furthermore, we know that polymer has several repeating units of monomers in the molecules and hence we can say that average molecular weight always coincides to average number of repeating units called **number average molecular weight** or sometimes say average weight that is known as **"weight average molecular weight"**.

Suppose "n_i" say chain length which consisting number of repeated units of in any polymer molecules which also including units at chain ends aswellas at branch points, then molecular weight of polymer for single repeated unit can be given as,

$$M_M = n_i \cdot M_S$$

(1)

$$M_M = n \cdot M_S$$

(2)

For single repeated unit

Where,

MM = Molecular weight of polymer molecule

n = Chain length in molecule

MS = Molecular weight of single repeated unit

2.19. Calculation of Number Average Molecular Weight

Let be consider that "n_i" is a macromolecules for each degree of polymerization of Pi. For obtaining accumulated degree of polymerization we will sum up the product of number of polymer molecules and degree of polymerization at each molecule that is given below.

$$= \sum_{i=1}^{\infty} n_i P_i$$

(3)

Dividing the equation -3 (which is a summation) by total number of monomers (M_N). Then we have a mathematical relationship that can be used to determine number average molecular weight.

$$W_{na} = \frac{\sum_{i=1}^{r} n_i \, p_i}{M_N} \tag{4}$$

Where,

Wna – Number average molecular weight

MN – Number of monomers units

2.20. Calculation of Weight Average Molecular Weight

Let be consider that Wi is the weight of a macromolecules or species for that degree of polymerization for each macromolecules is n_i, then total weight Wt can be calculated by the following relationship.

$$\text{Total weight of polymer} = Wt = \sum_{i=1}^{x} Wi \tag{5}$$

Where,

Wt = Total weight of polymer

Wi = Weight of each macromolecules or species

For accumulated weight of degree of polymerization for "n_i", we will have to put I = 1, 2, 3….

Further, molecular weight of chain length that is equal to "**n**". For obtaining accumulated degree of polymerization we will sum up the product of weight of polymer molecules and degree of polymerization at each molecule that is given below.

$$= \sum_{i=1}^{x} Wi\, Mi \tag{6}$$

Finally, weight average molecular weight cab be given by the following relationship,

$$\text{Weight average molecular weight} = W_{am} = \frac{\sum_{i=1}^{r} Wi\, Mi}{Wt} \tag{7}$$

Where,

Wam = Weight average molecular weight

Wt = Total weight of polymer

Wi = Weight of each macromolecules or species

2.21. Calculation of Weight Distribution in Polymer

Consider following mathematical relationship.

$$Wn = [\ Wi\ /\ Wt\ ]$$

Where, Wn is defined as the ratio of mass of polymer to degree of polymerization (DP) that is equal to **n** or molecular weight = n. Ms.

2.22. Calculation of Weight Average Chain Length (WACL)

Weight average chain length can be calculated by the following relationship.

$$W_{acl} = \frac{W_{am}}{M_s} = \frac{\text{Weight average molecular weight}}{\text{Molecular weight of single repeated unit}} \tag{1}$$

It must be remembered that if molecular weight distribution is very narrow then the number average and weight average molecular weights must be equal on the other hand when the distribution is broad then the molecular weight distribution is given as,

Molecular weight distribution =

$$\frac{W_{am}}{M_s} = \frac{\text{Weight average molecular weight}}{\text{Number average molecular weight}} \tag{2}$$

Note

- It has been observed that tensile & impact strength increases with rise of molecular weight of polymer.
- If polymer has high molecular weight then melt viscosity rises linearly with respect to low molecular weight polymer.
- Molecular weight distribution also a play a vital role because it definitely influences the property of polymer material or product.
- Molecular weight of polymer is a sum of the molecular weight of **mers** in the molecules, that can be given as, Molecular weight of polymer = n X [Molecular weight of each repeating unit of polymer] Where, n = Chain length
- At room temperature, short chain polymer & intermediate chain polymer have molecular weight between 10^4 to 10^7 g/mol.
- Consider following table that shows degree of polymerization (DP) & molecular weight of polymer (MW) for some common polymer material.

Polymer material	Degree of polymerization (n)	Molecular weight
Polyethylene	10,000	30,000
Polyvinylchloride (PVC)	1500	10,000
Nylon	120	15,000
Polycarbonate	200	40,000

2.23. End-group Analysis

A very simple method of measuring polymer molecular weight is to count the number of molecules in a given polymer sample. The product of the sample weight and Avogadro's number when divided by the total number of molecules gives the number-average molecular weight. This technique is best suitable for linear molecules which has two reactive end groups. Consequently, linear condensation polymers made by step-growth polymerization and possessing carboxyl, hydroxyl, or amine chain ends are logical candidates for end-group analysis.

2.24. Property of Polymer Material

Like metals, ceramic, composites and other relative materials polymer material also has been utilized in diverse field of application. Now a days these polymer products are being used in engineering aswellas non-engineering application both because they exhibit mechanical property, physical property, chemical property, optical property and electrical property etc. For example – low density between 4 to 5 (Less than metal & ceramics), high weight saving, high strength say plastic (Used in automobile, marine & aero-space application), high compressive strength, high strength to weight ratio, easy processability, corrosion resistance, low tensile strength, low creep, resistance to chemical & water, tough & high strength (Acetal, polycarbonate, nylon, polyester), performance ranges from low to high **[Commodity plastic (PE,PP,PS.PVC) < Intermediate plastic (PMMA, ABS) < Engineering plastic (PC, nylon, PPS) < Advance high performance plastic (LCP, PEEK, PES)].** Further good mechanical property (tensile strength, compressive strength, flexural strength, hardness, creep, fatigue & impact resistance), brittle behavior at room temperature or below it (say polystyrene, poly (methyl methacrylate) and many phenol-formaldehyde resins), hardness & toughness property (Cellulose acetate), rigidity (PVC & polystyrene polyblends), electrical insulator, excellent die-electric property (Work as good insulator), showing electrical conductivity in glassy state (Between 10^{-13} to 10^{-19} ohm^{-1}), insulator for high frequency application (Radar & microwave application), low thermal conductivity, low stiffness, low tensile strength in some polymer, low hardness than that of metals & ceramics, moderate ductility, high coefficient of thermal expansion, low melting point, high specific heat than that of metals & ceramics, flammability, indicate thermal & mechanical behaviour (Depending upon temperature, strain rate & morphology), stress & strain behaviour [brittle, plastic & highly elastic (Elastomer & rubber), brittle nature (High crystalline polymer), plastically deformed (Amorphous polymer), tensile modulus & tensile strength lower than metals, (hardness, fatigue resistance, % elongation & impact resistance) [Depending upon the molecular weight & crystallinity], visco-elastic nature (In elastomers due to cross linked structure), excellent chemical & electrical resistance, optical property (radiation absorbs), translucent & opaque (Crystalline polymer) and so forth. The above notable attributes are attracting our mind about some property. Further I will try to demonstrate some possible properties of polymer in brief.

2.24.1. Mechanical Property of Polymer

It has been observed that where ever, any material is used to those articles / parts which become a parts of any machine components then study of mechanical behaviour always on prime consideration. In similar manner when never polymer material employed for such an application then we also analyse its mechanical properties. If polymer apply as an engineering material then it should have tensile strength, compressive and flexural strength, hardness, creep resistance, fatigue resistance and impact resistance etc. Designer or design engineer if apply polymer material for any kind of engineering application they must have to assimilate their mechanical behaviour. Like metals or any other engineering material, we can also draw a stress & strain curve for polymer material by which we can calculate ultimate tensile strength, modulus of elasticity, yield point stress & other relative parameters. Consider following general diagram which is being drawn between stress & strain that is dedicated to polymeric material known as stress-strain diagram or curve.

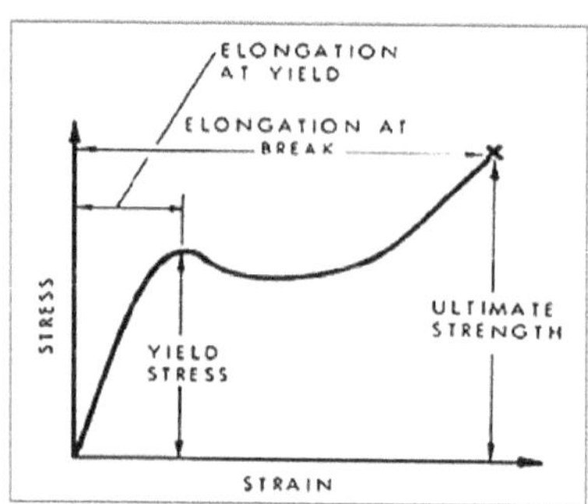

Diagram 2.31: Stress & Strain Curve for Some Common Polymeric Material

In such an analysis, we will have to take a test specimen upon which tensile load is applied until up to, it does not fracture or break. At initial stage there is no initial stress & strain in polymer specimen. when load is applied, we have straight line portion of curve which indicate stress is proportional to strain that is known as hook's law.

$$\text{Stress } \alpha \text{ strain}$$
$$\sigma \, \alpha \, \delta$$
$$\sigma = E. \, \delta$$

Where,

E is elastic modulus

σ is applied stress

δ is strain

Up to straight line portion, polymer remain elastic means after relieving the load it returns back to its original shape and we can calculate the value of young modulus or modulus of elasticity which is a measure of stiffness. Moreover, we also have an information about yield point stress of polymer (A point from which yielding of polymer is begin), maximum tensile strength or ultimate tensile strength at which breaking or fracture takes place. An area under the curve from origin to breaking point / fracturing point is known as toughness of the polymer material. Furthermore, it has been observed that each & every polymer material indicate different mechanical behavior whenever analyses is performed by strain & strain curve diagram, some important outcomes regarding to polymer is given below.

- Below transition temperature (T_g), all the amorphous polymer material showing hardness & brittleness. Initial slope of curve exhibit very high modulus whereas further portion of the curve showing moderate strength, low ultimate tensile strength, low elongation or breaking point & also low toughness.
- Usually some polymer material showing hard and brittle behavior at room temperature or below the room temperature. For example - Polystyrene, poly-methyl-methacrylate (PMMA) and many other phenol-formaldehyde resins.
- In case hard & strong polymers it has been observed that it has high modulus of elasticity, high strength and low elongation. Such a behaviour is identified by Poly vinyl chloride (PVC) & polystyrene-polyblends.
- Cellulose acetate, cellulose nitrate & nylon showing hardness & toughness both. During tensile testing of above, we have got high yield point stress, high modulus, high strength & large elongation.

- Soft & low toughness polymer showing low modulus, low yield point stress value, moderate strength & high elongation. Such a behavior is noticed in plasticized PVC, rubber and elastomers.

- Consider following stress & strain curve which is dedicated to hard & brittle, hard & strong, hard & tough and soft & tough polymer material. This curve shows various parameter like elastic modulus, strength, ultimate tensile strength and breaking points.

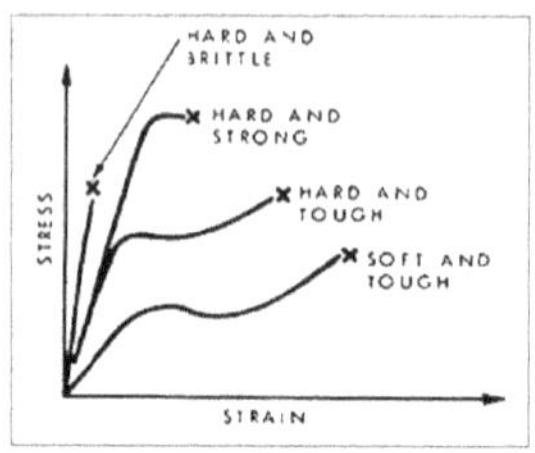

Diagram 2.32: Stress Strain Curve for Four Type of Polymer Material

In most cases, mechanical behavior or stress & strain behavior either brittle, plastics or highly elastic. It has also been observed that polymer mechanical behavior depending upon the temperature. For example - Some polymer material indicates brittle nature at low temperature on the other hand it represents rubber like behavior or characteristics at high temperature. In another example - High crystalline polymer indicates brittle nature whereas amorphous polymer indicates plastic deformation at low temperature. Moreover, cross-linked polymer has capacity to recover deformation even after very high strains, such a phenomenon also represented by the elastomer polymer. Consider the following stress & strain curve that is dedicated to polymer material such brittle plastic material, general plastic material and elastomeric materials.

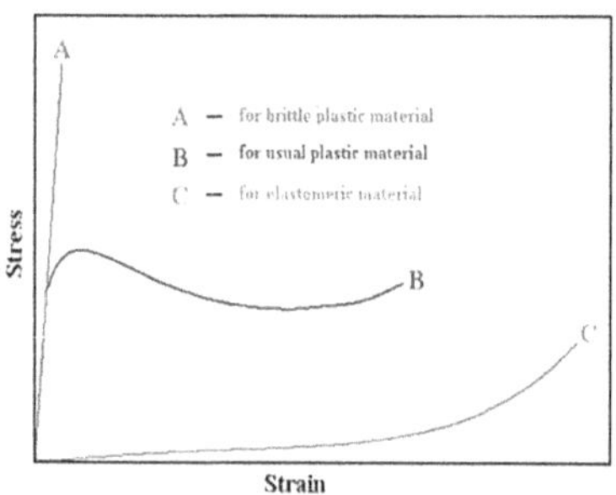

Diagram 2.33: Stress Strain Curve for Brittle Plastic, General Plastic & Elastomer Polymer Material

Conclusion

- For brittle plastic material it breaks at low stress value with little elongation.
- For general plastic material it shows parameter like – elastic modulus, yield point stress and ultimate tensile strength.
- In case of elastomeric material, it indicates high elongation, high modulus of elasticity and high ultimate tensile strength.

Consider stress & strain behavior of another glassy polystyrene that is dedicated to tension aswellas compression analyses.

The curve is generated from origin point say "O" at which no initial stress & strain is found in test specimen. Whenever polymer material is subjected to tensile load a little sloped portion of curve is straight line up to which it obey hooks law say stress is proportional to strain and further loading tends to sudden fracture or breaking of test specimen which will indicate brittle fracture and respective strength is known as ultimate strength of given polymer. To understand it considers following diagram.

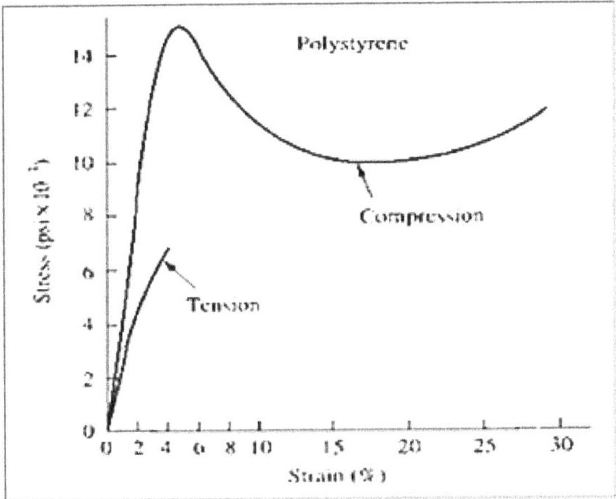

Diagram 2.34: Stress & Strain Curve for Dedicated to Glassy Polystyrene Under Tension & Compression

On the contaray, it has been found that in case of compression application, it indicate higher compressive strength. Furthermore, it also has been concluded that strength in compression always higher than that of tension. Eventually, we can say that glassy polystyrene is best suitable for compression rathar than tension.

Note

- With the rising of temperature, both rigidness & yield strength decreases where as elongation improves because temperature rise impart softening & ductility in polymer material.

- Molecular weight & crystallanity affecting the various mechanical properties of polymer material like – hardness, fatigue resistance, elongation & impact resistance.

- For augmenting brittle strength, it is imperative that molecular weight must be improved & reducing crystallinty.

- If cryastallinity decreases with temperature then stiffness, hardness and yield point strength decreases for polymer materials.

- Elastomers deform elstically & represent low modulus, high deformation, high reversibile elongation, at low value of stress due to their cross-linked structure.

- Elastic modulus of elastomers increases with rise in temperature.

- Most of the polymer indicate visco-elsticity means whenever external force is applied both elastic & plastic deformation takes place. Further visco-elastic behaviour of polymeric material always depending upon the temperature.

- We know that plastic polymer consisting weak secondary bonds. whenever tensile stress is applied upon polymer, it deform elstically due to elongation in chain molecules & exapansion in strong covalent bonds.

- Thermoplastic may be fractured in dutile & brittle manner and deforamation of plastic material may either elastic, plastic or combination of both and strength of thermoplastic strongly depending upon the, average molecular mass, degree of cryastallisation, presence of side group and presence of phenyl rings in main chain and sometimes addition of reinforcement.

- Degree of cryastallinity also affecting the mechanical behaviour of thermplastic. If cryastallinty is improve then strength modulus and density also rises.

- For manufacuring high strength engineering polymer, phenyl ring is introduced in main chain of thermplastic polymer.

- The mechanical strength of thermplastic material can be augmented by reinforcing either glass fiber, E – glass fiber, S- glass fiber, carbon fiber, kevlar fiber (kevlar-29, kevlar-49).

- Young modulus or tensile modulus or modulus of elasticity for any polymer material is defined as the ratio of stress to strain which also indicate stiffness of material. It is given by the following relatinship.

$$E = [\text{ Tensile stress } (\sigma) \text{ / Tensile strain } (e)]$$

- Liquid like polymer indicate shear property rather than tensile property.

- A selection of polymer depending upon the application for example - structural application & non-structural application. For structural application property like – tensile strength, stiffness, impact strength and chemical resistance are always desirable on the other hand for non-structural application, surface finishing, painting facility, resistance to humidity, crack resistance and ultraviolet resistance ae imperative.

In addition to tensile & compressive property, polymer material also indicates creep resistance, hardness, abrasion resistance, impact resistance & fatigue resistance etc. Let be consider definition of above property.

- **Fatigue:-** Fatigue is nothing but it is a phenomenon in which test specimen subjected to fluctuating loading. During fatigue test of polymer test specimen subjected to tensile & compressive stress. Fatigue resistance depending upon the rigidity. if rigidity is higher, than fatigue resistance would be lowered.

- **Impact Strength:-** Impact strength indicate measurement of toughness or resistance to breakage/ damage when test specimen is subjected to high velocity impact loading, for impact strength polymer material should have toughness. Polystyrene, poly-methyl-methacrylate (PMMA) and non-plasticized PVC showing impact strength or impact resistance. Furthermore, toughness of polymer material also depending the temperature & respective deformation during rate of deformation. Basically, impact resistance of polymer & other plastics governs to glass transition temperature (Tg) & crystallization. For example – Below the glass transition temperature, amorphous polymer break in brittle manner when it is subjected to impact loading on the other hand it become tougher & showing impact strength at the level of transition temperature. Above the transition temperature, amorphous polymer indicates rubbery state at which no impact resistance or impact strength is achieved because as temperature increases then rigidity decreases.

- **Hardness:-** Hardness is nothing but it is a resistance to scratching, abrasion and indentation. Further we can also define as a resistance of polymer material to plastic deformation or permanent deformation say bending, breaking and shape changing when load is applied.

- **Micro-hardness:-** A resistance to deformation, scratching, abrasion and indentation when polymer material to subjected to low load. It is generally evaluated with the help of Vickers & Knoop hardness test. Hardness is only identified in number, during micro-hardness test we will use load of 160 gram to 5 KG. Hardness depending upon the temperature & crystallinity, if degree of crystallinity decreases with temperature than hardness reduces.

Mechanical behavior of polymer material is different for metals, ceramics and other relative material due to following reasons.

- Difference in structure.

- Size of entities, force among molecules, force between monomers are also responsible for deformation during the evaluation of mechanical behavior.

2.24.2. Electrical Property of Polymer

We exactly know that conduction of charge always considered as an electrical property but the opposite notion may also be considered as an electrical behavior of any material. This property is recognized as a "Non-conducting behavior of material". Like glass, wood, bio-composites, ceramics material, "Polymer material also indicating non-conducting behavior" and behave as an insulating material. Except the electrical insulating material characteristics, polymer material also exhibits some other property like – Die-electric property, Die-electric polarization, electrical conductivity, permittivity, dissipation factor, die-electric loss and so forth. Further, all those parameters regarding to electrical property are briefly described under here,

1. **Die-electric Property:-** It has been observed that polymer material is insulator but they have capacity to store electrical energy. Whenever polymer non-conducting material is kept in electric field then they act as or work as storage container of electrical charge in such a condition it is known as dielectric and this characteristic is known as die-electric strength. Further die-electric strength is defined as,

"It measures the value of highest current that can be applied to a plastic before it allows current to pass".

Now a day's polymer like PTFE, PE, PVC, EP and MF are widely used as die-electric material in various type of applications. These can be manufactured at low cost and also indicate easy processability.

Notes

- Die-electric strength is measured in voltage.
- Polymer showing excellent die-electric strength property and it is noticed that die-electric material always having a good insulating property.
- Die-electric strength depending upon the temperature & thickness of polymer material and voltage application.
- Polymer material is used as insulating material by which we can insulate wires, electrical cables and other electrical & electronics equipment's.
- Polymer are promising candidates for many applications because of ease in processability, physical property, cost effectiveness and manufacturing of product according to design specification.
- Polymer material also indicate excellent chemical resistance.
- Polymer material also represent good optical properties & hence it can be used in other engineering applications.
- Polymer material die-electric strength may be as high as 1000 MV/m and the upper limit of die-electric strength is maintain by ionization energy of electrons in covalent bonds that exists throughout the polymer structure.

2. **Die-electric Constant:-** Like mica, inorganic substance and silicon dioxides, polymer material having a capacity to store electrical charge or current thereby we can develop an electrical field without any loss of energy or sometimes negligible loss. Further die-electric constant is a dimensionless parameter that is defined as,

"It is a ratio between applied external electrical field to stored die-electric charge/field inside the any polymer material".

Mathematically we can write as,

$$\varepsilon = [\, V_a / V_s \,]$$

Where,

"ε" = Die-electric constant

V_a = Applied electrical filed

V_s = Stored charge

Furthermore, die-electric constant may also be under stand as,

"How polymer or other plastic material can be polarized relative to vacuum".

For any capacitor die electric constant may also be defined in following manner.

"It is a ratio between capacity of electric capacitor filled to capacity of same capacitor in vacuum at definite external field frequency".

$$\text{Die electric constant} = \frac{\text{Capacity of electric capacitor in ambient condition (Filled with polymer)}}{\text{Capacity of electric capacitor in vacuum condition (Filled with polymer)}}$$

Note

- Die-electric constant is a dimensionless parameter.
- Die-electric material are widely used in high frequency application where performance & efficiency affected by temperature, moisture & frequency aswellas it is also depending upon the thickness.

3. **Electrical Conduction:-** We know that polymer material is insulator material however it can also be used as a conductor. It is possible with those polymers which has low molecular weight & having certain impurities. Further we also know that for charge conduction presence of free electrons & ions are imperative throughout the material. Further electrical conductivity can be mathematically expressed as,

$$X = A \cdot e^{-\Delta U / R.T}$$

Where,

X = Electrical conductivity of polymer

A = Constant (Depending upon the temperature)

R = Universal gas constant

ΔU = Activation energy

Note

- In glassy state, conductivity of polymer ranges between 10^{-13} to 10^{-19} ohm^{-1}.
- If temperature rises, then conductivity of polymer increases exponentially.

4. **Dissipation Factor:-** Dissipation factor is nothing but it indicates amount of energy dissipation when current is flow through polymer material. It is defined as the ratio between energy loss (As heat) to current flow.

$$\textbf{Dissipation factor} = \frac{\text{Energy loss as heat}}{\text{Amount of current flow}}$$

Note

- For polymer material dissipation factor must be low in high frequency application (For example – Radar & micro-wave) where it is used as insulator.
- Dissipation factor measured in megahertz (MHz).

5. **Die-electric Loss:-** It is a measurement of consumption of electrical energy in the form of heat which dissipated & could not be regain. Basically, such a loss in polymeric material comes on the front line due to displacement of chain which has plenty of monomers unit.

6. **Volume Resistivity:-** It is a measurement of conductivity when direct current is applied across a polymer material. Further it can also be measured as,

"How a plastic surface offer resistance to current flow when surface having contamination or moisture".

7. **Polymer Behavior in a Steady DC Electric Field:-** Electrical property of polymer or any other material is defined as, its response when they experienced an electric field of various strength & frequencies. Similar to other insulator material, polymer offers resistance to electric current and follow to ohm's law. The value of current can be calculated as,

$$\text{Current flow (I)} = \frac{A.E}{R}$$

Where,

I = Current flow in ampere,

E= Electric field strength

A = Cross section area of conductor

R = Resistivity in ohm

Insulating property or resistivity exists in polymer due to valance electrons among covalent bonds between pairs of atoms. However, it has been noticed that small current flow may occurs because of structural defects, impurities or other contamination involvements. Furthermore, resistivity can be reduced if concentration of impurities increases, in this condition resistivity also depending upon the temperature, if temperature rises then it will reduce. Moreover, when polymer material interacts either to ionizing radiation, absorption of water or inclusion of plasticizers then we have got enhancement in charge carriers' concentration & consequently conductivity may improve. But in actual practice, no steady current flow occurs in insulator in static electric field however energy storage takes place due to die-electric polarization. Polarization occurs in all the materials achieved either by movement of electrons with in the atoms. But permittivity plays a dominating role which regulates the force that acting between pairs of electric charge.

In case of DC electric field, permittivity depending upon the bonds in structure. Polymer such as polyethylene, polypropylene and poly-tetra-fluor-ethylene has no di-poles and bonding & non-bonding electrons are very tightly arranged and hence slight die-electric polarization occurs as a result of which low permittivity represent nonetheless polar polymer such as poly-methyl-methacrylate (PMMA), polyvinylchloride (PVC), and polyvinyl fluoride (PVDF) exhibits high value of permittivity.

2.24.3. Physical Property of Polymer

Like mechanical, thermal & electrical property, polymer physical property also play a significant role in the class of industrial / engineering material for manufacturing various type of product or articles. These physical properties also help to identifying their utility & usefulness for different kind of applications. Molecular weight, melting point temperature, glass transition temperature, density, specific gravity, thermal expansion, specific heat, thermal conductivity, strength, resistance to plastic flow, hardness, etc., are some important physical property of polymer materials. Molecular weight of polymer is nothing but it is a product of the degree of polymerization & the molecular weight of repeating units. Degree of polymerization in polymer molecules indicate the number of repeating units in the polymer chains. Molecular weight always calculates as average molecular weight rather than confirm fixed molecular value of weight. Basically, a reason is that polymer molecules are composition of many species which has variation in degree of polymerization and hence polymer may either high molecular weight polymer or low molecular weight polymer that has different crystal structure & consequently showing different physical properties and hence evidently it can be used in diverse field of application. One of the most important property comes on the front line is that high molecular weight polymer indicates high melt viscosity rather than low molecular weight polymer. The decreasing order of molecular weight of some common polymer material is given below.

Poly-carbonate > Polyethylene > Nylon > Polyvinyl chloride (PVC)
(40,000) (30,000) (15000) (10000)

Furthermore, consider some high molecular & low molecular weight polymer.

- **Rubber:-** Rubber is nothing but it is an elastomer which has high molecular weight whose structure consisting long flexible chain & low intensity molecular forces. For example - Natural & Synthetic rubber. Rubber are widely used to manufacturing vehicle, bike, bi-cylices tires. Elastomer are amorphous polymer material, whenever elastomer subjected to tensile force then elongation takes place as a result of which amorphous regions become straight & up to elastic limit it will touch the semi-crystalline state. And after releasing tension contraction occurs & elastomer return back to original state. Means we can say that it does not occupy the plastic state. Such a case obtained in natural rubber like – [Cis-poly-2-mehyl-1], [3- butadiene] and [Cis – polyisoprene].

- **Plastics:-** Plastic is tough & strong polymer material which has high molecular weight. Under this category polyethylene, polypropylene, PVC, polystyrene is most common.

We know that plastics are organic materials of high molecular weight, which can be molded into any desired shape by the application of heat and pressure. Now a days plastics gained popularity because of some specific property that is attributed as,

  - Lightness in weight & specific gravity ranges between 1 to 2.4.
  - Good thermal and superior electrical insulation (They possess very low thermal conductivity and high electrical resistance).
  - Corrosion-resistance (Plastic has high resistant to corrosion).
  - Easy workability (Like casting, molding, drilling, sawing, machining, etc. of plastics can perform easily).
  - Chemical Inertness (Plastics indicate inertness to lights, oils, acids and dampness).
  - Transparency (Some plastics are highly transparent and translucent. They can be ground and used as optical lenses).
  - Low softening point (Most of the plastics have low softening points, even as low as 50°C).

- **Polyvinylchloride (PVC):-** It is a high molecular weight plastic polymer.

Properties: PVC is a colourless, odourless, no inflammable and chemically inert, resistant to light, oxygen, inorganic acids and alkalis, but soluble in hot chlorinated hydrocarbons like ethyl chloride. Pure resin possesses a high softening point (148°C) and greater stiffness as compared to polyethylene. It is the third most widely used synthetic plastic after polythene and polypropylene.

Uses: PVC is used for making sheets, light fittings, safety helmets, cycle and motorcycle mudguards, pipes and pipe-fittings, chemical container, frames for doors and windows, debit and credit cards, etc.

- **Polystyrene:-** It is a high molecular weight plastic polymer.

Properties:- Polystyrene is a transparent, light, light stable, excellent moisture resistant polymer. It is highly electric insulating, highly resistant to acids and good chemical resistant. It's softening point ranges between 90°-100°C and showing brittle nature. It has a unique property of transmitting light through curved sections.

Uses: Polystyrene is used in protective packaging and containers as Styrofoam, toys, disposable food container. They are also used as insulators. They are used in making lenses, disposable cutlery (like spoons, forks, plates, trays), making radio, tv & refrigerator parts, CD cases, etc.

- **Nylon-6,6:-** It is a high molecular weight plastic polymer.

Properties: Nylons are translucent, showing high melting point temperature (160° to 264°C) polymers. They possess high temperature stability and good abrasion-resistance aswellas having self-lubricating property.

Uses: It is used for ball bearing cages, electro insulating elements, pipes, profiles and various machine parts. Other applications include carpet-fibers, apparel, airbags, tyres, zip ties, ropes, conveyor belts, under-garments, dresses, carpets, hoses and the outer layer of turnout blankets.

Consider following table which shows some common high molecular & low molecular weight polymer with their uses.

Table 2.17: Some Common High Molecular Weight Polymer & their Uses

High molecular weight polymer	Uses
Poly-carbonate	Safety helmets, lenses, bullet resistance window, electrical insulators, wind shields etc.
Polyethylene	Flexible bottles, toys, tumblers, battery parts, ice-trays
Nylon	Bearings, gears, cams, bushings, handles, jackets for wire & cables etc.
Ply-vinyl-chloride (PVC)	Sheet, light fitting, safety helmets, mud guards, pipes, chemical containers, debit & credit cards (ATM-cards), doors & window frames, foot ware, wire & cables coating
Poly-propylene	packaging films, TV cabinmates etc.
Polystyrenes	Wall tiles, battery cases, toys, lighting panels, house appliances, radio-components, medical devices, pens, safety razor, house wares, pens etc.
Natural rubber	Vehicles tires, gaskets, hoses
Synthetic rubber	Vehicles tires, shock absorbers, seals & belts.

Furthermore, low physical property polymer (Like - plastic) are generally used to manufacturing consumable goods such as plastic forks, knives, plastic cups & plates, containers and trash bag etc. In this class polyethylene, polystyrene and polypropylene are very famous.

Moreover, physical property always governs to some important parameters like – crystallites & amorphous regions, degree of flexibility of chain, cross-links and force acting between chains (Dispersion force, hydrogen bonding etc.).

1. **Amorphous Polymer:-** An amorphous polymer is one which has no crystals. If the bonding between chains are weak then low tensile strength would be achieved beyond the low tensile strength limit, polymer adopt plastic flow during which chain slip occurrence take place.

2. **Un-oriented Crystalline Polymer:-** In such a polymer crystallites are randomly oriented throughout the structure as a result of which it requires high melting temperature for melting the crystallites at this high temperature polymer become amorphous & undergoes plastic flow and further facilitate to molding due to existence of viscosity.

3. **Oriented Crystalline Polymer:-** Oriented crystalline polymer is one in which crystallites are aligned or oriented in specific direction rather than random orientation. For example – polymer produced by cold drawing process. To accomplish such an orientation, polymer material is heated above the transition temperature (Tg) and then stress is applied in one specific direction so that plastic deformation may occurs as a result of which all the un-oriented crystallites become oriented along the direction of applied stress. In such a case we have high tensile strength. Consider an example of poly-propene, which is available in three forms. Among three, two are highly crystalline & one has elastic & amorphous nature. These polymers have three form isotactic, Sydio-tactic and atactic polymer.

   - Atactic material is soft and elastic in nature.
   - Isotactic polymer is hard, clear, strong and having a melting point of $175^{o}C$.

Physical property of polymer also depending upon the force which acts between the chains & hence we can say that polymer material may either be amorphous or semi-crystalline in nature. For example – polyethene that is made from molecules of 1000 to 2000 CH_2 group has continuous chain, exhibit crystalline characteristics in which chain of CH_2 grouped oriented with respect to one another likewise a chain in crystalline, low molecular weight hydrocarbons.

Such a crystalline regions known as crystallites. Further between the crystallite's regions of polyethene, amorphous & non-crystalline regions, exists throughout which polymer chains are randomly distributed.

Note

- Force among the crystallites chain of polymer may either be Vander Waals or dispersion force. These forces are relatively weak.
- The inter-molecular force which acts between pairs of hydrogen bond in the crystallites of polyethene ranges between 0.1 to 0.2 Kcal/mol/pair on the other hand crystalline segments of 1000 CH_2 / unit, the sum of these interaction must be greater than that of C-C bond strength.
- For high molecular weight polymer such as nylons, nylon-66, strong molecular force can be attained by hydrogen bonding.

Temperature always influenced the intensity of bond strength either it may be amorphous or semi-crystalline. The amorphous regions of polymer is achieved at low temperature throughout which polymer molecules only vibrate rather than any kind of displacement or movement of molecules. Such a state of polymer is known as glassy state. In such a conditions polymer indicate, brittleness, hardness and rigidness as like a glass on the other hand if it is heated then force between the polymer chain reduced as a result of which polymer become soft and flexible similar to rubber and such a state is called rubbery state. We know that temperature variation always affecting the physical property of polymer which further leads to affect their usefulness or use. At low temperature, polymer remain hard & showing glass like structure. at temperature below or equivalent temperature at which glass like structure is achieved is known as glass transition temperature (Tg). But whenever polymer heated then crystals become melt, this temperature referred as melting point temperature (Tm). At this temperature a physical property known as viscosity comes on the front line which also define to molding temperature or workable temperature. Another temperature comes on the front line, that is known as decomposition temperature at which thermal breakdown of polymer chain takes place. Decomposition may either be due to presence of impurities, inhibitor or antioxidants. Existence of such a contamination in polymer not only leads to decomposition but also create problem in plastic molding. In addition to the above physical property, polymer material also indicates lower density, specific gravity (about 1.2), high coefficient of thermal expansion (Say 5 times to metal & 10 times to ceramics), Specific heat [4 times than that of metals & ceramics] and high thermal stability (Flammability) etc.

- The glass transition temperature (Tg) is the temperature at which the glassy state makes the transition to the rubbery state. The glass transition temperature is the property of the amorphous region of the polymer, while the crystalline region is characterized by the melting point.

2.24.4. Optical Property of Polymer Material

Basically polymers do not reflect or absorb light (they are white in color or mostly colorless). However, few pure polymers absorb radiation in the visible spectrum, which is roughly between 380 and 760 nm, resulting in most polymers being colorless. They include some thermosets and elastomers, including Phenol formaldehyde resins (PF), some polyurethanes, epoxies and furan resins. They absorb strongly at the blue end of the spectrum and as a result of which appear brownish when viewed by reflected light. Even though several polymers show no absorption of radiation at visible wavelength while they regularly absorb strongly at certain frequencies in the infrared. Apart from color, the visual appearance and optical performance of a polymer material is dependent on the nature of its surface and its light transmission properties. Polymeric materials that are colorless range from highly transparent to opaque and a loss of transparency arises from light scattering processes within the material. When a ray of light strikes the plane surface of a colorless transparent material, the optical phenomena takes place which governed by the refractive index 'n'. In polymer materials, heterogeneity of refractive index can take place due to differences of density in amorphous and crystalline regions within the polymer itself, or from solid particles embodied as pigments or fillers, or from voids. The degree or intensity of scattering is strongly dependent on the variation of refractive index and additionally on the size of the heterogeneities. Crystalline polymers are normally translucent or opaque unless (as in poly-methyl-pentene) there is little to negligible difference in the refractive indices of crystalline and amorphous regions or (as in cellulose triacetate) that the spherulites are unusually small. It is readily possible to intensify the transparency of polymer materials by assisted nucleation or by high-speed cooling from the melt, both means of reducing the spherulite size. Additionally, stretching is also an effective way to increase transparency.

Note

- Polystyrene is a transparent to light on the other hand nylons is translucent.

2.25. Application of Polymer Material

In modern age polymer is a promising material for diverse filed of application from home appliances to industrial & other engineering field because it possess some specific remarkable property. Consider the following table which shows, type of polymer, their properties & possible uses / field of application.

Table 2.18: Application of Some Important Polymer Material

Type of polymer	Property	Uses/application
Acetals	Good strength & stiffness, Good resistance to creep, abrasion, moisture, heat and chemicals	Manufacturing of mechanical parts or components where high performance is required over prolong of period. For example- bearings, cams, gears, bushings, rollers; wear surfaces, pipes, valves, housing etc.
Acrylics (PMMA)	Indicate moderate strength, Good optical properties, Resistance to wear and chemicals, Transparent (Opaque) and have good electrical resistance	Lenses, lighted signs, displays, window glazing, automotive lenses, windshields, lighting fixtures and furniture
Acrylonitrile-Butadiene-Styrene (ABS)	Dimensionally stable and rigid, Good impact, abrasion and chemical resistance, Good strength, toughness and high electrical resistance, Low temperature resistance,	Cabinets for consumer and electronic goods, automotive components (wheel covers, head light bezels), boat hulls, telephones, refrigerator liners, helmets, tool handles, pipes, fittings, decorative panels etc.
Fluro-carbons	Good resistance to high temperature, chemicals, weather and electricity, Unique non adhesive properties and low friction	Linings for chemical process equipment, non-stick coatings for cookware, electrical insulation for high temperature wire and cables, gaskets, low-friction surfaces, bearings and seals
Nylon	Good mechanical property, Abrasion and chemical resistance, Self-lubricating	All nylons are hygroscopic, reduces mechanical properties and increases part dimension
Aramids (aromatic polyamides)	Have very high tensile strength, Good stiffness	Typical applications - Fibers for reinforced plastic, bullet proof vests, cables and radial tyres
Polycarbonates	Good Mechanical and Electrical properties, High Impact and chemical resistance	Safety helmets, optical lenses, bullet resistant window glazing, load bearing electrical components, wind shields, medical apparatus, electrical insulators, business machine components etc.

Polyesters (acid and alcohol)	Good mechanical property, Good electrical property, Good chemical Resistance, Abrasion resistance and low friction	Gears, cams, rollers, load bearing members, pumps, electromechanical components, PET bottles, PBT engineering components, Terylene dress materials, polyester foam cushions, polyester sun control films and tracing films
Polyethylene (Three major classes - LDPE, HDPE, UHMWPE)	Good electrical and chemical properties, Mechanical properties depend on composition and structure, Inert to chemicals	LDPE, HDPE – House wares, bottles, ducts, bumpers, garbage cans, toys, packaging materials, tubing etc. UHMWPE – Parts requiring high impact toughness and resistance to abrasive wear (artificial knee and hip joints)
Polystyrene	Exhibits average properties and good resistance to tearing	Automotive components, medical devices, appliance parts, packing trays, foam insulation, radio components, house wares, toys, furniture parts, pens, safety razor etc.
Polyvinyl Chloride (PVC)	Wide range of properties, inexpensive, water resistant and self-extinguishing **and** Not suitable for applications related to strength and heat resistance, *Rigid* PVC is tough and hard	Used in construction industry (pipes, turbine pumps), *Flexible* PVC - wire and cable coatings, in low pressure flexible tubing and hose, footwear, gaskets, seals, sheet coatings, ATM cards etc.
	Thermosetting Plastics	
Alkyds (from alkyl, - alcohol and acid)	Possess good electrical insulating properties, impact resistance and dimensional stability, Low water absorption	Typical applications in electrical and electronic components
Aminos (Urea and Melamine)	Properties depend upon composition, Aminos are hard and rigid., Resistant to abrasion, creep and electrical arcing	Small appliance housings, toilet tops, handles **Urea** – Electrical and electronic components, stove hardware, small housings etc. **Melamine** – Dinner ware, buttons, military uniforms, ashtrays etc.

Epoxies	Excellent mechanical and electrical properties, Good dimensional stability, Strong adhesive property, Good resistance to heat and chemicals, Electrical components requiring good mechanical strength and high insulation	Tools, dies and adhesives, Fiber Reinforced Epoxies – Excellent mechanical properties, Pressure vessels, rocket motor casings, tanks and similar structural components
Phenolic	Brittle, are rigid and dimensionally stable, Resistant to heat, water, chemicals and electricity,	Handles, knobs, laminated, panels, electronic components (insulators, wiring devices), Used as bonding materials to hold abrasive grains together in grinding wheels, Pressure cooker handles
Unsaturated Polyesters	Good mechanical strength, chemical and electrical properties, Polyesters are generally reinforced with glass (or other) fibers	Used as casting resins, Boats, luggage carriers, chairs, automotive body
Polyimides	Good mechanical, physical and electrical properties at elevated temperature, Good creep resistance, low friction and wear characteristics, Non melting characteristic (self-extinguisher) of thermosets, but structure of thermoplastic	Pump components (bearings, seals, valve seats, piston rings), electrical connectors for high temperature use, Aerospace and automobile parts, sports equipment and safety vests.
Silicones	Properties depend on composition, Posses excellent electrical property over wide range of temperature and humidity, Resistant to chemicals and heat	Electrical components requiring strength at high elevated temperatures, oven gaskets, heat seals, and water proof materials, A silicone-based material called super lightweight ablator, used as a heat shield on parts of the booster of space shuttle (Silicon rubber adhesive is used to fix ceramic tiles on space shuttle)
	Elastomers	
Natural Rubber		Tires, Gaskets and Hoses
Polyacrylate		Oil Rings, O-rings, footwear, belts
SBR Molded		Mechanical Goods and Disposable items.
Silicone		Electric Insulation, Seals and Gaskets

Material	Properties	Applications
Synthetic Rubbers (Modified natural rubber) Synthetic natural rubber, butyl, styrene-butadiene, and ethylene propylene	Better resistance to heat, gasoline and chemicals, and higher range of useful temperatures	Tires, shock absorbers, seals and belts
Polyurethane	This Elastomer has good overall properties of high strength, stiffness and hardness, Resistance to abrasion, cutting and tearing	Seals, gaskets, cushioning, diaphragms for rubber forming of sheet metals and auto body parts
Styrene butadiene rubber -- Copolymer.	Low cost.	Used in hoses, belts, gaskets and auto tires
Chloroprene (Neoprene) 1st commercial synthetic rubber	Weather resistant	Used in gaskets, shock mounts, seals, etc.
Nitrile rubber	Resists swelling in oils and gasoline	Used in O-rings, gasoline hoses, seals, etc.
EPDM -- Copolymer.	Exceptional weathering resistance, affected by oils.	Used in wire insulation, weathering tripping and auto door and decklid seals, etc. Often used in foamed seals
Rubber	High friction properties, Good resistance to abrasion and fatigue, unaffected to fuels, oils and chemicals, absorb vibration and noise	Work as insulator, Engine Mount, shock absorber, noise absorber, tires
Conducting Polymers	Electromagnetic interference (EMI) shielding, Electromagnetic pulse applications, Electrical signal transfer, Electrostatic painting of panels, Electrostatic discharge, Lightning Strike Protection, Electro-optical devices (photovoltaic cells)	Conductive Paints, Coatings, Caulks, Sealants, Adhesives, Sheets, Tubes, and Structural Components **Industrial use:-** Space, Aerospace, Electronics, Automotive, Chemical

Table 2.19: Application of Some Important Polymer Material

Some other common polymer prepared from substituted alkenes

Polymer	Applications
Polyethene	Machine parts, toys, buckets, containers, shrink wrap, bag, films, pipes, electrical insulation
Polypropylene	Food packaging, ropes, clothing, carpeting, molded plastic parts, labware, pipes, valves
Poly-chloro-ethylene, polyvinylchloride (PVC), Vinyl	Building materials, pipes, tubing, electrical insulation, signs, toys
Poly-chloro-pene , neoprene	High quality automobile & garden hoses, wet suits, gaskets
Plexi-glass, Lucite, acrylic glass, poly-methyl-methacrylate (PMMA)	Helmet visors, aquariums, hard contact lenses, dentures, office furniture, artificial fingernails
Teflon – poly-tetra-fluoroethylene (PTFE)	Non-stick coating, used as lubricant coating for mechanical parts
Polyacrylonitrile, Orlon, Acrilan, Acrylic fibers	Acrylic fabrics, tents, awnings, sails, carpet, carbon fibers (for bricks, tennis rackets, auto-bodies)
Polystyrene or expanded polystyrene (Styrofoam)	Packaging, insulation, modelling, bio-logical labware, toys.

2.26. Polymer Processing

Polymer processing is nothing but it is an art which indicate **"How polymer material can be synthesized so that by applying subsequent process we can transformed polymer into product of desired design specification or convert it into usable form"**. For example – polymer rods, pipes, molded product, fiber, tires, cup & trays, toys etc. Generally, polymer product may either be manufacture by single polymer (Pure polymer), blends, filled polymer, polymer with additives in which additive may work as either be dyes, plasticizers or anti-oxidants. Eventually, we can say that characteristics of any kind of polymer product wholly depending upon the polymer & their respective composition.

Polymeric materials processing technique depending upon the number of factors aswellas type of polymer material that is being processed.

(a) Processing of thermoplastic or thermoset

(b) Melting/degradation temperature

(c) Atmospheric stability

(d) Shape and intricacy of the product

Polymers are often formed at elevated temperatures under pressure. Thermoplastics are formed above the glass transition temperatures and final shape is accomplish by pressure application on the other hand thermosets are formed in two stages – At initial stage polymer liquid form is prepared & then in final stage molding is performed. Further following are some important type of processing technique which are mostly preferred for producing polymer articles or products. These are:

1. Extrusion process
2. Injection molding technique
3. Fiber spinning
4. Compression molding
5. Transfer molding
6. Blow molding
7. Vulcanization process

2.26.1. Extrusion Process

Extrusion molding process is a continuous molding process used for shaping & forming thermoplastic material, such a process are mostly employed to produce continuous wire or uniform cross section articles. For example - solid & hollow tubes, solid rods, strips & sheets or to insulate or jacketing electrical cables. In extrusion process, thermoplastics composition is heated up to plastic state and then feed into die with the help of screw conveyor and consequently it gets shape of die or mold cavity & simultaneously plastic articles cooling is performed either in atmosphere or by air jets. Consider following diagram which shows production of thermo-plastic product say bottles and sheets.

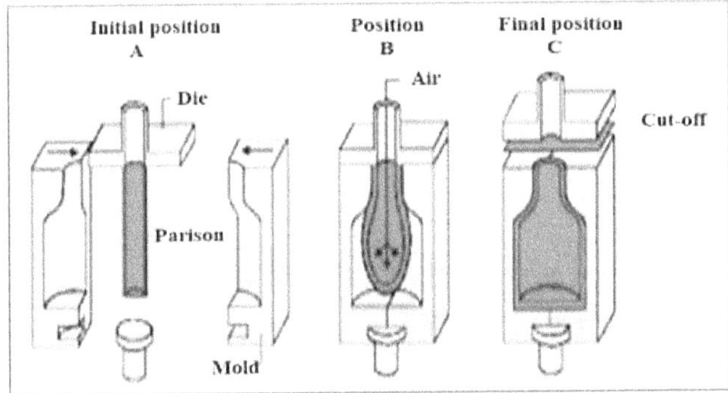

Diagram 2.35: Manufacturing of Thermo-plastic Bottle by Extrusion Below Molding Process

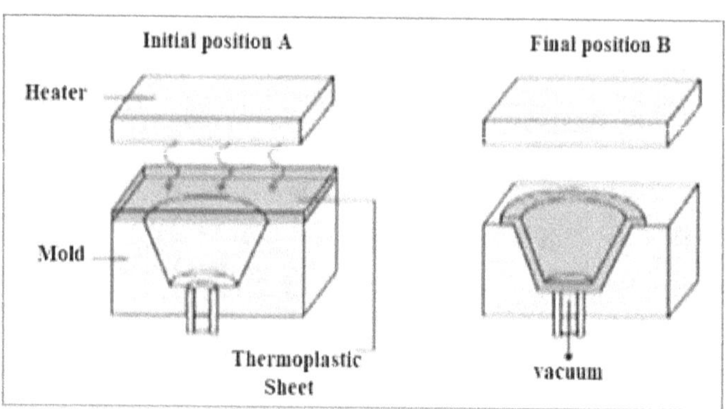

Diagram 2.36: Manufacturing of Thermo-plastic Sheets by Extrusion Below Molding Process

Furthermore, extrusion process is performed either by single screw extruder or multi screw extruder. Let be consider the case of single screw extruder, it consisting following components.

- Hopper for feeding thermoplastic polymer in dried state.
- Single screw extruder that is operated by pressure and plunger mechanism.
- Cylindrical barrel inside which single helical screw is free to rotate.

In order to manufacturing a thermoplastic polymer product, thermoplastic polymer or composition of thermoplastic polymer composition, fed into the extruder in the form of pallets which further fills the vacant space between of screw & barrel. During the feeding, the polymer material is in dried form as feeding has been completed then hopper is encapsulted by inert nitrogen gas. For melting & heating purpose barrel is heated upto melting temperature of polymer. A space nearer to hopper, always equipped with cooling facility (water cooling) whose function is to faciliate only melting the polymer in the barrel rathar than hopper so that solid plug may be formed which also inhibit surplus entering of polymer from extruder. Consider the following diagram which shows complelte over-view of single screw extruder.

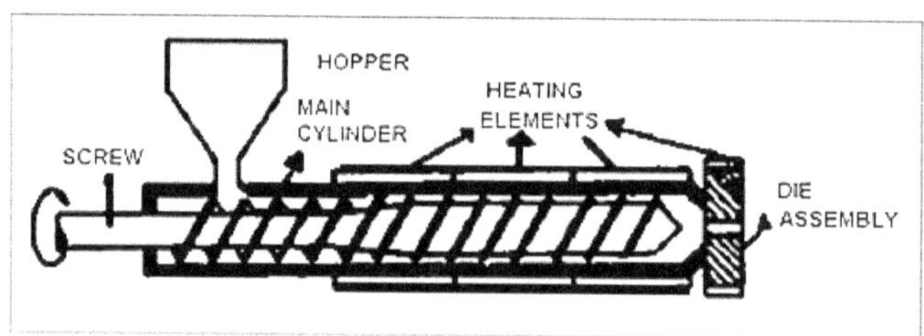

Diagram 2.37: Extrusion Machine or Extrusion Technique

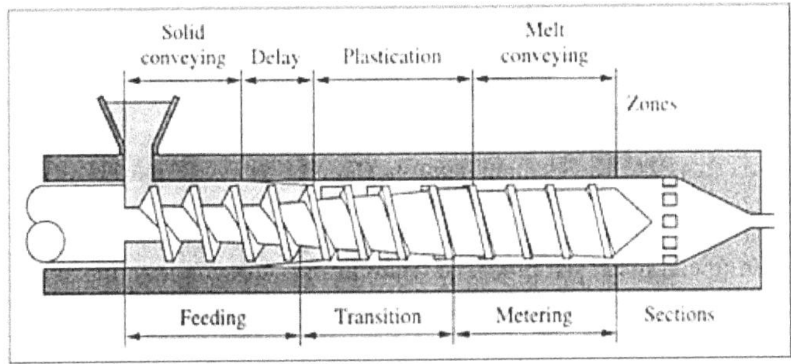

Diagram 2.38: Single Screw Extruder

It consisting following zones.

1. Solid conveying zone (Feeding zone)

2. Plastication (Transition zone)

3. Metering zone (Melt conveying zone)

4. Final section zone (After that final product is collected)

When screw rotates it will appply force upon the polymer material that has plastic state as a result of which it moves along the channels. During the processing, it has also been observed that channel vacant space reduces and at the final stage i.e. melt conveying zone, it will guide to melt & passing through the section of barrel and cosequently we have desired shape of product after passout of extrucsion die.

2.26.2. Injection Molding Technique

Injection molding is a versatile process that falls under the categeory of extrution technique which is being used to fabricate thermoplastics and thermosets both. Consider the following diagram which shows complete over-view of injection molding process.

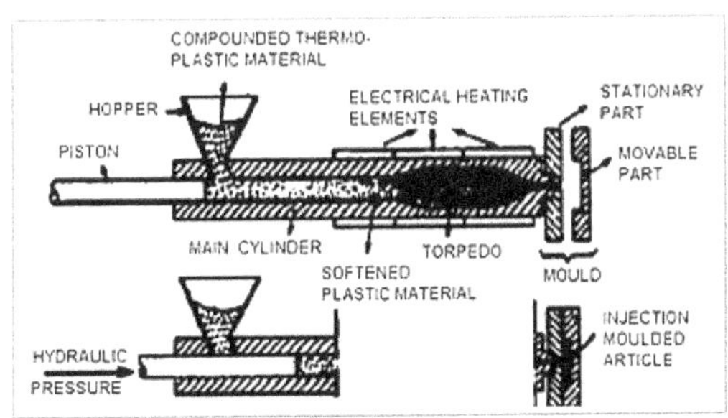

Diagram 2.39: Injection Molding Machine & Technique

It contains following components.

- Hopper, through which palletized or plastic powder is fed.
- Cylinder, in which polymer composition is heated upto plastic state that is known as melt.
- Nozzle, through which melt is injected inside the mold cavity.
- Cavity, which has two parts, one is stationary or fixed part and another is called movable part. A complete setup is known as mould cavity.
- Pistion that is operated by plunger mechanism. Plunger mechanism faciliates forward & backward motion to pistion. The basic function of piston is to build up desired degree of pressure force which is used to pushing the polymer material for filling the die cavity and then return back to its original position by plunger mechanism.

In this technique palletized material or plastic powder material is fed through hopper into a cylinder where it is heated up to plastic stage by means of oil or electricity. This melt is then injected at a predefined rate into mold cavity through the nozzle where it transformed or adopt a shape of mould cavity. After that mould is left for cooling as a result of which hot plastic become hard & rigid. after suffcent curing half movable part of mold is removed & collecting a finished product that is made from injection molding technique. The cycle is again repeated for next product or articles. This process can be repeated again & again between few second & minutes. Injection molding technique is also known as a recirocating screw-injection molding technique because inside the cylinder piston perform reciprocating motion to accomplish the operation.

2.26.3. Compression Molding

Compression molding process also applied to fabricating thermoplastics, thermosetting resins and elastomers. Consider the following diagram which shows the complete over view of compression molding technique that is dedicated to plastic & rubber.

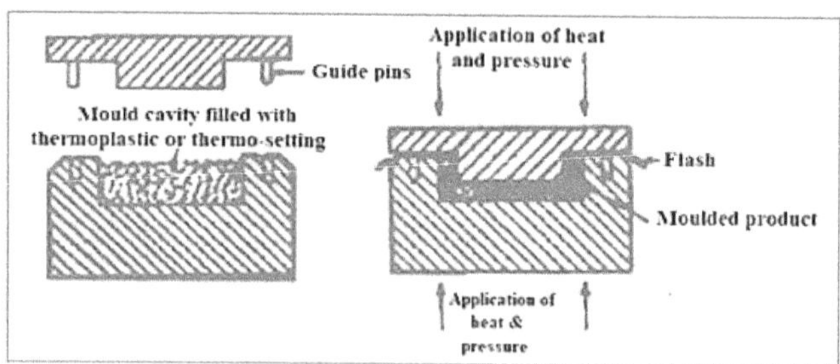

Diagram 2.40: Compression Moulding Technique for Thermosets / Thermoplastics

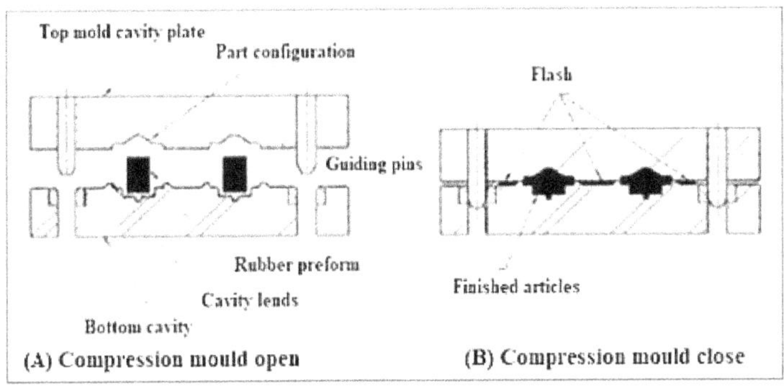

Diagram 2.41: Compression Molding Technique for Elastomer (Rubber)

It consists following components.

- Mould cavity, inside which thermoplastic or thermo-setting resins polymer material is filled manually in the form of powder or other form.

- Guiding pins, whose function is to align properly both the portion i.e. top portion & bottom portion of mould pattern.

- Heating & pressure force application device, whose function is to heat & melt the polymer up to plastic state and simultaneously pressure force is applied.

- Flash, an unwanted or surplus material of polymer composition that is release & disposed of after pressure application known as flash.

In compression molding technique, appropriate amount of polymer composition [(polymer + additives) are mixed in proper proportion] is kept inside the mold cavity manually. Mould cavity has two parts i.e. upper half portion & bottom half portion. After filling bottom half portion of mould cavity it is tightly closed by upper half portion and then heat & pressure is applied. During heating process, polymer composition transformed into melt & viscosity increases thereafter pressure force is applied as a result of which polymer composition takes a shape of cavity and then allow for cooling. As it cools, ejecting the article / product from mould cavity, we have got molded article according to design specification of mould cavity.

2.26.4. Transfer Molding

Transfer molding is different from both injection & compression molding. In injection molding technique, polymer composition is supplied through the hopper inside the cylinder where heating & melting is performed. Similarly, in case of compression molding we can put the polymer material, manually in mold cavity & then heat is supplied to melt it. But in case of transfer molding, polymer composition is kept & heated into a chamber outside the mold cavity. When mold is closed, a plunger mechanism is operated as a result of which pressure force would be acted upon the polymer melt & consequently cavity, filling take place. All the other mechanism is same but it has difference in melt filling procedure to cavity.

2.26.5. Blow Molding Technique

Below molding is employ to manufacturing glass article for instance - glass bottles. In similar manner, polymer material product may also be manufactured by blow molding techniques.

2.26.6. Vulcanization

Vulcanization process consists of heating the raw rubber at between temperature range of $100 - 140^oC$ with sulphur between time duration of 1 hours to 4 hours. As a result of which double bonds formation take place between different rubber molecules and hence cross-linking is achieved among the chains. This cross-linking not only impart stiffness to the rubber but also inhibit intermolecular displacement of rubber molecules. Stiffness of rubber always depending upon the sulphur contents. For example – Vehicle tyre consisting 3% – 5% sulphur contents and its maximum possible value as high as up to 30%. At the value of 30% sulphur content, hard and rigid rubber is formed that is known as "ebonite" rubber.

Following are Some Important Advantages of Vulcanization

- Vulcanized rubber has superior tensile strength, it is about 10 times the tensile strength of raw rubber.
- It has excellent resilience i.e., articles made from it returns to the original shape when the deforming load is removed.
- It has better resistance to moisture, oxidation, abrasion.
- It has much higher resistance to wear and tear compared to raw rubber.
- It has useful temperature range between -40° to 100^oC.
- It is a superior electrical insulator Ex: Ebonite
- It has resistance to Organic solvents [like petrol, benzene, CCl4], Fats and oils.
- It has low elasticity or sometimes has no elasticity like ebonite.

Note

- Usually extruder is required to melt, homogenize, and pump the thermoplastic material.
- Extrusion process is continuous process by which articles like-tubes, rods, and flat sheets are manufactured.

- In extrusion process, extruder is equipped with polymer processing technique.

- Injection molding is known as cyclic process which is widely used to manufacturing low & high technological product or articles for instance – For fabricating ceramic heat engine components that is employed for high temperature application.

- Injection molding machine setup is expensive but production costs are low.

- The process can be used to mold thermoplastics & thermosets. In addition, fillers can be added to make high-strength composite materials & foaming agents can be added to reduce the density of the molded article.

- During injection molding, inert gas like nitrogen also supplied into the mold so that molten polymer facing less viscosity. Consequently, we have got reduction in weight and also permitting to produce curved and hollow sections.

- Polyethylene, polypropylene, and polystyrene are processed by injection molding technique for making containers, toys, and housewares etc.

- Polyesters are used for manufacturing gears, bearings, electrical connectors, switches, sockets, appliance housings, and handles.

- Nylons are used for manufacturing high-temperature applications articles like - automobile radiator header tanks and for anticorrosion properties, acetals are preferred for making items such as gears, bearings, pump impellers, faucets, and pipe fittings. Other moldable polymers that are frequently encountered are polymethyl methacrylate for lenses and light covers, and polycarbonates and ABS for appliance housings and automobile parts.

- Injection-molding machine, specify by the screw diameter, the maximum shot size in ounces, and the force in tons with which the mold is clamped to the injection unit of the machine.

- Mold cavity for injection molding usually made from alloy steels that has internal coating of either chrome or nickel. The basic purpose of internal coating is to impart desire wear resistant, corrosion resistant & inhibit distortion during thermal cycling and sustain long service life.

- Mostly molds are water-cooled & can have multiple cavities.

- In case of single mold cavity, the injection of melt perform through a single nozzle on the contrary in case of multiple cavity molds, a runner system is demanded like a casting process to join the sprue to the gate or entry point of each cavity because it is imperative that all the cavity must be filled at the same time. Consider following diagram regarding to feed system of multiple cavity molds.

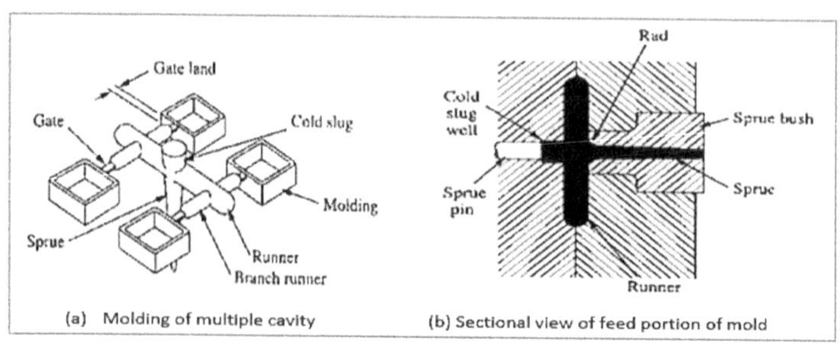

Diagram 2.42: Feed System

- Melt temperature, mold temperature, injection rate, gate pressure, packing time and the cooling time are considered as a machine variable for injection molding process which is mostly influencing the quality of polymer product.

- Injection molding showing incredible characteristic that has very short period of processing time.

Consider following table which Differentiate between both thermoplastics and thermo-setting plastics.

Table 2.20: Difference between Thermoplastics and Thermo-setting Plastics or Polymer

Thermo-plastics	Thermo-setting plastics
Thermoplastics are usually manufactured either by condensation or addition polymerization process.	Thermosetting polymers are manufactured by condensation polymerization process.
It consisting either linear chain or branched chain structure.	It has cross-linked or network structure
Polymer chain having Vander Waals force or by dipole-dipole force or by hydrogen bonds	Polymer chain have strong covalent bond with cross-links.
They facilitate recycling i.e. reheat & reshaped and remolding.	They cannot be recycled.
They become soft after heating & stiffness improve after cooling	They do not soften after heating instead it may burn
They can be soluble in some solvent if they have low molecular weight.	It indicates insolubility in any kind of solubility
During molding process chemical composition & structure remain unchanged.	They undergo chemical change if during polymerization perform & cross link structure is formed during molding process.
They have softness, flexibility and fusibility characteristics	They have rigidity, hardness & infusibility characteristics.
They can be recycled after reheating i.e. reversible	They do not recycle i.e. irreversible
They swell, dissolve in organic solvents	They don't swell, dissolve in organic solvents
It has toughness	They are hard & brittle in nature
Mould articles / product cooling is performed at room temperature after ejecting form mold	Mould articles / product cooling is performed after ejecting form mold still in hot condition.
Curing can be performed by cooling	Curing can be performed by applying heat & pressure
Acrylics, PVC, nylons, polypropylene, polystyrene, polymethyl methacrylate etc. are some common example of thermoplastics.	Epoxies, vulcanized rubbers, phenolics, unsaturated polyester resins, and amino resins (ureas and melamine) etc. are some examples of thermosetting plastic.

We have number of thermoplastic compounds, thermosetting compound and elastomer compound, consider the following table which shows some common polymer.

Table 2.21: Type of Thermoplastic, Thermosetting & Elastomer Compounds

Thermoplastic compound	Acrylics, PVC, nylons, polypropylene, polystyrene, polymethyl-methacrylate (PMMA), styrene acrylonitrile (SAN), ploy-ethylene, fluorocarbon, polycarbonate, polyamide (nylon-6 & nylon-66), acrylonitrile butadiene styrene (ABS), polyacetal, polycarbonate, Bakelite, polyethylene etc.
Thermosetting compound	Epoxies, vulcanized rubber, phenolics, unsaturated polyester resin, amin-resins, phenol formaldehyde, urea formaldehyde, melamine formaldehyde etc.
Elastomer & their compound	Polyurethane, buns-s, nitrile rubber, butyl rubber, Thiokol, poly-urethanes rubber, silicone resins etc.

Let be focus upon some polymer compound which is important from context point of view.

2.27. Thermoplastics Compounds

These plastics soften when heated and harden when cools – processes that are totally reversible and may be repeated. These materials are normally fabricated by the simultaneous application of heat and pressure. They are linear polymers without any cross-linking in structure where long molecular chains are bonded to each other by secondary bonds or inter-wined. They have the property of increasing plasticity with increasing temperature which breaks the secondary bonds between individual chains.

2.27.1. Polyethylene

Polyethylene is prepared by polymerization of ethylene in presence of oxygen.

Condition:- Polymerization is performed between 150°C to °C. Consider following reaction that id dedicated to manufacturing polyethylene.

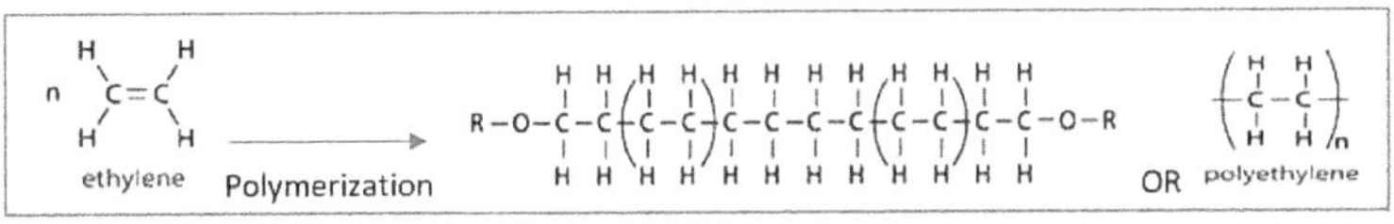

Diagram 2.43: Reaction of Formation of Polyethylene

Polyethylene have two forms low density polyethylene & high-density polyethylene. In case of low-density polyethylene density ranges between 0.91 to 0.925 g/cm^3 on the other hand high density polyethylene indicate better chemical resistance & density falls between range of 0.941 to 0.965 g/cm^3. Furthermore, LDPE is formed when free radicals are used as initiator while ionic catalyst are responsible for high density polyethylene (HDPE).

Characteristics:- Resistance to chemical & electricity, tough, low coefficient of friction, low strength, low weather resistance, rigid, translucent and showing easy crystallization.

Application:- It is generally used to manufacturing flexible bottles, toys, tumblers, battery parts, bottles cap, pipes etc. LDPE is used to manufacturing film & sheets, pipes [that is used for agriculture industries, irrigation field & domestic water supply pipes] on the other hand HDPE IS used for manufacturing toys & home appliances.

Note

- Polythene is a pure thermoplastic rather than compounds.

2.27.2. Nylon

Nylon is a compound of thermoplastic polymer which falls under the category of polyamides. It has two variety Nylon-6 & Nylon-66. Let be discuss in brief.

Nylon is manufactured from condensation process that falls under a compound of thermoplastic. It has following elements.

> **Composition** = Hexamethylenediamine (1,6- hexane diamine) + 5% aqueous solution of sebacoyl-chloride (Adipyl-chloride) + 5% solution of cyclohexane sodium hydroxide + 20% NaOH aqueous solution

Application:- Used for making parachutes & for making synthetic fibers.

Nylon – 6:- Nylon-6 is prepared by self-condensation of amino-caproic acid. It may also be prepared by ring opening polymerization of caprolactam.

Application:- It is used for manufacturing tyre cords.

2.27.3. Nylon – 66

Nylon – 66 is manufactured by condensation technique. For manufacturing nylon-66 we will have to take two components.

- Adipic acid (1,6- hex-dioic acid)
- Hexamethylenediamine (1,6- diamino-hexane)

Both the compounds are processed in presence of nitrogen at 280°C, processing reaction of nylon-66 is given below.

$$
\underset{\text{Adipoyl chloride}}{Cl-\overset{O}{\overset{\|}{C}}CH_2CH_2CH_2CH_2\overset{O}{\overset{\|}{C}}-Cl} \;+\; \underset{\text{Hexamethylenediamine}}{H-\overset{H}{\overset{|}{N}}CH_2CH_2CH_2CH_2CH_2CH_2\overset{H}{\overset{|}{N}}-H} \;\longrightarrow\; \underset{\text{Nylon 6-6}}{-\overset{O}{\overset{\|}{C}}CH_2CH_2CH_2CH_2\overset{O}{\overset{\|}{C}}-\overset{H}{\overset{|}{N}}CH_2CH_2CH_2CH_2CH_2CH_2\overset{H}{\overset{|}{N}}-}
$$

Diagram 2.44: Reaction for Manufacturing Nylon-66

The above reaction may also be rewrite as,

$$nHO-\overset{O}{\overset{\|}{C}}-(CH_2)_4-\overset{O}{\overset{\|}{C}}-OH + nH_2N-(CH_2)_6-NH_2 \longrightarrow \left[-\overset{O}{\overset{\|}{C}}-(CH_2)_4-\overset{O}{\overset{\|}{C}}-NH(CH_2)_6-NH\right]_n + 2nH_2O$$

Adipic acid   Hexamethylene diamine

Whenever reaction takes place between both the compounds then at the interface of both nylon – 66 is achieved.

Characteristics:- Nylon-66 is a linear polymer which has resistance to alkalis & mineral acid but they become degrade in presence of oxidizing agent like hydrogen peroxide & potassium permanganate etc. Furthermore, they have high temperature stability, good abrasion resistance, good self-lubricating property and translucent.

Applications:- It is usually employed for manufacturing ball bearings, cages, electrical insulating elements, pipes, machine parts, carpet fibers, apparel, air-bags, tyres, zip ties, ropes, gears, socks, bushes,, can, hair combs, for toughening electrical ware, for jacketing primary electrical insulation to impart abrasion resistance, for making tyre cords, to manufacturing bristles & brushes (By using mono-filaments), radiator parts of cars, relay coils formers, housing & casing of domestic appliances (Glass reinforced nylon plastics), for molding gear bearings, making vehicle tyres, used as fiber & fiber cloths.

2.27.4. Nylon – 11

Nylon-11 is prepared by self-condensation of W- amino-undecanoic acid.

Application:- It is used for manufacturing flexible tubes which is mostly used to conveying petrol.

Note

- Nylon – 66 indicate that it is made from the six-carbon diamine, 1,6- hexane diamine, six-carbon diacid and hexane-olioic acid.

- Nylon – 66, polymer material can be transformed into fibers by melting & extruding in the form of filaments or fiber and then cooling it. It is also used to manufacturing molded articles.

- Nylon -6 can be prepared by polymerization of 1-aza-2 cycloheptane (Caprolactam) that is obtained through the Beckmann rearrangement of cyclo-hexa-none-oxime.

2.27.5. *Polyvinyl Chloride (PVC)*

Polyvinylchloride is formed by heating a water emulsion of vinyl chloride in presence of somewhat benzoyl peroxide or H_2O_2 in auto-clave. Further vinyl chloride is generally prepared by heating hydrogen chloride with acetylene between temperature range of 60°C to 100°C in presence of metal chloride catalyst at the pressure that ranges between 1 to 1.5 atm. Consider following reaction which shows formation of vinyl chloride & subsequent reaction shows formation of polyvinylchloride.

$$CH = CH + HCl \longrightarrow CH_2 = CH\,Cl$$

Acetylene   Vinyl chloride

Diagram 2.45: Formation of Vinyl Chlorides

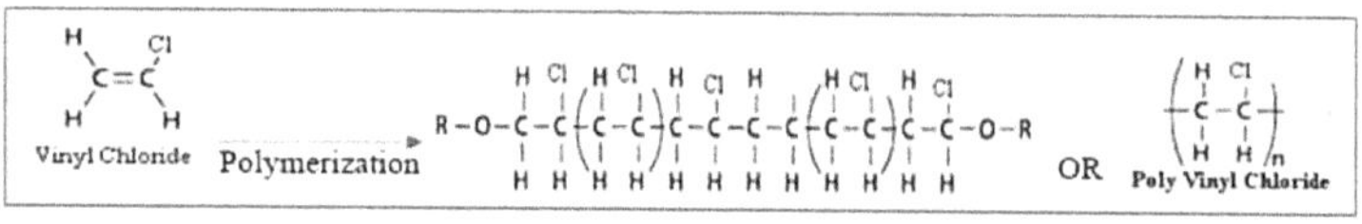

Diagram 2.46: Formation – Reaction of Poly Vinyl Chlorides (PVC)

Characteristics:- PVC is colorless, odorless, flammable, chemically inert, resistant to [light, oxygen, inorganic acids & alkalis} but soluble in hot chlorinated hydrocarbons say [ethyl chloride], high softening temperature point, high stiffness than that of polyethylene and high density etc.

Application:- Used for cable insulation, used for manufacturing [toys, leather cloths, film, sheets, floor covering, light fitting, safety helmets, cycle & bike mudguards, frame for window & doors, chemical containers, pipes & other pipe fittings]. Furthermore, a very common valuable application of PVC is used to manufacturing debit & credit cards in addition to it, PVC pipes are also used for transportation of corrosive chemical / liquid in petrochemical industries.

Note

- PVC has high stiffness & high softening temperature as a compare to polyethylene.
- PVC is a third synthetic thermoplastic which is widely used after polythene & polypropylene.

2.27.6. *Polyester*

Polyester is a condensation polymer that is also falls under the category of thermoplastic compound. Polyester may be of two types i.e. linear polyester and crosslinked polymer (Having cross-linked three-dimensional structure). Further variety of polyester may be formed, for example.

1. **Terylene:-** Terylene is a polymer fiber that is made from reaction between ethylene & terephthalic acid in presence of catalyst atmospheric oxidation of P-Xylene. Consider following reaction.

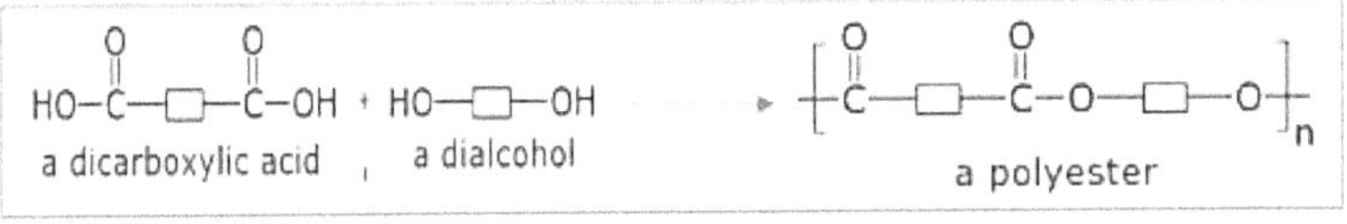

Diagram 2.47: Formation – Reaction of Polyester

Characteristics:- It is colorless, rigid & hard, indicate resistance to organic acids & other minerals, less resistance to alkalis, high melting point temperature [Due to aromatic rings] etc.

Application:- Used for making synthetic fibers, blending cab be performed with wool & cotton and showing wrinkle resistance.

2. **Linear Polyester:-** Linear polyester is aromatic which prepared by reaction between terephthalic acid & ethylene glycol, consider the following reaction that is dedicated to linear polyester.

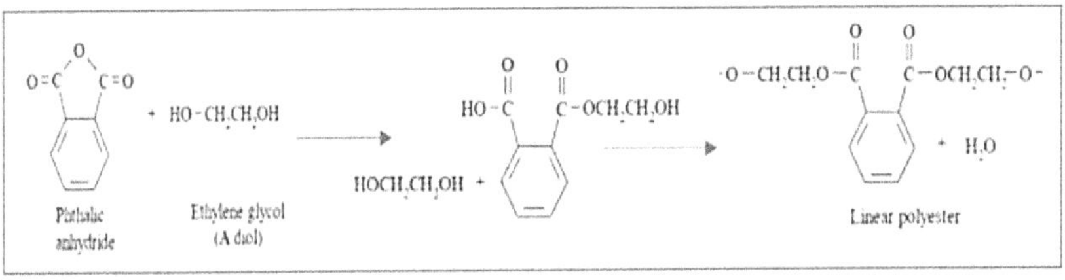

Diagram 2.48: Chain Reaction for Formation of Linear Polymer

Further if more than two functional group are existing in one monomer than cross-linked polymer chain would be formed which has three-dimensional network structure known as polyester glyptol. A reaction for that is given below.

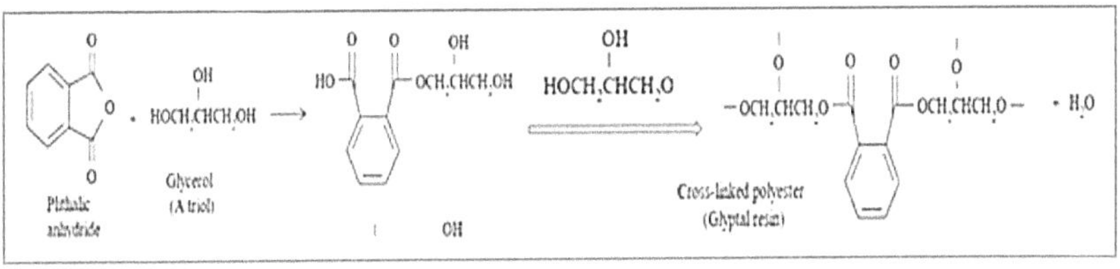

Diagram 2.49: Formation – Reaction for Cross Linked Polyester

Polyester glyptol is rigid than linear structure & used to manufacturing paints and coating material. Further, glyptol resin may also be formed by reaction between of following composition which gives highly cross linked resins also known as glyptol.

> [1,2,3-Propane-triol (Glycerol)] + [1,2-Benezene dicarboxylic anhydride (Phthalic anhydride)]
>
> { 2 Moles }   { 3- Moles }

The structure of highly cross linked glyptal resins is given below.

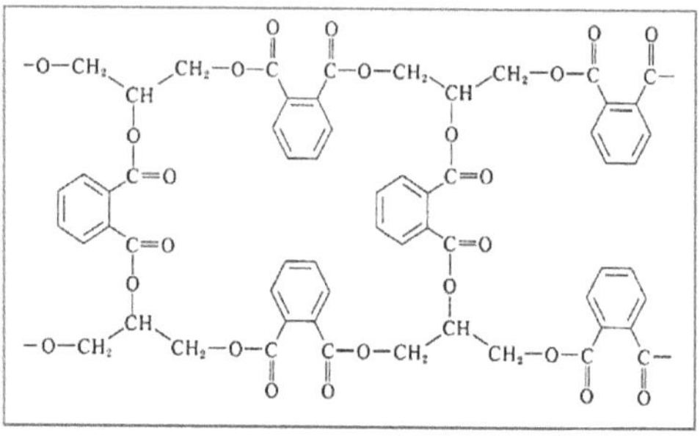

Diagram 2.50: Structure of Highly Cross-linked Glyptal Resin

3. **Dacron:-** Dacron is also a polyester that is made from condensation. In such a process reaction would be takes place in presence of metal oxide catalyst at 200°C between dimethyl 1,4 – benzene – dicarboxylate & 1,2-ethanediol (Ethylene glycol) as a result of which Dacron is formed. Consider the following reaction which shows formation of Dacron.

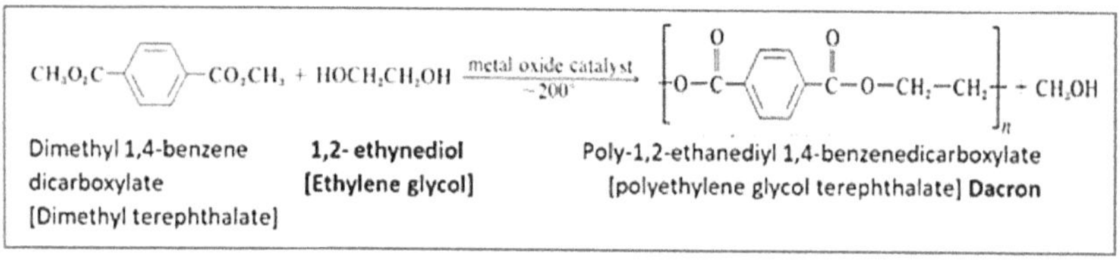

Dimethyl 1,4-benzene dicarboxylate [Dimethyl terephthalate]   1,2- ethynediol [Ethylene glycol]   Poly-1,2-ethanediyl 1,4-benzenedicarboxylate [polyethylene glycol terephthalate] **Dacron**

Diagram 2.51: Formation Reaction of Dacron

Characteristics:- Superior toughness, excellent fatigue, tear resistance, resistance to humidity, acids, greases, oil and solvents.

Application:- Used to manufacturing magnetic recording tapes, clothing, automobile tyre cords and beverage containers.

2.27.7. Teflon

Teflon is also known as polytetrafluoroethylene or fluon. It is manufactured by polymerization of water – emulsion tetrafluoroethylene in presence of benzoyl peroxide catalyst with the application of pressure. The reaction of Teflon manufacturing is given below.

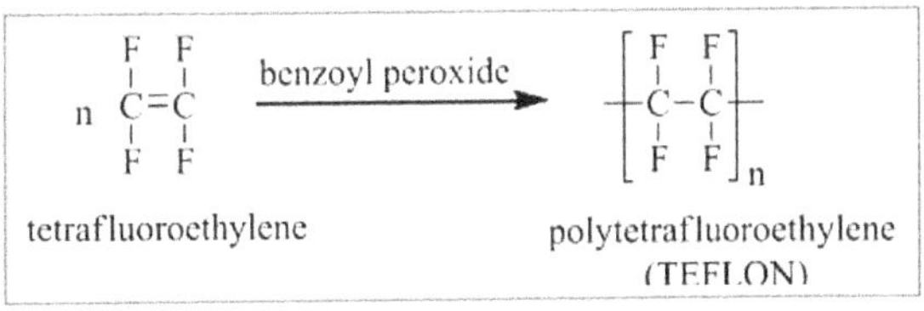

Diagram 2.52: Manufacturing Reaction of Teflon

Characteristics:- High toughness, high softening temperature (350°C), high chemical resistance, high density (2.1 to 2.3 g/cm^3), low coefficient of friction, behave as wax, good electrical property, good mechanical property (It can be machined, punched and drilled), It can be moulded in any form by applying high pressure after heating up to 350°C.

Application:- It can be used as insulating material for jacketing cables, wires and transformer, they are also used in making gaskets, stop cocks, pipes for chemical transportation, coating to high quality non-stick cooking panes and for manufacturing self-lubricating bearings.

2.27.8. Polystyrene

Polystyrene is formed by polymerization of styrene in presence of benzoyl peroxide as catalyst (Initiator). Consider following diagram which shows reaction of formation of polystyrene.

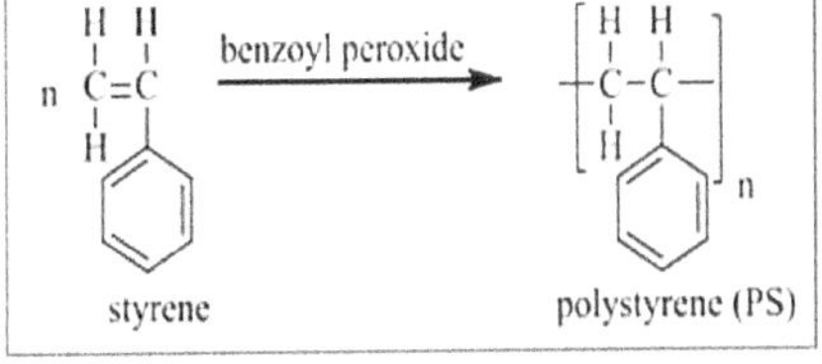

Diagram 2.53: Manufacturing Reaction of Polystyrene

Characteristics:- Polystyrene is transparent, light in weight, excellent electrical property (Insulator), optical property, good thermal property, dimensional stable, less expensive, superior resistance to [moisture, acids & chemical], brittle & hard, low softening point temperature [90°C to 100°C] and transmit light through curved surface or section.

Application:- Used for manufacturing containers, CD-Cases, jewel box, coffee cups, grocery store trays, building insulation, table ware [fork, knives & spoons], wall tiles, battery cases, toys, indoor lightning panels, home appliances, disposable food containers, lenses, plates, manufacturing parts of [radio, TV & Refrigerator] and Styrofoam toys etc.

2.27.9. Polypropylene

Polypropylene is a thermoplastic addition polymer which is produced from propene or propylene monomer. Consider the following diagram which show structure of polypropylene.

$$\left[\begin{array}{c} CH_3 \\ | \\ CH-CH_2 \end{array} \right]_n$$

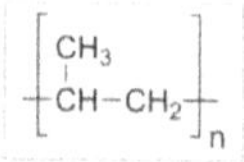

Diagram 2.54: Structure of Polypropylene

Characteristics:- It is rigid & hard in nature & showing chemical resistance, heat resistance, excellent electrical property & fatigue strength, inexpensive and poor resistance to UV LIGHT etc.

Application:- It is widely used to manufacturing packaging material, translucent part, house ware, toys, luggage, bumpers, trays, battery cases, film, tape, staple fibers, continuous filaments, chemical tanks, sheets, pipes, sterilizable bottles & TV cabinets etc.

2.27.10. Acrylics (Poly-methyl-methacrylate)

Acrylics is formed by free radical polymerization of methyl methacrylate by using vinyl. Consider the following diagram which shows the formation reaction of the Acrylics.

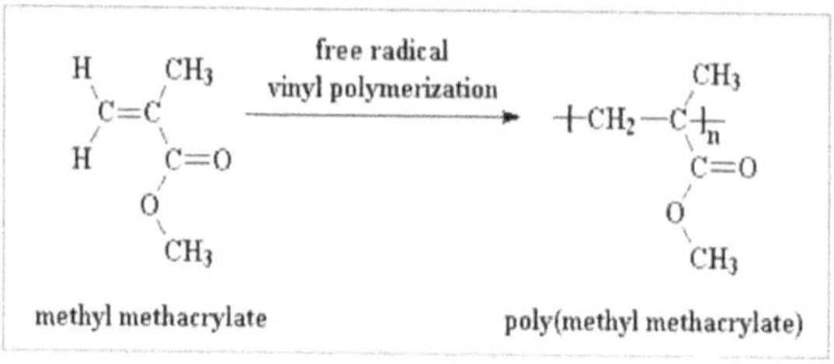

H   CH3
  C=C
H   C=O
         O
         CH3

free radical
vinyl polymerization

CH3
+CH2—C+n
         C=O
         O
         CH3

methyl methacrylate   poly(methyl methacrylate)

Diagram 2.55: Formation Reaction of Acrylics

Characteristics:- Outstanding light transmission and resistance to weathering, only fair mechanical properties.

Application:- Used to manufacturing lenses, transparent aircraft enclosure, drafting equipment's, outdoor sign etc.

2.27.11. Acrylonitrile-butadiene-styrene (ABS)

Acrylonitrile-butadiene-styrene is a thermoplastic compound which shows following structure.

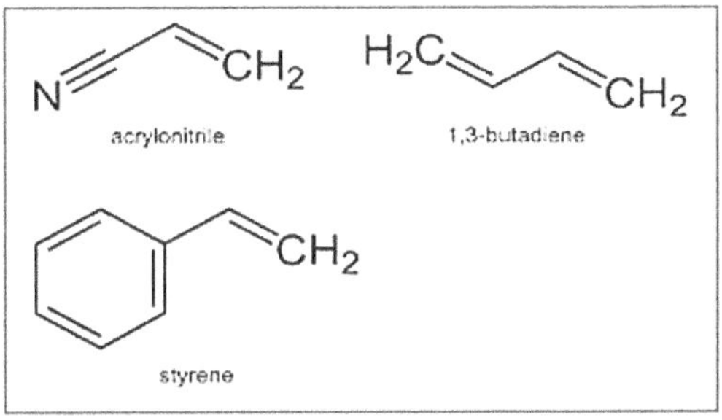

Diagram 2.56: Structure of Acrylonitrile-butadiene-styrene

Characteristics:- Superior strength and toughness, resistance to heat distortion, good electrical properties, flammable and soluble in some organic solvents.

Application:- Refrigerator lining, lawn and garden equipment, toys, highway safety devices.

2.27.12. Polycarbonates

Polycarbonates showing following structure.

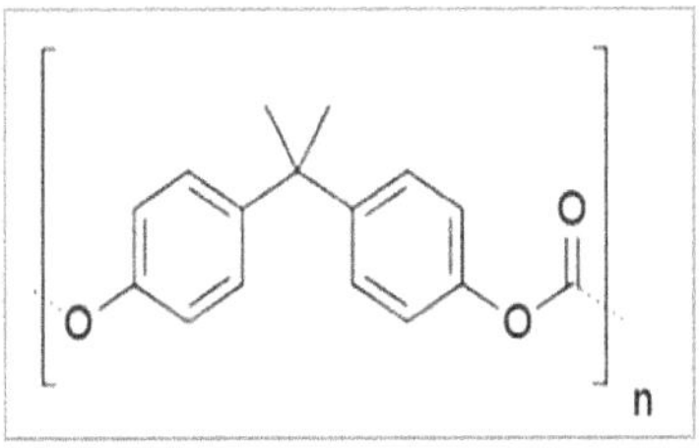

Diagram 2.57: Structure of Polycarbonate

Characteristics:- Dimensionally stable, high impact strength, high ductility & low water absorption etc.

Application:- Used to manufacturing safety helmet, lenses light globes, base of photographic film etc.

2.27.13. Fluorocarbons

Fluorocarbon also recognize by the acronym PTFE or TFE, It shows following structure.

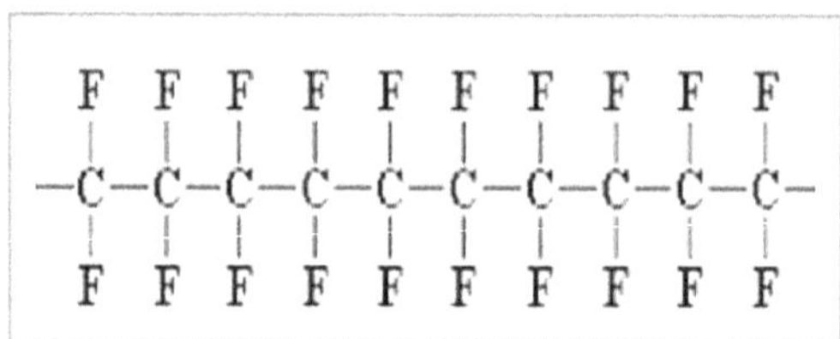

Diagram 2.58: Structure of Fluorocarbon

Characteristics:- Chemically inert, low coefficient of friction, high temperature stability [up to 260⁰C], weak in nature, excellent electrical property etc.

Application:- Used to manufacturing bearings, high temperature electronic parts, valves, anti-corrosive seals, chemical pipes and anti-adhesive coatings etc.

2.27.14. Bakelite

Bakelite is a first thermosetting synthetic polymer or we can say that it is a polymer compound. It is manufactured by condensation method in which phenol reacts with formaldehyde in presence of acid or alkalis catalyst. The formation of Bakelite consisting three steps.

1. **Step- 1st:-** In 1st step reaction between benzenol & methanol takes place which gives [4-hydroxphenyl] methanol. Consider following reaction, which gives complete information about the process.

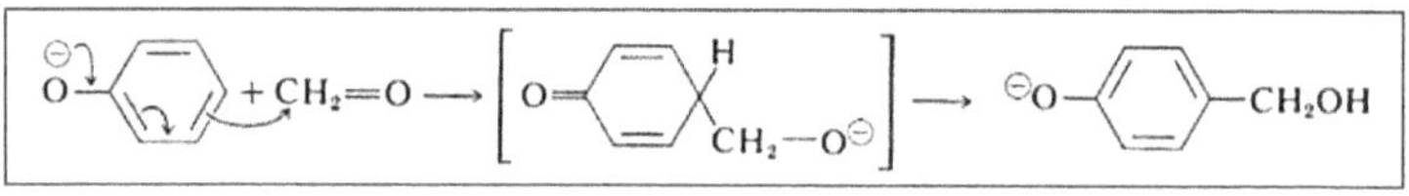

Diagram 2.59: Formation Reaction of [4-hydroxphenyl] Methanol

2. **Step - 2nd:-** In 2nd step we will perform dehydration of [4-hydroxphenyl]methanol by addition of base. Consider the following reaction which shows dehydration.

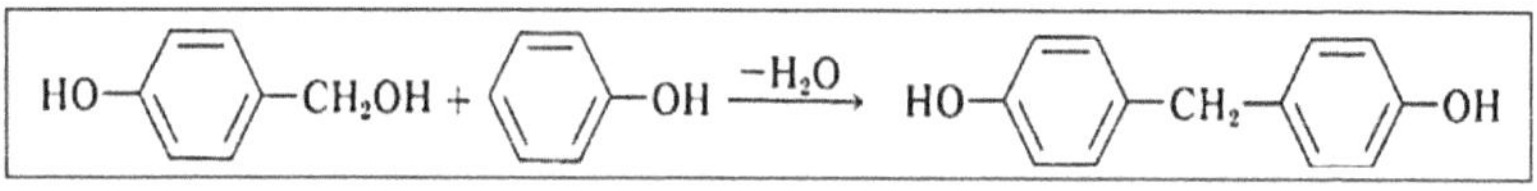

Diagram 2.60: Formation Reaction for bis[hydroxyphenyl] Methane

3. **Step – 3rd:-** If these reaction will be continued then 2,4 and 6 position of the benzenol form cross linked three dimensional Bakelite resin whose structure is given below.

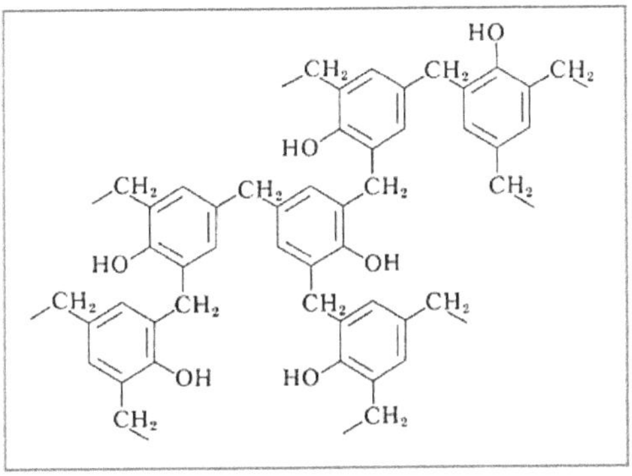

Diagram 2.61: Structure of Cross-linked Three-dimensional Bakelite Resin

A complete over view of manufacturing reaction of Bakelite [crossl-inked] is given below.

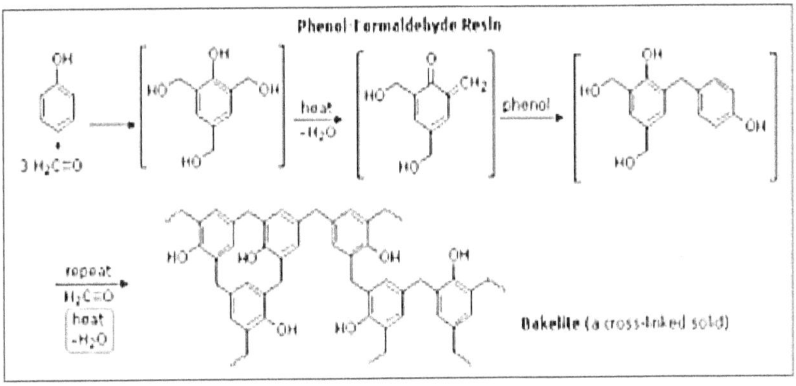

Diagram 2.62: Complete Reaction of Bakelite Formation

Characteristics:- It is hard, rigid & strong, exhibit electrical insulation, indicate resistance to [moisture, chemical & abrasion], excellent bonding strength & adhesive property [low molecular weight], infusible, insoluble, resistance to {acids, salt & organic solvents}, affected by alkalis and showing scratch resistance.

Application:- Used for making electrical insulator parts [switches, plug], switch board, heater handles, telephone parts, radio & TV cabinets, valve parts, knobs, buttons, knife handles, electronic instruments, handle of cookware & pans, distributor heads of cars, grinding wheel grains adhesives, brake lining, varnishes, protective coating and making ion exchange resins etc.

Note

- Bakelite is available in two forms these are nova lacks & resole.
- **Nova lacks** are phenol formaldehyde resins obtained by condensation of phenol & formaldehyde in presence of acid as catalyst when phenol to formaldehyde ratio is greater than one.
- Resole is a low melting polymer when it is heated in mould it transformed into insoluble & infusible polymer.

2.28. Thermosets Compound

These plastics require heat and pressure to mould them into shape. They are formed into a permanent shape and cured or 'set' by chemical reactions such as extensive cross-linking. They cannot be re-melted or reformed into another shape but decompose upon being heated to too high a temperature. Thus, thermosets cannot be recycled, whereas thermoplastic can be recycled. The term thermoset implies that heat is required to permanently set the plastic. Most thermosets composed of long chains that are strongly cross-linked (Covalently bonded) to one another to form 3-D network structures to form a rigid solid. Thermosets are generally stronger, but more brittle than thermoplastic. Advantages of thermosets for engineering design applications include one or more of the following: high thermal stability, high dimensional stability, high rigidity, light weight, high electrical and thermal insulating properties and resistance to creep and deformation under load. Consider some compound that falls under the category of thermosets,

2.28.1. Melamine

Melamine is a compound that is formed when 2,4,6 – tri-amino – 1,3,5 – triazine reacts with methanol. The structure of melamine is given below.

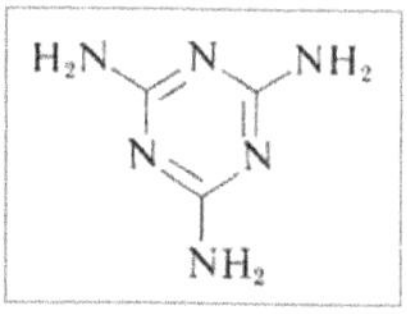

Diagram 2.63: Structure of 2,4,6 – tri-amino – 1,3,5 – triazine [Melamine]

Application:- It is used for manufacturing plastic dishes which is recognized as a Melmac.

Similarly, whenever urea reacts with methanol then urea methanol resin would be formed, consider following diagram which shows three-dimensional structure of urea-methanol.

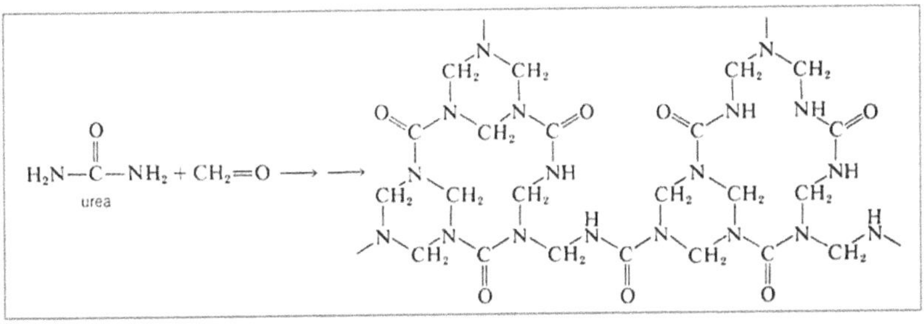

Diagram 2.64: Reaction between Urea & Methanol Showing Formation of Urea-methanol Resin

2.28.2. *Epoxy Resins*

Epoxy resins falls under the category of plastics & adhesives that is prepared by condensations polymerization which takes place between bisphenol & epichlorohydrin in the presence of alkalis at 60°C in which one monomer must be contained epoxy groups. Consider the following reaction which shows formation of epoxy resins, further next diagram shows structure of epoxies.

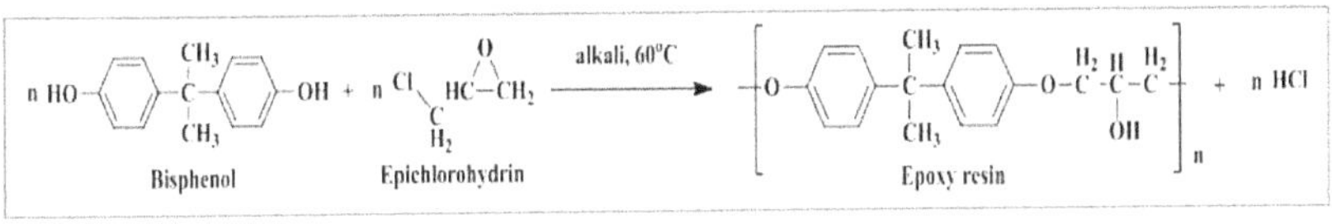

Diagram 2.65: Formation Reaction of Epoxy Resin

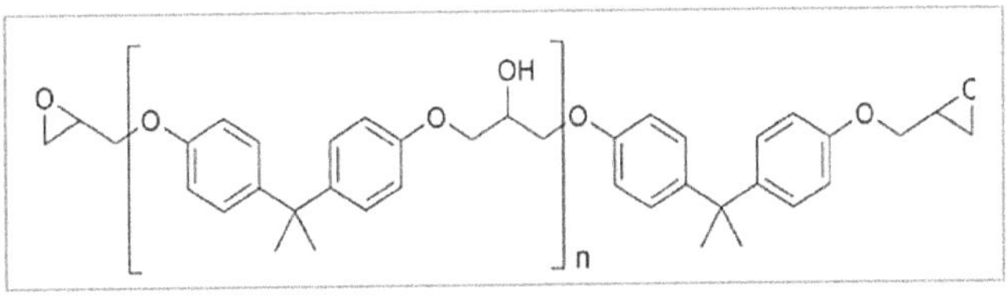

Diagram 2.66: Structure of Epoxy

At initial stage, condensation polymerization of the sodium salt of bisphenol takes place with the epoxies, when reaction is going on, we have two reactions.

- Sn – 2 type displacements
- Oxide – ring product

Consider following diagram which shows complete reaction of the process, we have got pre-polymer epoxy resins and then curing is performed which converted into three-dimensional network structure of cross-link structure.

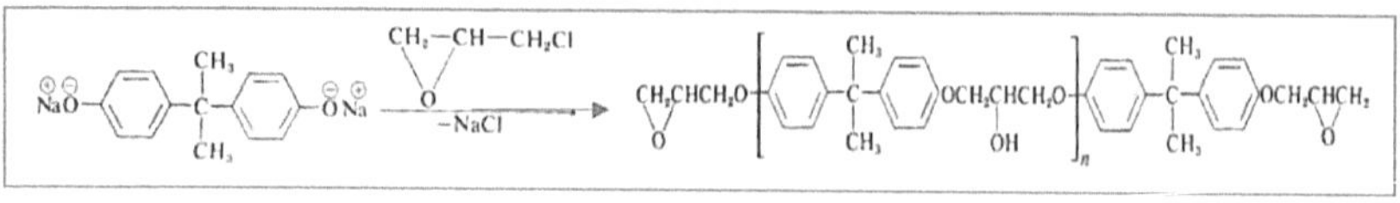

Diagram 2.67: Pre Polymer for Epoxy Resins

Furthermore, three-dimensional cross-linked structure may also be formed by the following reaction, by mixing the above product with trifunctional amine as a result of which we have extension in polymer china & formation of cross-linked structure would be formed, consider following reaction.

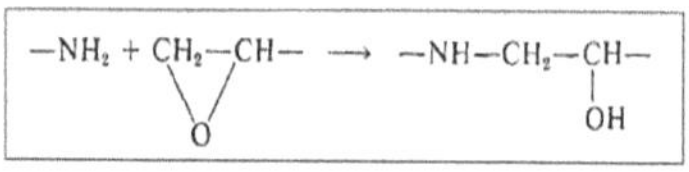

Further, a poly basic acid anhydride can be used to link the chain with secondary alcohol & then which subsequently link to oxide rings.

Characteristics:- Epoxy resin indicate high resistance to corrosion [chemical, water, solvents, acids & alkalis etc.], dimensionally stable, high flexibility, heat resistance, toughness, superior adhesion property, inexpensive and having electrical property etc.

Application:- Used to apply surface protective coating, apply as adhesives, skid resistance [Apply on highways surface], mould manufacturing [Epoxy resins] which is used to produce aircraft & automobiles components, electrical moulding, sinks and employed for fiber glass laminations.

2.28.3. Phenolic

Phenolic is obtained by the reaction of phenols [carbolic acid] with formaldehyde which catalysing by an acid or base. Consider the following diagram which shows structure of phenolic.

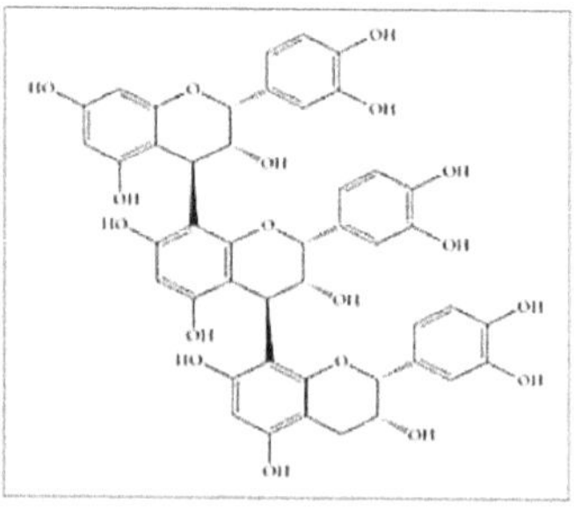

Diagram 2.68: Structure of Phenolics

Characteristics:- Thermally stable [Up to 150°C], may be compounded with resins & fillers, inexpensive, flame resistance etc.

Application:- Used for making moulded products, used as coating & adhesives, aircraft interiors, stow bins, motor housing, telephones, auto-distributors and electrical fixtures etc.

2.29. Elastomers

Also known as rubbers, these are polymers which can undergo large elongations under load, at room temperature, and return to their original shape when the load is released. There are number of man-made elastomers in addition to natural rubber. These consist of coil-like polymer chains those can reversibly stretch by applying a force. Processing of polymers mainly involves preparing a particular polymer by synthesis of available raw materials, followed by forming into various shapes. Raw materials for polymerization are usually derived from coal and petroleum products. The large molecules of many commercially useful polymers must be synthesized from substances having smaller molecules. The synthesis of the large molecule polymers is known as polymerization in which monomer units are joined over and over to become a large molecule. More upon, properties of a polymer can be enhanced or modified with the addition of special materials. This is followed by forming operation. Addition polymerization and condensation polymerization are the two main ways of polymerization. Addition polymerization, also known as chain reaction polymerization, is a process in which multi-functional monomer units are attached one at a time in chainlike fashion to form linear/3-D macromolecules. The composition of the macro-molecule is an exact multiple of for that of the original reactant monomer. This kind of polymerization involves three distinct stages – initiation, propagation and termination. To initiate the process, an initiator is added to the monomer. This forms free radicals with a reactive site that attracts one of the carbon atoms of the monomer. When this occurs, the reactive site is transferred to the other carbon atom in the monomer and a chain begins to form in propagation stage. A common initiator is benzoyl peroxide. When polymerization is nearly complete, remaining monomers must diffuse a long distance to reach reactive site, thus the growth rate decreases. Let be discussed here some elastomers,

2.29.1. Buna – S

Buna – S, sometimes also known as styrene rubber. It is a synthetic rubber, which is produced by copolymerization of butadiene (About 75% by weight) and styrene (25% by weight). Consider following terms which shows manufacturing reaction of Buna – s.

$$nCH_2=CH-CH=CH_2 \ + \ n \ CH_2=CH-Ph \longrightarrow \ -(-H_2C\ -CH=CH-CH-CH_2-CH-Ph-)n-$$

[1,3-butadiene (75%)]   **[Styrene(25%)]**   **[Buna – S]**

Diagram 2.69: Manufacturing Reaction of Buna – s Rubber

Characteristics:- Good quality in finished product, high abrasion-resistance, high load-bearing capacity and resilience, vulcanization is possible by using sulphur or sulphur monochloride (S_2Cl_2) but it has some draw backs i.e. oxidized [In presence of ozone in the atmosphere], become swells when interacts with either oils or solvents.

Application:- Used for manufacturing of motor tyres, floor tiles, shoe soles, gaskets, foot-wear components, wire and cable insulations, carpet backing, adhesives, tank-linings, etc.

2.29.2. Nitrile Rubber

Nitrile Rubber sometimes also referred as Buna – N or NBR.

Preparation:- It is prepared by the copolymerization of butadiene and acrylonitrile in emulsion system. Consider following reaction.

Co-polymerization

$$mCH_2=CH-CH=CH_2 + nCH_2=CH-CN \longrightarrow -(-CH_2-CH=CH-CH_2-)m-(CH_2-CH(CN)-)n-$$

[1,3-butadiene]   **[Acrylonitrile]**   **[Nitrile Rubber]**

Diagram 2.70: Manufacturing Reaction of Nitrile Rubber

Properties:- Nitrile rubber is less resistance to alkalis than natural rubber, Excellent resistance to [oils, chemicals, aging (Sun light)], If % value of acrylonitrile increases in nitrile rubber then resistance to acids, salts, oils, solvents etc. improves, vulcanized nitrile rubber showing high temperature stability, good abrasion resistance, good abrasion resistance also in presence of gasoline or oils.

Application:- It is used for making Conveyor belts, Lining of tanks, Gaskets, Printing rollers, Oil-resistance foams, Automobile parts and high altitude air-craft components Hoses and adhesives.

2.29.3. Butyl Rubber

Butyl rubber is prepared by the aluminium chloride initiated cationic copolymerisation of isobutene with small amount (1-5%) of isoprene.

Characteristics:- Amorphous in nature however it crystallizes on stretching, resistant to oxidation [Due to low degree of unsaturation], vulcanized but impossible to achieve hardness, become degraded due to heat or light to sticky low-molecular weight products, good electrical insulation properties, low permeability to air and other gases, flexibility [Up to temperature of -50°C], Soluble in hydrocarbon solvents [Benzene] but indicates excellent resistance to polar solvents [like-alcohol, acetone and ageing chemicals (ex: HCl, HF, HNO_3, H_2SO_4 etc.)].

Application:- Used for Insulation of high voltage wires and cables, Inner tubes of automobile tyres. For manufacturing conveyor belts for food and other materials and lining of tanks Hoses etc.

2.29.4. Thiokol

It is also known as polysulphide rubber (or Gr-P). It can be prepared by the condensation polymerization of sodium polysulphide (Na_2Sx) and ethylene dichloride. The reaction of formation is given below.

$$Cl-CH_2-CH_2-Cl + Na-\overset{\overset{\textstyle S}{|}}{S}-\overset{\overset{\textstyle S}{|}}{S}-Na + Cl-CH_2-CH_2-Cl \longrightarrow \ -CH_2-CH_2-\overset{\overset{\textstyle S}{|}}{S}-\overset{\overset{\textstyle S}{|}}{S}-CH_2-CH_2-$$

Ethylene dichloride   Sodium polysulphide   Ethylene dichloride   Thiokol

Condensation

Diagram 2.71: Formation Reaction of Thiokol

Application:- Used for manufacturing of oils hoses, chemically resistant tubing and engine gaskets, Diaphragms and seals which remain in contact with solvents and Printing rolls, Containers for transporting solvents and Solid propellant fuels for rockets, etc.

2.29.5. Polyurethane

Polyurethane falls under the class of elastomeric material which has linear block copolymer that consisting of alternating soft and hard segments with phase separation due to thermodynamic segmental incompatibility. Such kind of block copolymer can be synthesized with two distinct blocks that may differ in their propensity to crystallize. Segment may either be soft or hard segments which further can be amorphous or semi crystalline.

2.29.5.1. Poly Urethanes Rubbers

Polyurethane or isocyanate rubber is prepared by reacting polyalcohol with di-isocyanates. During the manufacturing of polyurethane following reaction is achieved.

n[OH-(CH2)2-OH+O=C=N-(CH2)2-N=C=O]--→[-O-(CH2)2-O-(CO)-NH-(CH2)2-NH-(CO)-]n

Polyalcohol   Di-isocyanates   Polyurethane

Diagram 2.72: Formation Reaction of Polyurethane

Furthermore, polyurethane commercially manufactured by the reaction between di-isocyanate & diol. For example-when methylene diphenyl di-isocyanate [Diphenylmethane di-iso-cyanate] reacts with ethane-1, 2-diol then we have got polyurethane polymers, consider following reaction.

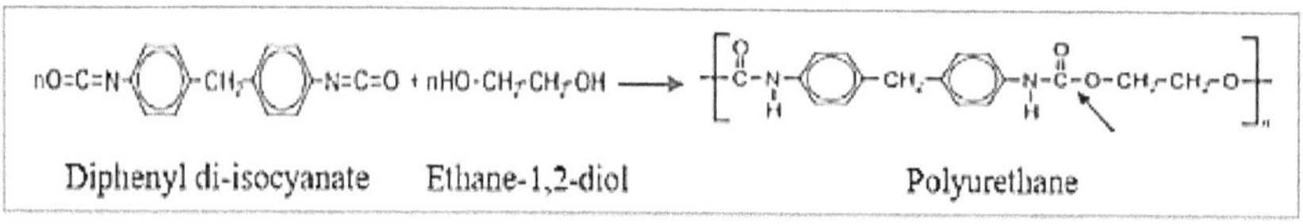

Diagram 2.73: Formation Reaction of Polyurethane

Polyurethane may also be formed when toluene diisocyanate reacts with glycerol. Consider the following reaction.

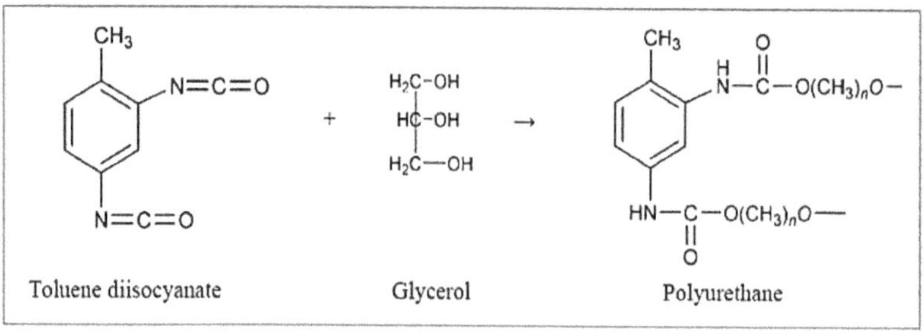

Diagram 2.74: Formation Reaction of Polyurethane

Characteristics:- Light in weight, showing resistance to [oxidation, heat, abrasion, chemical & weathering], toughness, resistance to organic solvents, unsuitable for acids & alkalis.

Application:- Used as surface coating, used to manufacturing other foam, spandex fibers, used as leather substitute as corfoam, to manufacturing gaskets & seals, films adhesives and elastomers.

2.29.5.2. Polyurethane (Foam Condensation Polymer)

Polyurethane foam is manufactured by condensation technique in which two liquid are mixed in equal proportion, For example.

> Part 1st + Part 2nd = Polyurethane foam
>
> Both the parts in liquid form

- Part 1st:- Part 1st consisting polymeric diol or triol [Glycerol] + Blowing agent + Silicone surfactants + Catalyst.
- Part 2nd:- Part 2nd consisting poly-isocyanate [Di-phenyl-methane-diisocyanate].

When polymerization reaction takes place between both the parts then large molecules are rigidly arranged in three-dimensional structure and at the same time somewhat water is responsible for decomposition of some amount of isocyanate as a result of which formation of carbon dioxide is achieved that transformed into foam. The above process can be easily understood by the following two reactions.

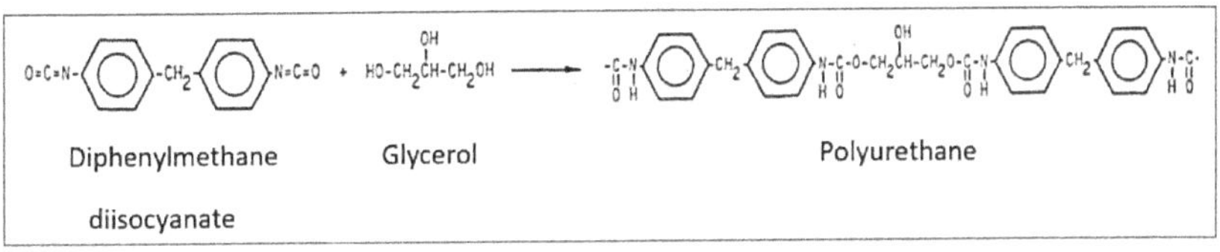

Diphenylmethane diisocyanate   Glycerol   Polyurethane

Diagram 2.75: Formation Reaction of Polyurethane Foam

$$R\text{-}N{=}C{=}O + H\text{-}O\text{-}H \rightarrow R\text{-}\underset{O}{\overset{H}{N}}\text{-}\overset{\|}{C}\text{-}OH \rightarrow R\text{-}N\text{-}H + CO_2$$

Diagram 2.76: Formation Reaction of Carbon Dioxide

Application:- Polyurethane foam is used to manufacturing casting objects, use as sound proofing material and used as insulator.

2.29.6. *Silicones*

Silicones is a family member of inorganic polymer that can not only be used as a elastomers (Rubber) but also use as a fluids, oil & resins. Silicones having Si-O bonds in back bones of structure. Consider the following diagram which shows structure of silicones.

$$\left[-O-\underset{R}{\overset{R}{Si}}-O- \right]_n$$

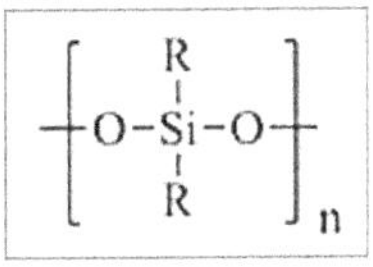

Diagram 2.77: Structure of Silicones

Where,

R is an alkyl or phenyl radical Silicones having two types i.e. linear silicones & tightly cross-linked silicones.

2.29.6.1. Formation of Linear Silicones

Whenever hydrolyses of bifunctional dimethyl-di-chlorosilane [$(CH_3)_2SiCl_2$] is performed then linear chain silicones would be formed, consider the following reaction which is dedicated to formation of linear silicones.

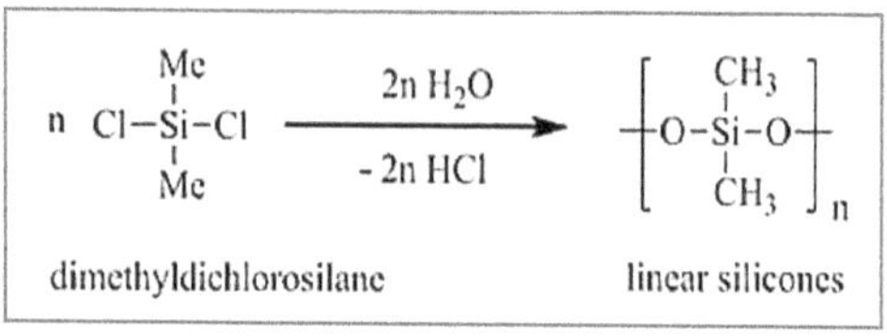

Diagram 2.78: Formation Reaction of Linear Chain Silicones

2.29.6.2. Formation of Cross-linked Silicones

Whenever hydrolyses of methyl-trichlorosilane is performed then cross-linked chain silicones would be formed, consider the following reaction which is dedicated to formation of cross linked silicones.

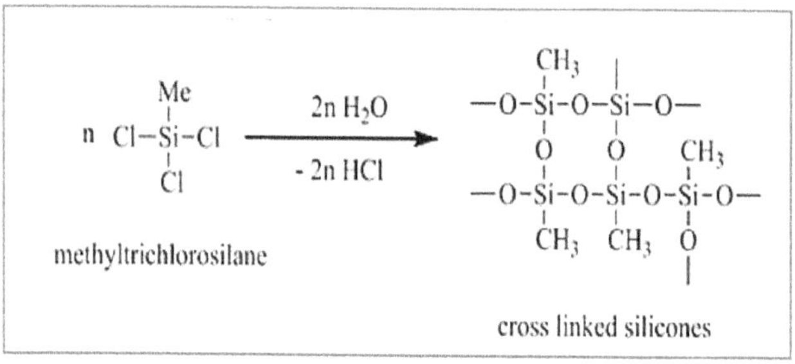

Diagram 2.79: Formation Reaction of Cross-linked Chain Silicones

Further, the intensity of cross-linked is always depending up on the proportion of methyl-trichlorosilane & dimethyl-di-chlorosilane in composition.

2.29.7. Manufacturing of Silicone Rubber

Silicon rubber is also a valuable product of silicone. It is manufactured by the following composition.

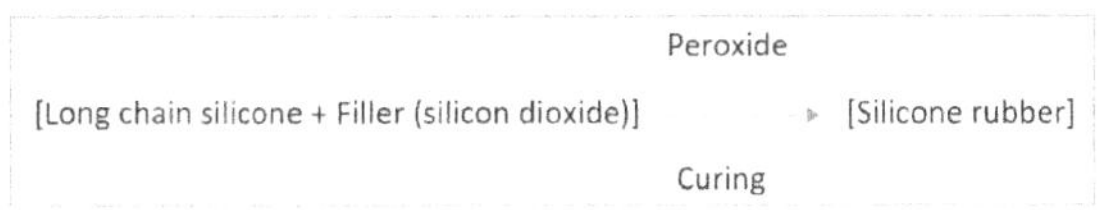

Diagram 2.80: Formation Reaction of Silicon Rubber

Silicone rubber consisting 6000 to 600000 Si units and elasticity ranges between temperature of -90°C to 250°C. These rubbers are generally used to manufacturing tyres of racing cars & aircraft in addition to it, it is also used to manufacturing sealing material of air-craft engines, lubricants, paints, boots and protective coating materials.

Characteristics:- Silicone exists in [liquid, viscous fluid, semi-solid (grease), rubber & solids], indicate high temperature stability, shows water resistance property and non-toxic in nature, excellent high temperatures, good oxidation-stability, low chemical resistance, specific gravity ranges between 1.03 to 2.1.

Application:- Following are some important application of silicone & their respective products.

1. Straight chain silicone polymer available in liquid form up to 500 Si units, they are used as water proof for building, glass ware and fabrics. also used in car & apply as a shoe polish.

2. Due to nontoxic & low surface tension, they are use in sewage disposed, textile dying, beer making industries, frothing of cooking oil in making potato Crips & chips.

3. Silicone oil are used as die-electric insulating material in high voltage transformer & also used as hydraulic fluids.

4. Silicone grease is applied for maintaining temperature variation [high to low or vise-versa] which is formed by blending silicone oil & lithium stearate soaps.

5. Silicone resin are highly cross-linked silicone, they are generally used to manufacturing high voltage insulator, printed circuit board, for encapsulating IC-chips & resistors, for providing non-stick coating on pans, for mould preparation which is being used for making car tyres etc.

6. Liquid silicones or silicone oils are relatively low molecular-weight silicones, generally made from dimethyl silicones. They possess great wetting-power for metals, Low surface tension and show very small changes in viscosity with temperature.

7. They are used as high temperature lubricants, antifoaming agents, water-repellent finishes for leather and textiles, heat transfer media, as damping and hydraulic fluids. They are also used in cosmetics and polishes.

Note

- Silicones & their relative products are very expensive but their fascinating characteristics & application making it a promising material for modern age.
- Linear chain of silicone can be terminated during the reaction by adding somewhat $(CH_3)_2SiCl$ with $(CH_3)_2SiCl_2$ & further subsequent hydrolyses.
- During the reaction between $(CH_3)_2SiCl$ & $(CH_3)_2SiCl_2$, chain length is evaluated as average chain length which is completely depending upon the mixture of composition.

In addition to the above, some other polymer compounds are given blow.

2.29.8. Polyphosphazines

Polyphosphaines is an inorganic polymer that is formed when phosphorous pentachloride reacts with ammonium chloride in presence of 1,1,2,2-tetra chloroethane. Consider the following diagram which shows structure & respective reaction of formation of polyphosphazines.

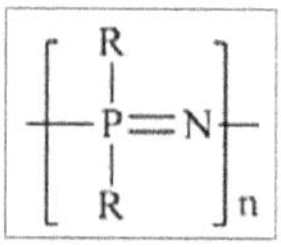

Diagram 2.81: Structure of Polyphosphazines

Where,

R – Cl is a Poly-phosphonitrillic chloride

R = OCH3 is a polydimethoxy phosphazines

R = OC2H5 is a polydiethoxy phosphazines

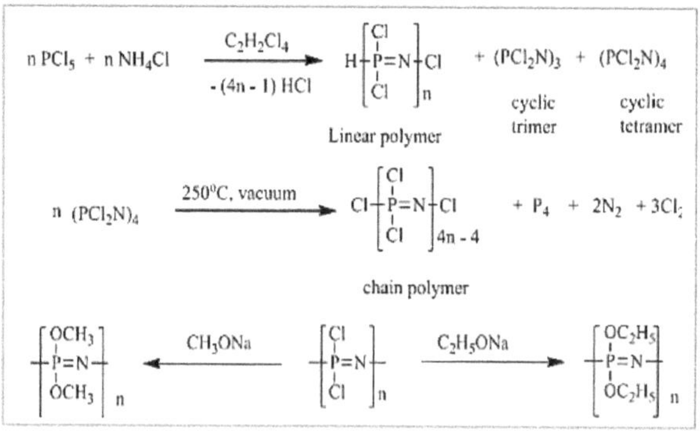

Diagram 2.82: Manufacturing Reaction of Polyphosphazines

Outcomes of Reaction:- we have following results.

- In presence of vacuum, at the temperature of 250°C, cyclic trimer & tetramer would be formed. It is a combination of linear chain & some cross-linked polymer that has minimum solubility in solvent.

- Poly-dimethoxy & poly-diethoxy phosphazines formation takes place when polyphosphonitrilic chloride reacts with sodium methoxide & sodium ehoxide respectively.

Characteristics of Polyphosphonitrilic Chlorides Polymer;- High elasticity in air but elasticity vanishes in moisture, gel formation achieved due to cross linking, fresh product is soluble in chloroform but insoluble in hexane.

2.29.9. *Sodium Polyacrylate – Copolymer*

The basic purpose of sodium polyacrylates is to absorb water and such a material is known as super absorbents. It is manufactured when starch – hydrolyzed polyacrylonitrile super absorber is mixed with either glycerin or ethylene glycol.

Application:- Used as pampers & other disposable diapers etc.

Note

- Super absorbent material is also known as water grabber or water lock.

2.29.10. *Sulphur based Inorganic Polymer*

In this class following two types are important.

1. **Polymeric Sulphur (PS):-** Polymeric Sulphur is prepared by heating rhombic Sulphur at the temperature of 170°C and then quenched the molten form into ice bath, after that product is washed with carbon di-sulfide to remove surplus amount of Sulphur.

2. **Polymeric Sulphur nitride (SN)n:-** Polymeric Sulphur nitride is also known as polythiazyl. It is a golden color polymer with metallic luster & showing electrical conductivity. It falls under the category of conductive inorganic polymer.

Manufacturing:- It is prepared by polymerization of di-sulfur dinitride (S_2N_2) which is a dimer.

- (S_2N_2) dimer is formed by synthesis of tetra sulfur (S_4N_4 – tetramer) in presence of catalyst hot silver wool. Consider following reaction which shows formation of polythiazyl.

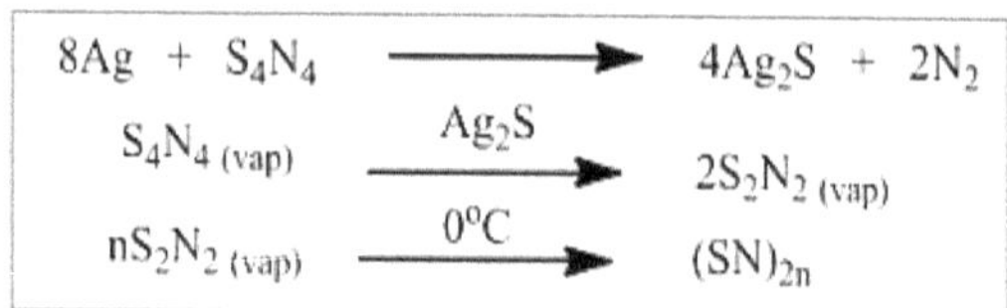

Diagram 2.83: Formation Reaction of Polymeric Sulphur Nitride

Characteristics:- Indicate electrical conductivity, Used as super conductor [non-metallic super conductor] etc.

Application:- Used in LED'S, transistors, battery cathodes and in solar cells etc.

2.29.11. *Chalcogenide Glass (As_2S_3)n*

Chalcogenide glass belongs to the family of amorphous cross-linked polymer. It is also a compound of polymer. Chalcogenide glass consisting following composition.

Composition = (Compound of chalcogens) + (One or more than one polyvalent elements)

- Compound of chalcogens may either Sulphur, selenium, or tellurium.
- Polyvalent elements may either phosphorus, thallium, arsenic, antimony, bismuth, indium, tin or germanium.

Chalcogenide glass may be prepared by following two method.

1. Heating the composition of chalcogenide glass up to molten state & then followed to quenching.
2. Heating the Chalcogenide compound up to state of vaporization & then flowing this vapor over the polyvalent elements.

Characteristics:- Chalcogenide glass having low softening temperature & low tensile strength and high refractive indices than that of boro-silicate, resistance to acid, high oxidation resistance [up to 3000°C], indicate optical property [Transmit infrared radiation], exhibit electrical property [Conductivity depending up on voltage] etc.

Application:- To manufacturing rewritable optical discs & phase change memory devices, used as infrared detectors, lenses and infrared optical fibers.

Question for Discussion

1. Define polymers? Explain in brief about natural & synthetic polymer.
2. How many types of polymer exists? indicate structure of any five-polymer material.

3. Write down name & application of thermoplastic polymer?

4. Is polymer may be used in civil engineering construction work? if yes then write down all the possible polymer with their application.

5. What do you understand by monomers? explain & gives an information about monomers structure of polyethylene, Polyvinylchloride (PVC) & polypropylene (PP).

6. Is polymer falls under the category of engineering material? if yes then explain.

7. Write short notes on followings.

    1. Polymer as a mechanical engineering material

    2. Polymer as a civil engineering material

    3. Polymer as an electrical engineering material

    4. Elastomer application in mechanical engineering field

8. What do you understand by polymer concrete & polymer modified concrete? Explain in brief.

9. Write down all the factor which limits or influence the application of polymer material? Explain.

10. Explain the contribution of polymer material in the field of medical application? Explain their applications.

11. Compare, metals, ceramics & polymer with suitable explanation?

12. What are the various modes of polymer classifications? Explain any two among them.

13. Write short notes on followings.

    1. Natural polymer & synthetic polymer

    2. Branched chain polymer & cross-linked polymer

    3. Elastomers & fibers

14. Explain about thermo-plastic polymer & thermo-setting polymer?

15. Explain about addition & condensation polymerization technique? Give some example regarding to above.

16. What is addition polymerization? Explain about initiation stage, propagation stage and termination stage.

17. Write down the application of addition polymer? Describes their characteristics.

18. Explain in brief about free radical polymerization, ionic polymerization & cationic polymerization.

19. What do you understand by condensation polymer? give some example with suitable explanation of condensation polymerization.

20. Describe about co-ordination polymerization & co-polymerization polymer?

21. Writ down the application of condensation polymer? describes characteristics of condensation polymerization technique?

22. What do you understand by molecular weight & molecular weight distribution? what is meant by number average molecular weight & weight average molecular weight.

23. Write short notes on followings.

    1. End group analyses
    2. Mechanical property of polymer
    3. Electrical property of polymer
    4. Physical property of polymer
    5. Amorphous, oriented & un-oriented crystalline polymer

24. What do you understand by higher molecular weight polymer? Write down name & describe their uses.

25. Describe in brief about optical property of polymer material?

26. Write down application of thermoplastics, thermo-setting & elastomer polymeric materials?

27. What is meant by polymer processing? Explain with suitable sketch about extrusion process & injection molding process.

28. Write short notes on followings.

    1. Compression molding technique
    2. Transfer molding technique
    3. Blow molding technique
    4. Vulcanization process

References

[1]   V.R. Gowariker, N.V. Viswanathan, and Jayadev Sreedhar, Polymer Science, New Age International (P) Limited publishers, Bangalore, 2001.

[2]   C.A. Harper, Handbook of Plastics Elastomers and Composites, Third Edition, McGraw Hill Professional Book Group, New York, 1996.

[3]   William D. Callister, Jr, Materials Science and Engineering – An introduction, sixth edition, John Wiley & Sons, Inc. 2004.

[4]   Calhoun, A., and Peacock, A., Polymer Chemistry – Properties and Applications, Hanser Publishers, Munich (2006).

[5]   Termonia, Y., and Smith, P., High Modulus Polymers, A.E. Zachariades, and R.S. Porter, Eds., Marcel Dekker Inc., New York, (1988).

[6]   Ehrenstein, G. W., Faserverbund-Kunststoffe, Hanser Publishers, Munich (2006).

[7]   Naranjo, A, Noriega, M., Osswald, T.A., Roldán, A. and Sierra, J., Plastics Testing and Characterization – Industrial Applications, Hanser Verlag, München, (2006).

[8]   Sánchez, K., Stöckl, K., Perdomo, A., and Osswald, T.A., Journal of Plastics Technology, accepted for publication (2012).

[9]   Gowariker, V.R., Viswanathan, N.V. and Shreedhar, J. Polymer Science, New Age International, New Delhi, 2005.

[10]   Chanda M., Introduction to Polymer Science and Chemistry, CRC Press, Taylor and Francis Group, FL, USA, 2006.

[11]   Cowie, J.M.G., Polymers: Chemistry and Physics of Modern Materials, Blackie Academic & Professional, Glasgow, 1991.

[12]   Mohanty A.K., Misra M., Drzal L.T., Natural fibers, Biopolymers, and Biocomposites, Chapter 1, CRC Press, Taylor & Francis Group, FL, USA, 2005.

[13]   Catia Bastioli, Handbook of Biodegradable Polymers, Chapter 1,3,6,8, Rapra Technology Limited, UK, 2005.

[14]   C Ricbard Brundle, Charles A. Evans, Jr. Sbaun Wihon, Encyclopaedia of Materials Characterization, Chapter 1,8, Manning Publications, Greenwich 1992.

[15]   Odian George, Principles of Polymerization, Fourth Edition, Wiley International, Jersey, USA, 2004.

[16]   Flory P.J., Principles of Polymer Chemistry, Sixteenth Printing, Cornell University Press, New York, 1995.

[17]   Piringer O.G., Baner A.L., Plastic Packaging: interaction with food and pharmaceutical, 2nd Edition, Wiley-VCH Verlag GmbH, Weinheim. Furuta, M, 1977.

[18]   Nunes, R.W., Martin, J.R. and Johnson, J.F., "Influence of Molecular Weight Distributions on Mechanical Properties of Polymers" Polymer Eng. Sci., 22(4), 193, 1982.

[19]   Ehrenstein G.W., Theriault R.P., Polymeric materials: structure, properties, applications, Hanser Gardner Publications, 2001.

[20]   Günter R., Jens-Uwe S., Polymer crystallization: observations, concepts, and interpretations, Lecture Notes in Physics, Vol. 606, Springer-Verlag Berlin Heidelberg, 7, 2003.

[21]   Mittal Vikas, Optimization of Polymer Nanocomposite Properties, Wiley, 13.1, 2010.

[22]   Takashi Yamamoto, Molecular Dynamics Modeling of Crystal-Melt Interfaces and Growth of Chain folded Lamellae, Adv. Polym. Sci., 2005.

[23]   Scheirs J., Compositional and Failure Analysis of Polymer, John Wiley & Sons, USA, 2000.

[24] Nicholson J.W., The Chemistry of Polymer, RSC Publishing, Cambridge, UK, 2012.

[25] Feughelman M., Mechanical properties and structure of alpha-keratin fibres: wool, human, UNSW Press, Sydney, 1997.

[26] Balta Calleja, F.J., Basanowska, J., Rueda, D.R., J. Mater. Sci., 28, 6074, 1993.

[27] Balta Calleja, F.J., Salazer, J.M., Rueda, D.R., "Encyclopedia of polymer science and engineering" Wiley, New York, 6, 614, 1986.

[28] Ania, F., Salazer, J.M., Balta calleja, F.J., J. Mater. Sci., 24, 1989.

[29] Deslandes, Y., Rossa, E.A., Briss, F., Menehini, T., J. Mater. Sci., 26, 2769, 1991.

[30] Balta Calleja, F.J., Cruze, C.S., Chen, D., Zachmann, H.G., Polym., 32, 2252, 1991.

[31] Ion, R., Pollock, H.M., Cames, C.R., J. Mater. Sci., 25, 1444, 1990.

[32] L.G. Petors and F. Knoop; Metals and Alloys, 13, 292, 1940.

[33] P. Agrawal, R. Bajpai and S.C. Dutt; Polym. Int., 249-251, 1994.

[34] Diego, J.A., J. Phys. D: Appl. Phys., 40, 1138, 2007.

[35] Migahed, M.D., Ahmed, M.T., Kotp, A.E., J. Macromole. Sci. B Phys., 44, 43, 2005.

[36] Caserta, G., Serra, A., J. Appl. Phys., 42, 3774, 1971.

[37] Palli, P.K.C., Phys. Stat. Sol., 17, 221, 1973.

[38] Singh, R., Datt, S.C., Thin Solid Films, 70, 325, 1980.

[39] Mahendru, P.C., J. Phys. D. Phys., 8, 305, 1975.

[40] Latour, M., Murphy, P.V., J. Electrostate, 3, 16, 1977.

[41] Jain, K., Phys. Stat. Sol., 21, 685, 1974.

[42] Talwar, M., Sharma, D.L., J. Electrochem. Soc., 125, 434, 1978.

[43] Singh, R., "3rd Inter. C of. On properties and application of dielectric materials", Tokyo, Japan, 1991.

[44] Datt, S.C., Keller, J.M., Indian Journal Pure & Appl. Phy., 29, 150, 1991.

[45] Kojima, K., Marda, T., J. Appl. Phys., 17, 1735, 1978.

[46] Takai, Y., J. Appl. Phys., 16, 1937, 1977.

[47] Shrivastava, A.P., Mathur, O.N., Indian J. Phys., 53, 91, 1979.

[48] Keller, J.M., Datt, S.C., Phys. Stat. Sol., 91, 205, 1985.

[49] Tanaka, T., J. Appl. Phys., 49, 784, 1978.

[50] Iqbal, T., Hogarth, C.A., Thin Solid Films, 61, 23, 1979.

[51] Eatah, A.I., Tawfik, A., Indian J. Phys. A I62A, 8, 88, 1988.

[52] Partridge, R.M., Polym. Letts., 5, 205, 1967.

[53] Rychkov, A.A., J. Phys. D. Phys., 25, 986, 1992.

[54] Lewandowshi, A.C., Phys. Revi. B, 49, 8029, 1994.

[55]   Thielen, A., J. Appl. Phys., 75, 8, 1994.

[56]   Christodoulides, C., Phys. Stat. Sol., 11, 325, 1998.

[57]   Gun'ko, V.M., Goncharuk, E.V., Adv. Collo. Inter. Sci., 22, 2006.

[58]   Catia Bastioli, Handbook of Biodegradable Polymers, Chapter 1,3,6,8, Rapra Technology Limited, Crewe, UK, 2005.

[59]   C Richard Brundle, Charles A. Evans, Jr. Shaun Wihon, Encyclopedia of Materials Characterization, Chapter 1,8, Boston, London, 1992.

[60]   Paul A Fowler, J Mark Hughes and Robert M Elias, J Sci Food Agric., 86: 1781–1789, (2006).

[61]   Xu Fenglan, Li Yubao, Wang Xuejiang, Journal of materials science 39, 5669–5672, 2004.

[62]   Ji-Guang Gu, Ji-Dong Gu, Journal of Polymer and Environment Vol. 13(1), 65-74, 2005.

[63]   D. Briassoulis, Journal of Polymers and the Environment, Vol. 12(2), 65-81, 2004.

[64]   Randal Shogren, Journal of Environmental Polymer Degradation, Vol. 5(2), 91- 95, 1997.

[65]   Randal L. Shogren, Zoran Petrovic, Zengshe Liu, and Sevim Z. Erhan, Journal of Polymers and the Environment, Vol. 12(3), 173-178, 2004.

[66]   Yeon-Hum Yun, Young-Ho Na and Soon-Do Yoon, Journal of Polymers and the Environment, Vol. 14(1), 2006.

[67]   VS Sangawar, RJ Dhokne, AU Ubale, PS Chikhalikar and SD Meshram, Bull. Mater. Sci., Vol. 30(2), 2007.

[68]   VS Sangawar, PS Chikhalikar, RJ Dhokne, AU Ubale and SD Meshram, Bull. Mater. Sci., Vol. 29(4), 2006.

[69]   Elias H.G. An Introduction to Polymer Science. Weinheim: VCH; 1997.

[70]   Belgacem MN, Gandini A. Monomers, Polymers and Composites from Renewable Resources. Elsevier; 2011.

[71]   Billmeyer FW. Textbook of Polymer Science. New York: Wiley–Interscience; 1971.

[72]   Karak N. Fundamentals of Polymers: Raw Materials to Finish Products. PHI Learning Pvt Ltd; 2009.

[73]   Chanda M, Roy SK. Industrial Polymers, Specialty Polymers, and Their Applications. Boca Raton: CRC Press; 2008.

[74]   G. Odian, Principles of Polymerization, McGraw-Hill Book Co., New York, 1970.

[75]   Polymer Engineering Science and Viscoelasticity, An Introduction, by Hal F. Brinson and L. Catherine Brinson Pages 55-97, Springer Science + Business Media, LLC 2008.

[76]   Alfred Kofi GAND, Tak-Ming CHAN, James Toby MOTTRAM, Front. Struct. Civ. Eng., 2013, 7(3): 227–244.

[77]   Advances in Civil Engineering Asaad, M.H. Kadhim, Hesham A. Numan, and Mustafa Özakça, Volume 2019.

[78]   Esmaeeli, Esmaeel & Barros, Joaquim. (2015). Flexural Strengthening of RC Beams using Hybrid Composite Plate (HCP).

[79]   F. Humphreys, Matthew. (2019). Research gate, The use of polymer composites in construction.

[80]   International Research Journal of Engineering and Technology (IRJET) e-ISSN: 2395-0056 Volume: 05 Issue: 12, Dec 2018 ISSN: 2395-0072.

[81]   Fuchs ER, Field FR, Roth R, Kirchain RE (2008) Strategic materials selection in the automobile body: Economic opportunities for polymer composite design, Composites science and technology, 68: 1989-2002.

[82]   Victor SP, Muthu J (2014) Bioactive mechanically favorable, and biodegradable copolymer nanocomposites for orthopaedic applications. Mater Sci Eng C Mater Biol Appl., 39: 150-160.

[83]   Ma Z, Mao Z, Gao C (2007) Surface modification and property analysis of biomedical polymers used for tissue engineering. Colloids Surf B Bio interfaces 60: 137-157.

[84]   Pertici G (2016) Introduction to bioresorbable polymers for biomedical, Bioresorbable Polymers for Biomedical Applications: From Fundamental to Translational Medicine p. 1.

[85]   Zare Y, Shabani I (2016) Polymer/metal nanocomposites for biomedical applications. Mater Sci Eng C Mater Biol Appl., 60 195-203.

[86]   Bassas-Galia M, Follonier S, Pusnik M, Zinn M (2016) 2-Natural polymers: A source of inspiration, Bioresorbable Polymers for Biomedical Applications. Perale G Hilborn J Eds., (2016): 31-64.

[87]   Thakur VK, Kessler MR (2015) Self-healing polymer nanocomposite materials: A review. Polymer 69 (2015): 369-383.

[88]   Cowie J.M.G. (1991). Polymers: Chemistry & Phisics of Modern Materials, Blackie, Glasgow (UK).

[89]   Flory P. (1953). Principles of Polymer Chemistry, Cornell Univ. Press, Ithaca.

[90]   Jenkins A.J., Loening K.L. (1988). Nomenclature, in "Comprehensive Polymer Science", G. Allen Ed., vol. 1, pp. 13-54.

[91]   Morawetz H. (1985). Polymers: The Origin and Growth of a Science. J. Wiley & Sons. Inc., New York.

[92]   Pino P., Porri L., Giannini U. (1987). Insertion Polymerisation in "Enciclopedia of Polymer Science and Engineering" J.I. Kroschnitz Ed., J. Wiley & Sons, New York, pp. 147-220.

[93]   Rudin A. (1999). The Elements of Polymer Science and Engineering. Academic Press, San Diego (USA).

[94]   V.R. Gowariker, N.V. Viswanathan, and Jayadev Sreedhar, Polymer Science, New Age International (P) Limited publishers, Bangalore, 2001.

[95]   C.A. Harper, Handbook of Plastics Elastomers and Composites, Third Edition, McGraw Hill Professional Book Group, New York, 1996.

[96]   William D. Callister, Jr, Materials Science and Engineering – An introduction, sixth edition, John Wiley & Sons, Inc. 2004.

[97]   Cichosz, Stefan, et al. "Polymer-based sensors: A review." Polymer Testing, vol. 67, 2018, pp. 342-348.

[98]   Fried, Joel R. Polymer Science and Technology. Pearson Education, 2014.

[99]   Hakim, Hussein, et al. "study the optical properties of polyvinylpyrrolidone (PVP) doped with KBR." European Scientific Journal, vol. 3, Dec. 2013.

[100]   Hall, Christopher. "Electrical and Optical Properties." Polymer Materials, 1981, pp. 92-112.

Index